Optimization in HPLC

Optimization in HPLC

Concepts and Strategies

Edited by

Stavros Kromidas

Editor

Dr. Stavros Kromidas
Consultant
Breslauer Str. 3
66440 Blieskastel
Germany

All books published by **WILEY-VCH** are carefully produced. Nevertheless, authors, editors, and publisher do not warrant the information contained in these books, including this book, to be free of errors. Readers are advised to keep in mind that statements, data, illustrations, procedural details or other items may inadvertently be inaccurate.

Library of Congress Card No.:
applied for

British Library Cataloguing-in-Publication Data
A catalogue record for this book is available from the British Library.

Bibliographic information published by the Deutsche Nationalbibliothek
The Deutsche Nationalbibliothek lists this publication in the Deutsche Nationalbibliografie; detailed bibliographic data are available on the Internet at <http://dnb.d-nb.de>.

© 2021 WILEY-VCH GmbH, Boschstr. 12, 69469 Weinheim, Germany

All rights reserved (including those of translation into other languages). No part of this book may be reproduced in any form – by photoprinting, microfilm, or any other means – nor transmitted or translated into a machine language without written permission from the publishers. Registered names, trademarks, etc. used in this book, even when not specifically marked as such, are not to be considered unprotected by law.

Print ISBN: 978-3-527-34789-6
ePDF ISBN: 978-3-527-82850-0
ePub ISBN: 978-3-527-82851-7
Obook ISBN: 978-3-527-83748-9

Cover Design Formgeber, Mannheim, Germany
Typesetting Straive, Chennai, India
Druck und bindung CPI Group (UK) Ltd, Croydon, CR0 4YY

Printed on acid-free paper

Contents

Preface *xv*
About the Book *xvii*

Part I Optimization Strategies for Different Modes and Uses of HPLC *1*

1.1 2D-HPLC – Method Development for Successful Separations *3*
Dwight R. Stoll, Ph.D.
1.1.1 Motivations for Two-Dimensional Separation *3*
1.1.1.1 Difficult-to-Separate Samples *3*
1.1.1.2 Complex Samples *4*
1.1.1.3 Separation Goals *4*
1.1.2 Choosing a Two-Dimensional Separation Mode *4*
1.1.2.1 Analytical Goals Dictate Choice of Mode *5*
1.1.2.2 Survey of Four 2D Separation Modes *5*
1.1.2.3 Hybrid Modes Provide Flexibility *7*
1.1.3 Choosing Separation Types/Mechanisms *8*
1.1.3.1 Complementarity as a Guiding Principle *8*
1.1.3.2 Pirok Compatibility Table *9*
1.1.3.3 Measuring the Complementarity of Separation Types *9*
1.1.4 Choosing Separation Conditions *11*
1.1.4.1 Starting with Fixed First-Dimension Conditions *11*
1.1.4.2 Starting from Scratch – Flexible First-Dimension Conditions *13*
1.1.4.3 Special Considerations for Comprehensive 2D-LC Methods *13*
1.1.4.4 Rules of Thumb *13*
1.1.5 Method Development Examples *14*
1.1.5.1 Example 1 – Use of LC–LC to Identify an Impurity in a Synthetic Oligonucleotide *14*
1.1.5.2 Example 2 – Comprehensive 2D-LC Separation of Surfactants *14*
1.1.6 Outlook for the Future *17*
Acknowledgment *18*
References *18*

1.2	**Do you HILIC? With Mass Spectrometry? Then do it Systematically** *23*	
	Thomas Letzel	
1.2.1	Initial Situation and Optimal Use of Stationary HILIC Phases *25*	
1.2.2	Initial Situation and Optimal Use of the "Mobile" HILIC Phase *28*	
1.2.2.1	Organic Solvent *28*	
1.2.2.2	Salts *31*	
1.2.2.3	pH Value *33*	
1.2.3	Further Settings and Conditions Specific to Mass Spectrometric Detection *35*	
1.2.4	Short Summary on Method Optimization in HILIC *36*	
	References *36*	
1.3	**Optimization Strategies in LC–MS Method Development** *39*	
	Markus M. Martin	
1.3.1	Introduction *39*	
1.3.2	Developing New Methods for HPLC–MS Separations *39*	
1.3.2.1	Optimizing the LC Separation *40*	
1.3.2.1.1	Optimizing for Sensitivity and Limit of Detection – Which Column to Take? *40*	
1.3.2.1.2	Optimizing Resolution vs. Sample Throughput *41*	
1.3.2.1.3	MS-Compatible Eluent Compositions and Additives *43*	
1.3.2.2	Optimizing Ion Source Conditions *44*	
1.3.2.3	Optimizing MS Detection *47*	
1.3.2.4	Verifying the Hyphenated Method *48*	
1.3.2.5	Method Development Supported by Software-based Parameter Variation *49*	
1.3.3	Transferring Established HPLC Methods to Mass spectrometry *50*	
1.3.3.1	Transfer of an Entire HPLC Method to a Mass Spectrometer *51*	
1.3.3.2	Selected Analysis of an Unknown Impurity – Solvent Change by Single-/Multi-Heartcut Techniques *52*	
	Abbreviations *54*	
	References *55*	
1.4	**Chromatographic Strategies for the Successful Characterization of Protein Biopharmaceuticals** *57*	
	Szabolcs Fekete, Valentina D'Atri, and Davy Guillarme	
1.4.1	Introduction to Protein Biopharmaceuticals *57*	
1.4.2	From Standard to High-Performance Chromatography of Protein Biopharmaceuticals *58*	
1.4.3	Online Coupling of Nondenaturing LC Modes with MS *62*	
1.4.4	Multidimensional LC Approaches for Protein Biopharmaceuticals *64*	
1.4.5	Conclusion and Future Trends in Protein Biopharmaceuticals Analysis *66*	
	References *67*	

1.5	**Optimization Strategies in HPLC for the Separation of Biomolecules** *73*
	Lisa Strasser, Florian Füssl, and Jonathan Bones
1.5.1	Optimizing a Chromatographic Separation *73*
1.5.2	Optimizing the Speed of an HPLC Method *77*
1.5.3	Optimizing the Sensitivity of an HPLC Method *79*
1.5.4	Multidimensional Separations (See also Chapter 1.1) *80*
1.5.5	Considerations for MS Detection (See also Chapter 1.3) *81*
1.5.6	Conclusions and Future Prospects *83*
	References *84*
1.6	**Optimization Strategies in Packed-Column Supercritical Fluid Chromatography (SFC)** *87*
	Caroline West
1.6.1	Selecting a Stationary Phase Allowing for Adequate Retention and Desired Selectivity *88*
1.6.1.1	Selecting a Stationary Phase for Chiral Separations *88*
1.6.1.2	Selecting a Stationary Phase for Achiral Separations *90*
1.6.2	Optimizing Mobile Phase to Elute all Analytes *93*
1.6.2.1	Nature of the Cosolvent *93*
1.6.2.2	Proportion of Cosolvent *94*
1.6.2.3	Use of Additives *96*
1.6.2.4	Sample Diluent *97*
1.6.3	Optimizing Temperature, Pressure, and Flow Rate *97*
1.6.3.1	Understanding the Effects of Temperature, Pressure, and Flow Rate on your Chromatograms *97*
1.6.3.2	Optimizing Temperature, Pressure, and Flow Rate Concomitantly *99*
1.6.4	Considerations on SFC–MS Coupling *100*
1.6.5	Summary of Method Optimization *101*
1.6.6	SFC as a Second Dimension in Two-Dimensional Chromatography *102*
1.6.7	Further Reading *102*
	References *103*
1.7	**Strategies for Enantioselective (Chiral) Separations** *107*
	Markus Juza
1.7.1	How to Start? *108*
1.7.2	Particle Size *109*
1.7.3	Chiral Polysaccharide Stationary Phases as First Choice *110*
1.7.4	Screening Coated and Immobilized Polysaccharide CSPs in Normal-Phase and Polar Organic Mode *113*
1.7.5	Screening Coated and Immobilized Polysaccharide CSPs in Reversed-Phase Mode *116*
1.7.6	Screening Immobilized Polysaccharide CSPs in Medium-Polarity Mode *119*

1.7.7	Screening Coated and Immobilized Polysaccharide CSPs under Polar Organic Supercritical Fluid Chromatography Conditions *120*	
1.7.8	Screening Immobilized Polysaccharide CSPs in Medium-Polarity Supercritical Fluid Chromatography Conditions *125*	
1.7.9	SFC First? *127*	
1.7.10	Are There Rules for Predicting Which CSP Is Suited for My Separation Problem? *127*	
1.7.11	Which Are the Most Promising Polysaccharide CSPs? *127*	
1.7.12	Are some CSPs Comparable? *129*	
1.7.13	"No-Go's," Pitfalls, and Peculiarities in Chiral HPLC and SFC *132*	
1.7.14	Gradients in Chiral Chromatography *133*	
1.7.15	Alternative Strategies to Chiral HPLC and SFC on Polysaccharide CSPs *133*	
1.7.16	How Can I Solve Enantiomer Separation Problems Without Going to the Laboratory? *135*	
1.7.17	The Future of Chiral Separations – Fast Chiral Separations (cUHPLC and cSFC)? *136*	
	References *138*	

1.8 **Optimization Strategies Based on the Structure of the Analytes** *141*
Christoph A. Fleckenstein

1.8.1	Introduction *141*	
1.8.2	The Impact of Functional Moieties *142*	
1.8.3	Hydrogen Bonds *143*	
1.8.4	Influence of Water Solubility by Hydrate Formation of Aldehydes and Ketones *146*	
1.8.5	Does "Polar" Equal "Hydrophilic"? *148*	
1.8.6	Peroxide Formation of Ethers *150*	
1.8.7	The pH Value in HPLC *151*	
1.8.7.1	Acidic Functional Groups *152*	
1.8.7.2	Basic Functional Groups *153*	
1.8.8	General Assessment and Estimation of Solubility of Complex Molecules *155*	
1.8.9	Octanol–Water Coefficient *157*	
1.8.10	Hansen Solubility Parameters *160*	
1.8.11	Conclusion and Outlook *162*	
	Acknowledgments *163*	
	References *163*	

1.9 **Optimization Opportunities in a Regulated Environment** *165*
Stavros Kromidas

1.9.1	Introduction *165*	
1.9.2	Preliminary Remark *165*	
1.9.3	Resolution *167*	

1.9.3.1	Hardware Changes	*167*
1.9.3.1.1	Preliminary Remark	*167*
1.9.3.1.2	UHPLC Systems	*168*
1.9.3.1.3	Column Oven	*168*
1.9.3.2	Improving the Peak Shape	*169*
1.9.4	Peak-to-Noise Ratio	*171*
1.9.4.1	Noise Reduction	*171*
1.9.5	Coefficient of Variation, VC (Relative Standard Deviation, RSD)	*171*
	References	*176*

Part II   Computer-aided Optimization *177*

2.1   Strategy for Automated Development of Reversed-Phase HPLC Methods for Domain-Specific Characterization of Monoclonal Antibodies *179*
Jennifer La, Mark Condina, Leexin Chong, Craig Kyngdon, Matthias Zimmermann, and Sergey Galushko

2.1.1	Introduction	*179*
2.1.2	Interaction with Instruments	*181*
2.1.3	Columns	*182*
2.1.4	Sample Preparation and HPLC Analysis	*183*
2.1.5	Automated Method Development	*184*
2.1.5.1	Columns Screening	*185*
2.1.5.2	Rapid Optimization	*186*
2.1.5.3	Fine Optimization and Sample Profiling	*188*
2.1.6	Robustness Tests	*188*
2.1.6.1	Selection of the Variables	*189*
2.1.6.2	Selection of the experimental design	*190*
2.1.6.3	Definition of the Different Levels for the Factors	*191*
2.1.6.4	Creation of the Experimental Set-up	*191*
2.1.6.5	Execution of Experiments	*192*
2.1.6.6	Calculation of Effects and Response and Numerical and Graphical Analysis of the Effects	*192*
2.1.6.7	Improving the Performance of the Method	*194*
2.1.7	Conclusions	*196*
	References	*196*

2.2   Fusion QbD® Software Implementation of APLM Best Practices for Analytical Method Development, Validation, and Transfer *199*
Richard Verseput

2.2.1	Introduction	*199*
2.2.1.1	Application to Chromatographic Separation Modes	*200*
2.2.1.2	Small- and Large-Molecule Applications	*200*

2.2.1.3	Use for Non-LC Method Development Procedures *200*
2.2.2	Overview – Experimental Design and Data Modeling in Fusion QbD *201*
2.2.3	Analytical Target Profile *201*
2.2.4	APLM Stage 1 – Procedure Design and Development *202*
2.2.4.1	Initial Sample Workup *202*
2.2.5	Chemistry System Screening *204*
2.2.5.1	Starting Points Based on Molecular Structure and Chemistry Considerations *205*
2.2.5.2	Trend Responses and Data Modeling *205*
2.2.6	Method Optimization *207*
2.2.6.1	Optimizing Mean Performance *207*
2.2.6.2	Optimizing Robustness In Silico – Monte Carlo Simulation *210*
2.2.6.3	A Few Words About Segmented (Multistep) Gradients and Robustness *213*
2.2.7	APLM Stage 2 – Procedure Performance Verification *214*
2.2.7.1	Replication Strategy *214*
2.2.8	The USP <1210> Tolerance Interval in Support of Method Transfer *214*
2.2.9	What is Coming – Expectations for 2021 and Beyond *216*
	References *217*

Part III Current Challenges for HPLC Users in Industry *219*

3.1 Modern HPLC Method Development *221*
Stefan Lamotte

3.1.1	Robust Approaches to Practice *222*
3.1.1.1	Generic Systems for all Tasks *222*
3.1.2	The Classic Reverse-phase System *225*
3.1.3	A System that Primarily Separates According to π–π Interactions *227*
3.1.4	A system that Primarily Separates According to Cation Exchange and Hydrogen Bridge Bonding Selectivity *227*
3.1.5	System for Nonpolar Analytes *228*
3.1.6	System for Polar Analytes *228*
3.1.7	Conclusion *230*
3.1.8	The Maximum Peak Capacity *230*
3.1.9	Outlook *231*
	References *231*

3.2 Optimization Strategies in HPLC from the Perspective of an Industrial Service Provider *233*
Juri Leonhardt and Michael Haustein

3.2.1	Introduction *233*
3.2.2	Research and Development *233*
3.2.3	Quality Control *234*

3.2.4	Process Control Analytics	235
3.2.5	Decision Tree for the Optimization Strategy Depending on the Final Application Field	237

3.3 Optimization Strategies in HPLC from the Perspective of a Service Provider – The UNTIE® Process of the CUP Laboratories 239
Dirk Freitag-Stechl and Melanie Janich

3.3.1	Common Challenges for a Service Provider	239
3.3.2	A Typical, Lengthy Project – How it Usually Goes and How it Should not be Done!	239
3.3.3	How Do We Make It Better? - The UNTIE® Process of the CUP Laboratories	241
3.3.4	Understanding Customer Needs	241
3.3.5	The Test of an Existing Method	242
3.3.6	Method Development and Optimization	243
3.3.7	Execution of the Validation	245
3.3.8	Summary	248
	Acknowledgments	249
	References	249

3.4 Optimization Strategies in HPLC 251
Bernard Burn

3.4.1	Definition of the Task	252
3.4.2	Relevant Data for the HPLC Analysis of a Substance (see also Chapter 1.8)	252
3.4.2.1	Solubility	252
3.4.2.2	Acidity Constants (pK_a)	257
3.4.2.2.1	Polarity of Acidic or Alkaline Substances (see also Chapter 1.8)	257
3.4.2.2.2	UV Spectra	259
3.4.2.2.3	Influence on the Peak Shape	259
3.4.2.2.4	Acid Constant Estimation	263
3.4.2.3	Octanol–Water Partition Coefficient	263
3.4.2.4	UV Absorption	270
3.4.2.5	Stability of the Dissolved Analyte	272
3.4.3	Generic Methods	278
3.4.3.1	General Method for the Analysis of Active Pharmaceutical Ingredients	278
3.4.3.2	Extensions of the Range of Application	279
3.4.3.3	Limits of this General Method	279
3.4.3.4	Example, Determination of Butamirate Dihydrogen Citrate in a Cough Syrup	279
3.4.3.4.1	Basic Data	279
3.4.3.4.2	Expected Difficulties	279
3.4.3.4.3	HPLC Method	279

3.4.3.4.4	Example Chromatogram *279*	
3.4.4	General Tips for Optimizing HPLC Methods *279*	
3.4.4.1	Production of Mobile Phases *284*	
3.4.4.1.1	Reagents *284*	
3.4.4.1.2	Vessels and Bottles *285*	
3.4.4.1.3	Measurement of Reagents and Solvent *285*	
3.4.4.1.4	Preparation of Buffer Solutions *286*	
3.4.4.1.5	Filtration of Solvents and Buffer *286*	
3.4.4.1.6	Degassing of Mobile Phases *287*	
3.4.4.2	Blank Samples *287*	
3.4.4.3	Defining Measurement Wavelengths for UV Detection *288*	
3.4.4.4	UV Detection at Low Wavelengths *288*	
3.4.4.4.1	Solvents *291*	
3.4.4.4.2	Acids and Buffer Additives *292*	
3.4.4.4.3	Drift at Solvent Gradients *294*	
3.4.4.5	Avoidance of Peak Tailing *295*	
3.4.4.6	Measurement Uncertainty and Method Design *302*	
3.4.4.6.1	Weighing in or Measuring *302*	
3.4.4.6.2	Dilutions *303*	
3.4.4.6.3	HPLC Analysis *304*	
3.4.4.6.4	Internal Standards *305*	
3.4.4.7	Column Dimension and Particle Sizes *305*	
	Reference *309*	

Part IV Current Challenges for HPLC Equipment Suppliers *311*

4.1 Optimization Strategies with your HPLC – Agilent Technologies *313*
Jens Trafkowski

4.1.1	Increase the Absolute Separation Performance: Zero Dead-Volume Fittings *314*
4.1.2	Separation Performance: Minimizing the Dispersion *314*
4.1.3	Increasing the Throughput – Different Ways to Lower the Turnaround Time *316*
4.1.4	Minimum Carryover for Trace Analysis: Multiwash *317*
4.1.5	Increase the Performance of What you have got – Modular or Stepwise Upgrade of Existing Systems *318*
4.1.6	Increase Automation, Ease of Use, and Reproducibility with the Features of a High-End Quaternary UHPLC Pump *319*
4.1.7	Increase Automation: Let your Autosampler do the Job *321*
4.1.8	Use Your System for Multiple Purposes: Multimethod and Method Development Systems *321*
4.1.9	Combine Sample Preparation with LC Analysis: Online SPE *322*

4.1.10	Boost Performance with a Second Chromatographic Dimension: 2D-LC (see also Chapter 1.1) *323*	
4.1.11	Think Different, Work with Supercritical CO$_2$ as Eluent: SFC – Supercritical Fluid Chromatography (see also Chapter 1.6) *324*	
4.1.12	Determine Different Concentration Ranges in One System: High-Definition Range (HDR) HPLC *325*	
4.1.13	Automize Even Your Method Transfer from other LC Systems: Intelligent System Emulation Technology (ISET) *326*	
4.1.14	Conclusion *327*	
	References *328*	
4.2	**To Empower the Customer – Optimization Through Individualization** *329*	
	Kristin Folmert and Kathryn Monks	
4.2.1	Introduction *329*	
4.2.2	Define Your Own Requirements *329*	
4.2.2.1	Specification Sheet, Timetable, or Catalogue of Measures *329*	
4.2.2.2	Personnel Optimization Helps to make Better Use of HPLC *331*	
4.2.2.3	Mastering Time-Consuming Method Optimizations in a Planned Manner *332*	
4.2.2.4	Optimizations at Device Level do not Always have to Mean an Investment *332*	
4.2.3	An Assistant Opens Up Many New Possibilities *333*	
4.2.3.1	If the HPLC System must Simply be able to do more in the Future *333*	
4.2.3.2	Individual Optimizations with an Assistant *333*	
4.2.3.3	Automatic Method Optimization and Column Screening *334*	
4.2.3.4	A New Perspective at Fractionation, Sample Preparation, and Peak Recycling *335*	
4.2.3.5	Continuous Chromatography, a New Level of Purification *336*	
4.2.4	The Used Materials in the Focus of the Optimization *337*	
4.2.4.1	Wetted vs. Dry Components of the HPLC *337*	
4.2.4.2	Chemical Resistance of Wetted Components *338*	
4.2.4.3	Bioinert Components *340*	
4.2.4.3.1	Material Certification *340*	
4.2.5	Software Optimization Requires Open-Mindedness *340*	
4.2.6	Outlook *341*	
4.3	**(U)HPLC Basics and Beyond** *343*	
	Gesa Schad, Brigitte Bollig, and Kyoko Watanabe	
4.3.1	An Evaluation of (U)HPLC-operating Parameters and their Effect on Chromatographic Performance *343*	
4.3.1.1	Compressibility Settings *343*	
4.3.1.2	Solvent Composition and Injection Volume *346*	
4.3.1.3	Photodiode Array Detector: Slit Width *348*	

4.3.2	"Analytical Intelligence" – AI, M2M, IoT – How Modern Technology can Simplify the Lab Routine *349*	
4.3.2.1	Auto-Diagnostics and Auto-Recovery to Maximize Reliability and Uptime *349*	
4.3.2.2	Advanced Peak Processing to Improve Resolution *350*	
4.3.2.3	Predictive Maintenance to Minimize System Downtime *353*	
	References *354*	
4.4	**Addressing Analytical Challenges in a Modern HPLC Laboratory** *355*	
	Frank Steiner and Soo Hyun Park	
4.4.1	Vanquish Core, Flex, and Horizon – Three Different Tiers, all Dedicated to Specific Requirements *356*	
4.4.2	Intelligent and Self-Contained HPLC Devices *362*	
4.4.3	2D-LC for Analyzing Complex Samples and Further Automation Capabilities (see also Chapter 1.1) *363*	
4.4.3.1	Loop-based Single-Heart-Cut 2D-LC *364*	
4.4.3.2	Loop-based Multi-Heart-Cut 2D-LC *364*	
4.4.3.3	Trap-based Single-Heart-Cut 2D-LC for Eluent Strength Reduction *366*	
4.4.3.4	Trap-based Single-Heart-Cut 2D LC–MS Using Vanquish Dual Split Sampler *367*	
4.4.4	Software-Assisted Automated Method Development *368*	
	Abbreviations *374*	
	References *374*	
4.5	**Systematic Method Development with an Analytical Quality-by-Design Approach Supported by Fusion QbD and UPLC–MS** *375*	
	Falk-Thilo Ferse, Detlev Kurth, Tran N. Pham, Fadi L. Alkhateeb, and Paul Rainville	
	References *384*	

Index *385*

Preface

The "HPLC world" is a diverse one – a lucky chance and challenging at the same time. Successful strategies for a "good" result can therefore look completely different.

The aim of this book is to provide interested colleagues successful strategies and proven ways for method development and optimization for all important areas in the field of HPLC and UHPLC. With this goal in mind, experts were invited to present their knowledge and experience in a practical and compact manner.

It was important to take both into account: Different challenges of a chromatographic nature, but also different framework conditions in everyday life. Only this enables a differentiated perspective and consequently a target-oriented approach: Hence, the authors are researchers or employees of well-known manufacturers, are service providers in industrial companies or private laboratories, or they have developed tools themselves.

Readers may find inspiration in the book for developing their individual optimization strategy.

I would like to thank my fellow authors for their time and commitment as well as WILEY-VCH, who made the realization of this project possible.

Blieskastel, June 2021 *Stavros Kromidas*

About the Book

The book is designed as a guide and does not have to be read in a linear fashion. The individual chapters represent self-contained modules; it is possible to "jump" at any time. In this way, we have tried to do justice to the book's character as a reference and hope that readers may benefit from this.

The book consists of four parts:

Part I: Optimization Strategies for Individual Problems

In the first part, optimization strategies for different analytes are discussed, from small molecules and chiral substances to biomolecules. Different modes of operation are also covered: LC–MS, 2D-HPLC, HILIC, SFC. Finally, optimization strategies based on structural info of the analytes are presented, and optimization possibilities in a regulated environment are discussed.

Part II: Computer-Aided Strategies (In silico Applications)

In Part II, concepts for computational method development for small molecules and biomolecules are presented, based on specific problems.

Part III: Users' Report

Service providers from two industrial companies and two private laboratories present their concepts for method development in Part III, based on the specifications and requests of internal and/or external customers.

Part IV: Manufacturers' Report

Employees of 5 well-known HPLC manufacturers show how the design of HPLC instruments, different tools, and the underlying philosophy support HPLC users in establishing the most efficient HPLC method possible, adapted to the problem at hand.

Part I

Optimization Strategies for Different Modes and Uses of HPLC

1.1

2D-HPLC – Method Development for Successful Separations
Dwight R. Stoll, Ph.D.

Gustavus Adolphus College, Department of Chemistry, 800 West College Avenue, St. Peter, MN 56082, USA

1.1.1 Motivations for Two-Dimensional Separation

Historically, much of the research devoted to multidimensional separations and their application to real analytical problems has been focused on dealing with complex samples. These have traditionally been described as containing hundreds or thousands of compounds and are often derived from natural sources such as plant extracts or body fluids (e.g. blood or urine). Increasingly, however, we observe that multidimensional separation can be exquisitely effective for dealing with samples containing analytes that are difficult to separate but are not complex by the traditional definition. Since this distinction can have a big impact on how one approaches method development, we start here by explicitly differentiating the two cases.

1.1.1.1 Difficult-to-Separate Samples

The difficulty associated with a separating a particular sample may originate from its sheer complexity (i.e. thousands of compounds). In this case relying on chromatographic separation alone will not be enough to fully separate the mixture, and some other source of selectivity will be needed (e.g. sample preparation, and/or selective detection such as mass spectrometry). However, it is now common to encounter samples that contain only a few compounds but are difficult to separate simply due to the high degree of similarity of the compounds in the mixture. For example, a mixture may only contain six compounds, but if two of those six compounds are enantiomers (**1a** and **1b**), then fully separating the mixture using a single column may be difficult even if the separation of compounds 2–5 from **1a**/**1b** is straightforward. Such situations are encountered more frequently now compared to the past, in part due to the development of small-molecule drugs with multiple chiral centers [1], and the increasing recognition of the importance of both the D- and L- enantiomers of amino acids ([2], see Chapter 1.7), for example.

Optimization in HPLC: Concepts and Strategies, First Edition. Edited by Stavros Kromidas.
© 2021 WILEY-VCH GmbH. Published 2021 by WILEY-VCH GmbH.

1.1.1.2 Complex Samples

As stated above, traditionally complex samples have been thought of as containing hundreds or thousands of different compounds. These samples often come from nature, but not always. For example, surfactants and polymers produced by chemical synthesis can result in highly heterogeneous mixtures of thousands of different compounds. Historically, the analysis of such samples by multidimensional chromatography has been mainly focused on so-called comprehensive methods of separation that yield a kind of global profile or "fingerprint" of the contents of the sample. However, in cases where only one or a few particular molecules in the sample are of importance to the analysis, simpler multidimensional separation methods such as heartcutting can be adequate, and even preferred.

1.1.1.3 Separation Goals

As is often discussed in the multidimensional separation literature, and below, the process of developing a multidimensional separation method is one full of compromises. For example, conditions that favor shorter analysis times do not lead to the best detection sensitivity, and vice versa. Therefore, it is important for the analyst to identify – at the very beginning of method development – what are the characteristics or performance metrics for the method that are most important to him/her. For example, if achieving baseline resolution of six critical pairs of analytes is critically important for the method to be successfully applied, then method development decisions should support this objective, even if it comes at the cost of increased analysis time, and/or lower detection sensitivity.

1.1.2 Choosing a Two-Dimensional Separation Mode

All two-dimensional separations can be executed either "offline" or "online." In the offline mode, one or more fractions of ^{1}D effluent are collected in some kind of storage device such as a set of vials or a wellplate. These fractions are then injected at some later time (minutes to years) into another LC system (i.e. the same LC system running different conditions from the ^{1}D separation, or a different LC system altogether), either with or without intermediate processing of these fractions. For example, in proteomics applications of 2D-LC, it is common to desalt the fractions, or dry them down by evaporation to remove organic solvent, before analysis by the ^{2}D separation [3]. In the online mode, fractions collected from the ^{1}D column are either processed immediately by direct injection into the ^{2}D column, or stored for a short time (seconds to hours) in some kind of device (typically capillary loops or sorbent-based traps) that is internal to the instrument. An example of an instrument configuration commonly used for this purpose is shown in Figure 1.1.1. In this case, the interface valve situated between the ^{1}D and ^{2}D columns has two positions. Switching between them changes the roles of loops 1 and 2 between collecting ^{1}D effluent and introducing the fraction of the ^{1}D effluent into the ^{2}D flow stream, effectively injecting that material into the ^{2}D column.

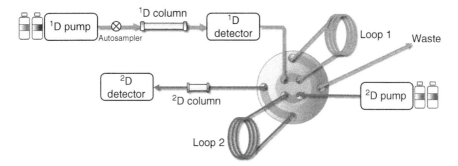

Figure 1.1.1 Illustration of an instrument configuration typically used for 2D-LC. *Source*: Dr. Gabriel Leme.

As commercially available equipment for 2D-LC separation has become more sophisticated and reliable, the trend in the industry has been to move away from offline separations because of challenges associated with implementation of offline separations for large numbers of samples, and with degradation and contamination of ^{1}D effluent fractions when they are handled external to the instrument [4]. Given this trend, I have chosen to focus entirely on online 2D-LC for the rest of this chapter. Readers interested in learning more about offline 2D-LC are referred to review articles dedicated to this topic [5, 6].

1.1.2.1 Analytical Goals Dictate Choice of Mode

Starting in the late 1970s, different groups began developing the modes of 2D-LC separation we have come to know as "heartcutting" and "comprehensive" [4, 7]. In the most recent decade, two additional modes have been developed, which are now known as "multiple heartcutting" and "selective comprehensive" 2D separations. Each of these four modes will be discussed in some detail in Section 1.1.2.2. At this point, though, I want to emphasize that choosing which separation mode you will use should be driven by the overall goals of the analysis. For example, if you have a complex sample and you want to learn as much as you can about that sample (i.e. identify hundreds of compounds), then the comprehensive mode of 2D separation will almost always be the best choice. However, if you are only interested in a few target compounds in the sample – even if the sample matrix is highly complex – then a more targeted mode of 2D separation such as heartcutting or multiple heartcutting will likely be the best approach. In practice, time spent on each ^{2}D separation is one of the most precious resources of the 2D-LC instrument, and allocating effort to ^{2}D separations that are not necessary to achieve the overall analytical goals of the analysis is costly (in terms of both time and supplies), wasteful, and adds unnecessary complexity to the method.

1.1.2.2 Survey of Four 2D Separation Modes

The vast majority of 2D-LC applications being developed today fit into one of the four modes of 2D separation illustrated in Figure 1.1.2. In the single-heartcut mode

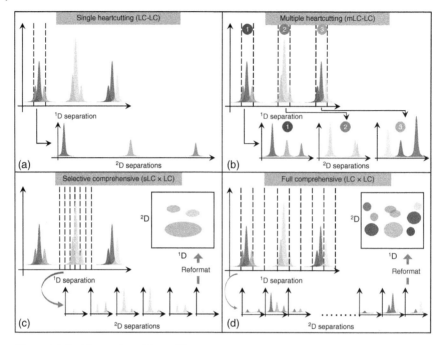

Figure 1.1.2 Illustration of four different modes of 2D-LC separation.

(A; LC–LC), a single fraction of ^{1}D effluent containing analytes of interest is captured at the outlet of the ^{1}D column and transferred to the ^{2}D column where this submixture of the original sample will be further separated if the separation mechanisms employed in the first and second dimensions are complementary. Perhaps the biggest advantage of the LC–LC mode is that the time that can be dedicated to separation of the ^{1}D effluent fraction in the second dimension is not strictly limited. This provides tremendous flexibility in terms of choosing parameters for the ^{2}D separation, including flow rate, column dimensions, and injection volume. The biggest disadvantage of LC–LC, however, is that the scope of the analysis is limited. We are restricted to the analysis of compounds that can be captured in a single fraction of ^{1}D effluent. Nevertheless, the LC–LC approach has been used to great effect in application areas ranging from identification of small-molecule pharmaceutical impurities [8] to the detection of drug metabolites in plasma [9].

The extreme opposite of LC–LC in terms of analytical scope is the comprehensive mode of 2D separation (D; LC×LC). As the illustration shows, in this case, fractions of ^{1}D effluent are collected and transferred – one at a time, in a regular, serial fashion – to the ^{2}D separation. Typically, this results in a long string of many (tens to hundreds) ^{2}D chromatograms collected in a single detector datafile. This long data string can then be parsed into pieces that correspond to individual ^{2}D separations and reformatted to produce a two-dimensional data array, which can then be viewed either as a contour map or a 3D surface rendering of the data. The advantages and disadvantages of the LC×LC approach are effectively the converse of those for the LC–LC approach. The main advantage is that the scope of the ^{2}D separation is as wide as the scope of the ^{1}D separation; the main disadvantage is that the time that can be dedicated to each ^{2}D separation is severely restricted because of the sheer number of fractions of ^{1}D effluent that must processed by the second dimension.

The two other modes illustrated in Figure 1.1.2 are hybrids of the LC–LC and LC × LC modes. In the case of multiple heartcutting (B; mLC–LC), one fraction of ^{1}D effluent is collected per region of the ^{1}D separation targeted for further separation, just like LC–LC, but this is repeated two or more times over the course of the 2D separation. Finally, in selective comprehensive separations (C; sLC × LC), multiple fractions of ^{1}D effluent are collected across a zone of interest in the ^{1}D separation, stored in loops or traps associated with the interface, and then injected one at a time into the ^{2}D column as in LC × LC separations. These hybrid modes are attractive in many situations because they capitalize on the strengths of LC–LC and LC × LC while mitigating their weaknesses. Specifically, mLC–LC and sLC × LC provide the analyst with a lot of flexibility in development and implementation of a 2D-LC method because they provide a means to decouple the process of collecting ^{1}D effluent fractions from the process of further separating those fractions in the second dimension [10].

1.1.2.3 Hybrid Modes Provide Flexibility

There are multiple ways that the added flexibility provided by mLC–LC and sLC × LC is practically useful, but I provide two examples for consideration here. First, sLC × LC is helpful for avoiding the so-called undersampling problem in 2D separations. Undersampling refers to the negative effect of collecting ^{1}D effluent fractions that are wider than about one-half of a ^{1}D peak width, whereby analytes eluting closely from the ^{1}D column are mixed back together in the sampling process. This effectively diminishes the performance of the first dimension of a 2D separation [11–13]. Overcoming this problem in the LC × LC mode is especially difficult when ^{1}D peaks are narrow (e.g. less than five seconds wide), but the sLC × LC mode can be used to manage this challenge by collecting several narrow (as low as one second or less) fractions over a particular region of interest in the ^{1}D separation. Second, sLC × LC can also be used to manage the volume of ^{1}D effluent that is injected into the ^{2}D column for each region of interest in the ^{1}D separation. A concrete example will make this benefit more clear. Suppose we have an existing 1D-LC separation running at 1 mL/min and we want to transfer a particular peak of interest to a ^{2}D column for further separation, and/or characterization by mass spectrometry. If the ^{1}D peak is 15 seconds wide, then the volume of the peak that has to be transferred is 250 μl. While it is certainly possible to transfer this volume in a single fraction, there are many cases where injecting such a large volume into the ^{2}D column will compromise the performance of the ^{2}D separation, especially when there is a mismatch between the mobile phases used in the ^{1}D and ^{2}D separations [14]. With sLC × LC, however, one could collect four fractions of the ^{1}D peak of interest instead of one, with each of the fractions being about 60 μl, and then these four fractions would be injected into the ^{2}D column one at a time [15]. This, of course, is likely to add time to the overall analysis and requires a more complex interface, but this kind of flexibility can be very valuable during method development.

1.1.3 Choosing Separation Types/Mechanisms

Once one has chosen which mode of 2D separation to use, the next most important decision involves choosing which two separation types will be used in the first and second dimensions of the 2D system.

1.1.3.1 Complementarity as a Guiding Principle

There has been much discussion in the 2D separations literature about the principle of "orthogonality" as it is related to choosing two separation types to use in a 2D separation. The reason for invoking orthogonality is that – from a purely theoretical standpoint – it is best if the retention patterns obtained from the ^{1}D and ^{2}D separations are not at all correlated [16]. However, I think it may be more practically relevant to think about the relationship between two separation types used in the 2D separation in terms of complementarity. To what extent does the separation type used in the second dimension complement the separation already used in the first dimension? A concrete chemical example will help make this point. Suppose we are separating a mixture of peptides that vary in both the total number of amino acids and the number of lysine residues such that the degree of positive charge on these peptides in solution varies as well (low pH). If we set up a 2D-LC system with reversed-phase C18 columns and low pH mobile phases in both dimensions, this will not yield an effective 2D separation because the ^{2}D separation does not add anything new to the separation in terms of selectivity. On the other hand, suppose we change the ^{1}D separation to cation-exchange (CEX) where peptides will elute mainly according to their degree of positive charge (low charge elutes first, high charge elutes last). Now, if we add a ^{2}D separation using a reversed-phase (RP) C18 column, this will nicely complement the ^{1}D separation because it will separate mainly according to the water solubility of the peptides (most soluble elutes first, least soluble elutes last). In this case, we can have two peptides that carry the same charge – and thus coelute from the CEX separation – but have very different water solubilities due to differences in the number and/or type of amino acids, and can be easily separated by the ^{2}D RP column.

Historically, a lot of effort has been dedicated to learning which separation types are most complementary for different sample types and applications. New users can use this prior research as a foundation for their own work. For some application areas, there are specific papers that illustrate the complementarity of different separation types for specific types of molecules such as peptides [17]. I encourage readers to consult databases of 2D-LC applications to quickly learn about which two separation types might be useful for their application (http://www.multidlc.org/literature/2DLC-Applications).

1.1.3.2 Pirok Compatibility Table

Unfortunately, we need to consider more than just the complementarity of the selectivities of two separation types used in a 2D-LC separation. Other factors such as the compatibility of the mobile phases used with each separation type are often important, and in fact can render useless a pairing of separation types that looks quite attractive from the point of view of selectivity. For example, pairing a normal-phase (NP) separation (i.e. bare silica stationary phase; hexane mobile phase) with an RP separation is attractive for some applications because the NP separation is dominated by adsorptive analyte–stationary phase interactions, whereas the RP separation is dominated by the partitioning of analytes into a bonded stationary phase. This difference in retention mechanisms can lead to highly complementary selectivities. However, we encounter a major practical difficulty in this case because the nonpolar organic solvent–rich mobile phases used for NP separations are not miscible with the water-rich mobile phases used for RP separations – at least not across a wide range of compositions. This difficulty has limited the use of some combinations of separation types such as NP–RP, although even in this case the miscibility problem can be managed by injecting very small volumes of ^{1}D effluent into large ^{2}D columns [18]. Pirok and Schoenmakers have summarized a lot of the knowledge in the 2D-LC field about which separation types work well together using the table shown in Figure 1.1.3. Combinations shaded with green colors are likely to work well, whereas combinations shaded with red colors present at least one major difficulty that will have to be managed if they are chosen for a 2D separation. Readers interested in a more detailed explanation of all of the information in this table are referred to the original paper of Pirok and Schoenmakers [19]. The table is also being updated as 2D-LC technology evolves; a current version can be found at our website (www.multidlc.org/megatable).

1.1.3.3 Measuring the Complementarity of Separation Types

Once we have made initial decisions about which two separation types to use in our 2D-LC separation, we need to assess the quality of the resulting separations. For more targeted separations, usually we are most interested in resolving one or more target compounds from the sample matrix, or from themselves. In this case, it is a straightforward matter to evaluate the extent to which the ^{2}D separation has resolved compounds that coeluted from the ^{1}D column, and thus the complementarity of the two separation types. For more comprehensive separations, we are usually interested in the extent to which the ^{1}D and ^{2}D separations – working together – spread the constituents of the sample out across the entire separation space. The need to assess this has led many groups to develop a variety of metrics, which have been critically discussed and compared in recent articles [20, 21]. In our own work, we have used the approach illustrated in Figure 1.1.4, which amounts to estimating the fraction of the available 2D separation space that is occupied by peaks by counting the number of bins that are occupied by peaks and dividing by the total number of available bins in the space. During method development, we adjust elution conditions in both dimensions to spread the peaks out as much as possible with the goal of reaching

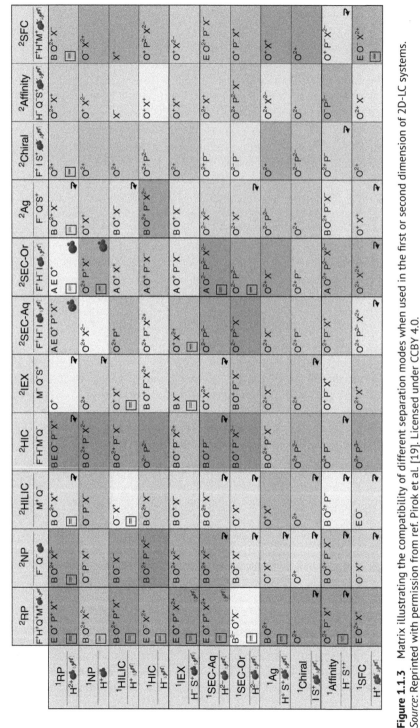

Figure 1.1.3 Matrix illustrating the compatibility of different separation modes when used in the first or second dimension of 2D-LC systems.
Source: Reprinted with permission from ref. Pirok et al. [19]. Licensed under CCBY 4.0.

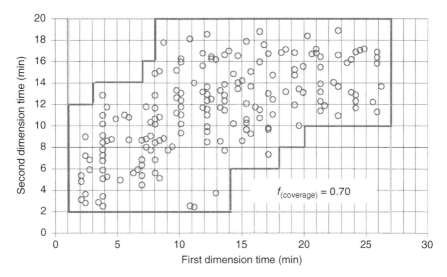

Figure 1.1.4 Illustration of a commonly used approach to estimate the fraction of 2D separation space that is occupied by peaks when analyzing real samples. The dark perimeter captures the bins where peaks can reasonably appear (outside of this perimeter lie the dead times and re-equilibration times associated with the first and second dimensions). *Source*: Adapted from Davis et al. [22].

100% coverage of the space. This is rarely achieved in practice, but some applications are able to achieve greater than 90% coverage [23].

1.1.4 Choosing Separation Conditions

Newcomers to multidimensional separations often describe being overwhelmed by the number of variables to consider when developing a 2D method, and the apparent complexity of 2D separations in general. While it certainly is true that there are more variables to consider in a 2D method than a 1D one, I also think it is possible to approach method development in a methodical way that should prevent being overwhelmed. Here, we consider method development from two distinct starting points: (i) cases where a 1D method already exists and we want to turn that into the ^{1}D separation of a 2D method; and (ii) cases where we want to start building a 2D method from scratch, such that we are not constrained by the parameters of an existing 1D method.

1.1.4.1 Starting with Fixed First-Dimension Conditions

Suppose we aim to develop a 2D method by adding a ^{2}D separation to an existing 1D method. This is common in cases where a 1D method is used to separate and determine the concentrations of several impurities in an active ingredient. If this method relies on UV detection, and suddenly a new peak appears, the first question will be – what is that new peak? This question can often be addressed quickly using a single-heartcut approach (LC–LC) and adding a ^{2}D separation that is compatible with MS detection. This enables capture of the new, unknown peak, transfer of that peak to the ^{2}D column, separation of the constituents of the fraction from themselves and/or the ^{1}D mobile phase buffer, and eventually elution and detection by mass spectrometry. Development of this type of method is straightforward in principle, but as soon as we start looking at the variables we quickly run into a problem. Suppose the existing 1D method runs at a flow rate of 1 ml/min, and the target unknown peak is five seconds wide at half-height. If we want to capture all of that peak, and add a little margin on the front and back sides of the peak to account for slight retention shifts from one analysis to the next, then we should collect a fraction of ^{1}D effluent that is about 20 seconds wide (8σ peak width, plus 10% on each side). Given the flow rate of 1000 µl/min, this corresponds to a fraction volume of about 333 µl. The volume of this fraction, which becomes the injection volume for the ^{2}D separation, is large compared to even the largest analytical columns (e.g. 150 mm × 4.6 mm i.d.) in common use today, and can have a serious negative impact on the quality of the ^{2}D separation. This is the first point where an important decision must be made. We have three main options; readers interested in more detail on each approach are referred to the recent literature [14, 24, 25]:

(1) Inject the entire fraction as it is into a large ^{2}D column, and hope for the best. This will work well in cases where the target analyte elutes from the ^{1}D column in effluent that is a weak solvent for the ^{2}D column (e.g. ^{1}D effluent that is 10/90 acetonitrile (ACN)/water injected into a ^{2}D mobile phase that is 20/80 ACN/water), but this is not always possible or convenient.

(2) Inject a smaller fraction of the ^{1}D peak to make the fraction volume more manageable (e.g. 40 µl). This approach will address the injection volume problem, but potentially introduces new problems: (i) one runs the risk of missing analytes that elute in the front or the tail of the target ^{1}D peak; and (ii) the repeatability of this approach will not be good because it will be sensitive to small shifts in ^{1}D retention time.

(3) Finally, the option that gives the analyst the most control is to use an active modulation technique to manage impact of the large fraction volume on the ^{2}D separation. The most common approaches are: (i) dilution of the ^{1}D effluent with weak solvent using an additional pump (e.g. diluting with water in the case of RP separation in the second dimension); (ii) Active solvent modulation using a valve to adjust the properties of the ^{1}D effluent fraction; and (iii) using a sorbent-based trap to separate the analytes of interest from the ^{1}D effluent matrix prior to injection into the ^{2}D column.

Once the ^{1}D effluent fraction volume and the approach to transfer this fraction to the second dimension are chosen, the dimensions and conditions for the ^{2}D column can be established. In the case of a LC–LC separation like the one described in this scenario, the analyst has a lot of flexibility in the second dimension, and most of the guidelines we use for choosing columns and conditions in 1D-LC apply here too. The column length, particle size, and flow rate are strongly dictated by the analysis time for the ^{2}D separation, and the theory to guide these choices is readily available and easily applied [26]. Second-dimension separations that are long and require high resolution will benefit from long columns (>100 mm). On the other hand, separations that are short, or in cases where only a crude ^{2}D separation is needed (e.g. desalting applications), short columns (<100 mm) will suffice. When using UV detection in the second dimension, larger diameter columns (>3.0 mm i.d.) and higher flow rates (>1 ml/min.) are desirable because these conditions will mitigate the effect of the large volume of ^{1}D effluent that is injected. On the other hand, when MS detection is used in the second-dimension, smaller-diameter columns (<2.1 mm i.d.) and lower flow rates (<1 ml/min) are desirable to avoid flooding the MS inlet with solvent.

1.1.4.2 Starting from Scratch – Flexible First-Dimension Conditions

When developing a new 2D-LC method from scratch, we don't have as many constraints to deal with as we do when adding a ^{2}D separation to an existing 1D-LC separation. In most cases, users leverage this freedom by reducing the flow rate of the ^{1}D separation so that the volume of ^{1}D effluent transferred to the ^{2}D column is not so large as in the scenario described above. This is especially helpful in the case of LC × LC separations, where ^{2}D columns tend to be small to facilitate fast ^{2}D separations. In this case, a typical ^{1}D flow rate is 50 μl/min. However, even for LC–LC and hybrid 2D-LC separations, dealing with a ^{1}D flow rate of 200 μl/min is a lot easier than dealing with 1000 μl/min.

1.1.4.3 Special Considerations for Comprehensive 2D-LC Methods

As illustrated in Figure 1.1.2d, in comprehensive 2D-LC separations, many fractions of ^{1}D effluent are collected and transferred to the ^{2}D column over the course of a single 2D-LC analysis. This means that all of the steps associated with a single ^{2}D separation happen tens or hundreds of times over the course of an LC × LC analysis, including one or two valve switches per fraction, and typically a change in mobile-phase composition to elute the injected components. Most useful LC × LC separations are on the order of 30 minutes to 3 hours in length, and the timescale of each ^{2}D separation is usually 15–120 seconds. This, in turn, requires that the second dimension is operated with short (<50 mm), narrow (2.1 mm i.d.) columns, and high (>1 ml/min) flow rates. Of course there are exceptions to these trends, but these conditions are typical. In our work, we have observed that both the design of the valve used to transfer ^{1}D effluent fractions to the second dimension and the pressure at which the ^{2}D column is operated can have a significant impact on the lifetime of ^{2}D columns, and therefore these parameters are worthy of serious consideration when developing LC × LC methods [23, 27].

1.1.4.4 Rules of Thumb

(1) When using RP separations in both dimensions, try to use the less retentive column in the first dimension. This will help mitigate the effect of the volume of ^{1}D effluent transferred to the ^{2}D column.

(2) Estimate the solvent strength of the ^{1}D effluent compared to the starting point in each ^{2}D separation. For RP separations, ensuring that the ^{1}D effluent contains about 10% less organic solvent will lead to good results more often than not. If this is not the case, then the organic content can be adjusted using active modulation approaches [24].

(3) A good starting point is to use columns of the same diameter for the ^{1}D and ^{2}D separations.

1.1.5 Method Development Examples

The scope of this chapter is such that there is not enough space to cover many aspects of method development in detail. I think the next best option is to talk through examples of 2D-LC methods and explain some of the key method development decisions. This will enable readers to use a similar thought process when developing their own methods.

1.1.5.1 Example 1 – Use of LC–LC to Identify an Impurity in a Synthetic Oligonucleotide

This first application example, which is from the work of Koshel et al., uses a single-heartcut 2D-LC method to identify an impurity in a synthetic oligonucleotide by mass spectrometry [28]. Historically, the identification of such impurities by MS has been difficult because the high concentrations of long-chain amine-based ion-pairing agents used for RP separations of oligonucleotides are not very compatible with MS detection. However, this 2D-LC method enables separation of the impurity from the target oligonucleotide by the ^{1}D column as shown in Figure 1.1.5a, and then separation of the impurity from the long-chain ion-pairing agent by the ^{2}D column prior to detection. In this case, the same RP column chemistry is used in both dimensions (i.e., ion-pairing reversed-phase (IPRP)). Although this is unusual in 2D-LC separations generally, it is effective here because the ion-pairing agents used in the first and second dimensions are chemically quite different (hexylamine vs. triethylamine plus hexafluoroisopropanol) and give rise to different selectivities. The important parameters for the separation are shown in Table 1.1.1. In this case the ^{1}D and ^{2}D flow rates are quite similar; this is possible because this is a single-heartcut method, and the speed of the ^{2}D separation is not critical. A single 50-μl fraction of ^{1}D effluent (0.5 minutes × 0.10 ml/min) is transferred to the ^{2}D column after dilution with aqueous eluent using the ^{2}D pump. This dilution step is important as it reduces the organic solvent content of the sample injected into the ^{2}D column, resulting in excellent peak shape in the ^{2}D chromatogram as shown in Figure 1.1.5b.

1.1.5.2 Example 2 – Comprehensive 2D-LC Separation of Surfactants

In this second example, we examine the use of LC × LC to separate Tween 20 (also known as polysorbate 20, or PS20), which is a complex mixture of fatty acid esters of ethoxylated sorbitan. The comprehensive mode of 2D-LC is clearly the best one

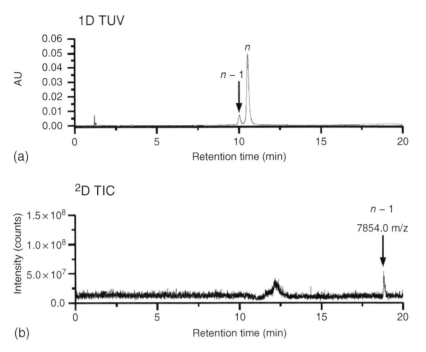

Figure 1.1.5 Identification of an impurity in a synthetic dye–labeled oligonucleotide using LC–LC with RP columns in both dimensions; a) UV detection; b) MS detection. The use of different ion-pairing conditions in the two dimensions provides complementary selectivities even though the column chemistry is the same, and makes the ^{2}D separation amenable to detection by mass spectrometry. *Source*: Koshel et al. [28]. © 2020, Elsevier.

to characterize a complex mixture like this. The separation shown in Figure 1.1.6, which is from the work of Vanhoenacker et al. [29], is a beautiful demonstration of how hydrophilic interaction (HILIC) and RP separations can nicely complement each other when used in a 2D separation format. Under the conditions of this separation, the ^{1}D HILIC column separates molecules primarily on the basis of the number of oxyethylene groups present, with more highly ethoxylated molecules eluting later in the first dimension. On the other hand, the ^{2}D RP column separates molecules primarily according to the length and number of fatty acid ester groups present. A summary of the conditions used for this separation is shown in Table 1.1.2. The ^{1}D column is operated at a flow rate that is considered low for a 2.1-mm i.d. column (30 μl/min), primarily so that the volume of ^{1}D effluent injected into the ^{2}D column is not too large (20 μl). This is particularly important here because each of these fractions will contain more than 75% organic solvent, whereas each ^{2}D separation starts with 60% organic solvent, and most of this organic is MeOH, which is a weaker solvent than ACN for RP separations. Increasing the ^{1}D flow rate, and therefore the volume of ^{1}D effluent injected into the ^{2}D column with each fraction, would cause severe broadening of ^{2}D peaks.

The ^{2}D column is relatively short (50 mm) and narrow (2.1 mm i.d.), and packed with small particles (1.8 μm), all of which favor the fast separations needed in the

Table 1.1.1 Summary of conditions for the IPRP–IPRP separation of dye-labeled oligonucleotides.

First dimension

Column: RP (C18); 50 mm × 2.1 mm i.d., 1.7 μm

Gradient elution from 66% to 76% B

    A solvent: 100 mM hexylamine acetate, pH 7

    B solvent: MeOH

Flow rate: 0.10 ml/min

Temperature: 60 °C

Injection volume: 1 μl

Analysis time: 30 min

Second dimension

Column: RP (C18); 50 mm × 2.1 mm i.d., 1.7 μm

Gradient elution from 20% to 70% B

    A solvent: 15 mM triethylamine, 400 mM hexafluoroisopropanol in water, pH 8

    B solvent: MeOH

Flow rate: 0.25 ml/min

Temperature: 60 °C

Analysis time (per fraction): 20 min

Interface

Dual six-port valves, with direct introduction of fraction into ^{2}D column

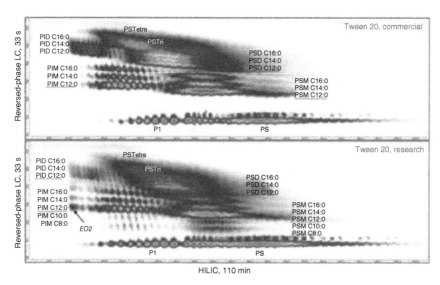

Figure 1.1.6 Separation of the constituents of commercial PS20 using LC × LC with HILIC and RP columns in the first and second dimensions, respectively. *Source*: Vanhoenacker et al. [29]. © 2020, MJH Life Sciences.

1.1.6 Outlook for the Future

Table 1.1.2 Summary of conditions for the HILIC × RP separation of PS20.

First dimension

Column: HILIC (bare silica); 100 mm × 2.1 mm i.d., 1.8 µm

Gradient elution from 98% to 0% B

    A solvent: 10 mM ammonium formate, pH 3 in 50/50 ACN/water

    B solvent: ACN

Flow rate: 30 µl/min

Temperature: 30 °C

Injection volume: 1 µl

Analysis time: 120 min

Second dimension

Column: RP (C18); 50 mm × 2.1 mm i.d., 1.8 µm

Gradient elution from 20% to 100% B

    A solvent: 5 mM ammonium formate, pH 3 in 50/50 MeOH/water

    B solvent: 5 mM ammonium formate, pH 3 in 50/50 ethylacetate/MeOH

Flow rate: 2 ml/min

Temperature: 80 °C

Analysis time (per fraction): 33 s

Interface

Dual-loop design with 20-µL loops, operated in cocurrent mode

---

second dimension. The column temperature is 80 °C, which facilitates the use of a flow rate that is considered high for this column (2 ml/min) [30]. Steep gradients are used so that the duration of each ^{2}D cycle is just 33 seconds, which helps minimize the undersampling effect and maintain separation of ^{1}D peaks provided by the ^{1}D column prior to sampling.

1.1.6 Outlook for the Future

At this point in time, we are observing a rapid proliferation of 2D-LC methods in application areas ranging from pharmaceutical analysis to food and beverage analysis. The community of 2D-LC users is broadening quickly in parallel with continuing advances in commercially available technology for 2D-LC and improved fundamental understanding of the technique. In the near future, I expect that we will see careful and critical evaluations of the robustness of the technology. This will be important to the wider use of the technique as various industries consider the possibility of implementing 2D-LC methods in regulated and quality control laboratories. The development of more streamlined approaches to method development (as opposed to approaches based on experience as discussed above), which will probably benefit substantially from the development of software tools for this purpose, is badly needed and will have a big impact if they can be realized in practice. Currently nearly all multidimensional LC separations are executed on a time basis, meaning that the ^{1}D separation and subsequent ^{2}D separation of ^{1}D effluent fractions are carried out serially. 2D separations executed spatially are very attractive and were explored decades ago [31], but faced serious engineering challenges at that time. These ideas are being revisited now primarily by the Schoenmakers group [32], and 3D separations in space could become practically useful for the most complex samples thanks to the explosive growth of 3D printing capabilities. Finally, one of the most active areas of research for 2D-LC will be focused on data analysis strategies as users look to improve the efficiency of extracting useful information from rich chromatograms like those shown in Figure 1.1.6.

Acknowledgment

I want to thank Dr. Gabriel Leme for the preparation of Figure 1.1.1.

References

1 Venkatramani, C.J., Al-Sayah, M., Li, G. et al. (2016). Simultaneous achiral-chiral analysis of pharmaceutical compounds using two-dimensional reversed phase liquid chromatography-supercritical fluid chromatography. *Talanta* 148 (February): 548–555. https://doi.org/10.1016/j.talanta.2015.10.054.

2 Hamase, K., Morikawa, A., Ohgusu, T. et al. (2007). Comprehensive analysis of branched aliphatic D-amino acids in mammals using an integrated multi-loop two-dimensional column-switching high-performance liquid chromatographic system combining reversed-phase and enantioselective columns. *Journal of Chromatography A* 1143 (1–2): 105–111. https://doi.org/10.1016/j.chroma.2006.12.078.

3 Wang, H., Sun, S., Zhang, Y. et al. (2015). An off-line high pH reversed-phase fractionation and nano-liquid chromatography–mass spectrometry method for global proteomic profiling of cell lines. *Journal of Chromatography B* 974 (January): 90–95. https://doi.org/10.1016/j.jchromb.2014.10.031.

4 Erni, F. and Frei, R. (1978a). Two-dimensional column liquid chromatographic technique for resolution of complex mixtures. *Journal of Chromatography A* 149 (February): 561–569. https://doi.org/10.1016/S0021-9673(00)81011-0.

5 Fairchild, J.N., Horváth, K., and Guiochon, G. (2009). Approaches to comprehensive multidimensional liquid chromatography systems. *Journal of Chromatography A* 1216 (9): 1363–1371. https://doi.org/10.1016/j.chroma.2008.12.073.

6 Horváth, K., Fairchild, J.N., and Guiochon, G. (2009). Optimization strategies for off-line two-dimensional liquid chromatography. *Journal of Chromatography A* 1216 (12): 2511–2518. https://doi.org/10.1016/j.chroma.2009.01.064.

7 Johnson, E.L., Gloor, R., and Majors, R.E. (1978). Coupled column chromatography employing exclusion and a reversed phase: a potential general approach to sequential analysis. *Journal of Chromatography* 149: 571–585. https://doi.org/10.1016/S0021-9673(00)81012-2.

8 Luo, H., Zhong, W., Yang, J. et al. (2017). 2D-LC as an on-line desalting tool allowing peptide identification directly from MS unfriendly HPLC methods. *Journal of Pharmaceutical and Biomedical Analysis* 137 (April): 139–145. https://doi.org/10.1016/j.jpba.2016.11.012.

9 Jayamanne, M., Granelli, I., Tjernberg, A., and Edlund, P.-O. (2010). Development of a two-dimensional liquid chromatography system for isolation of drug metabolites. *Journal of Pharmaceutical and Biomedical Analysis* 51 (3): 649–657. https://doi.org/10.1016/j.jpba.2009.09.007.

10 Groskreutz, S.R., Swenson, M.M., Secor, L.B., and Stoll, D.R. (2012). Selective comprehensive multi-dimensional separation for resolution enhancement in high performance liquid chromatography, Part I – principles and instrumentation. *Journal of Chromatography A* 1228: 31–40. https://doi.org/10.1016/j.chroma.2011.06.035.

11 Davis, J.M., Stoll, D.R., and Carr, P.W. (2008a). Effect of first-dimension undersampling on effective peak capacity in comprehensive two-dimensional separations. *Analytical Chemistry* 80 (2): 461–473. https://doi.org/10.1021/ac071504j.

12 Horie, K., Kimura, H., Ikegami, T. et al. (2007). Calculating optimal modulation periods to maximize the peak capacity in two-dimensional HPLC. *Analytical Chemistry* 79 (10): 3764–3770. https://doi.org/10.1021/ac062002t.

13 Murphy, R.E., Schure, M.R., and Foley, J.P. (1998). Effect of sampling rate on resolution in comprehensive two-dimensional liquid chromatography. *Analytical Chemistry* 70 (April): 1585–1594. https://doi.org/10.1021/ac971184b.

14 Stoll, D.R., Sajulga, R.W., Voigt, B.N. et al. (2017a). Simulation of elution profiles in liquid chromatography – II: investigation of injection volume overload under gradient elution conditions applied to second dimension separations in two-dimensional liquid chromatography. *Journal of Chromatography A* 1523 (July): 162–172. https://doi.org/10.1016/j.chroma.2017.07.041.

15 Pursch, M. and Buckenmaier, S. (2015). Loop-based multiple heart-cutting two-dimensional liquid chromatography for target analysis in complex matrices. *Analytical Chemistry* 87 (10): 5310–5317. https://doi.org/10.1021/acs.analchem.5b00492.

16 Giddings, J.C. (1984). Two-dimensional separations: concept and promise. *Analytical Chemistry* 56 (12): 1258A–1270A. https://doi.org/10.1021/ac00276a003.

17 Gilar, M., Olivova, P., Daly, A.E., and Gebler, J.C. (2005). Orthogonality of separation in two-dimensional liquid chromatography. *Analytical Chemistry* 77 (19): 6426–6434. https://doi.org/10.1021/ac050923i.

18 Dugo, P., Favoino, O., Luppino, R. et al. (2004). Comprehensive two-dimensional normal-phase (adsorption)–reversed-phase liquid chromatography. *Analytical Chemistry* 76 (9): 2525–2530. https://doi.org/10.1021/ac0352981.

19 Pirok, B.W.J., Gargano, A.F.G., and Schoenmakers, P.J. (2017). Optimizing separations in on-line comprehensive two-dimensional liquid chromatography.

Journal of Separation Science 41 (1): 68–98. https://doi.org/10.1002/jssc.201700863.

20 Gilar, M., Fridrich, J., Schure, M.R., and Jaworski, A. (2012). Comparison of orthogonality estimation methods for the two-dimensional separations of peptides. *Analytical Chemistry* 84 (20): 8722–8732. https://doi.org/10.1021/ac3020214.

21 Schure, M.R. and Davis, J.M. (2015). Orthogonal separations: comparison of orthogonality metrics by statistical analysis. *Journal of Chromatography A* 1414 (October): 60–76. https://doi.org/10.1016/j.chroma.2015.08.029.

22 Davis, J.M., Stoll, D.R., and Carr, P.W. (2008b). Dependence of effective peak capacity in comprehensive two-dimensional separations on the distribution of peak capacity between the two dimensions. *Analytical Chemistry* 80 (21): 8122–8134. https://doi.org/10.1021/ac800933z.

23 Stoll, D.R., Lhotka, H.R., Harmes, D.C. et al. (2019). High resolution two-dimensional liquid chromatography coupled with mass spectrometry for robust and sensitive characterization of therapeutic antibodies at the peptide level. *Journal of Chromatography B*: 121832. https://doi.org/10.1016/j.jchromb.2019.121832.

24 Pirok, B.W.J., Stoll, D.R., and Schoenmakers, P.J. (2019). Recent developments in two-dimensional liquid chromatography – fundamental improvements for practical applications. *Analytical Chemistry* 91 (1): 240–263. https://doi.org/10.1021/acs.analchem.8b04841.

25 Stoll, D.R., Shoykhet, K., Petersson, P., and Buckenmaier, S. (2017b). Active solvent modulation – a valve-based approach to improve separation compatibility in two-dimensional liquid chromatography. *Analytical Chemistry* 89 (17): 9260–9267. https://doi.org/10.1021/acs.analchem.7b02046.

26 Carr, P.W., Wang, X., and Stoll, D.R. (2009). Effect of pressure, particle size, and time on optimizing performance in liquid chromatography. *Analytical Chemistry* 81 (13): 5342–5353. https://doi.org/10.1021/ac9001244.

27 Talus, E.S., Witt, K.E., and Stoll, D.R. (2015). Effect of pressure pulses at the interface valve on the stability of second dimension columns in online comprehensive two-dimensional liquid chromatography. *Journal of Chromatography A* 1378 (January): 50–57. https://doi.org/10.1016/j.chroma.2014.12.019.

28 Koshel, B., Birdsall, R., and Chen, W. (2020). Two-dimensional liquid chromatography coupled to mass spectrometry for impurity analysis of dye-conjugated oligonucleotides. *Journal of Chromatography B* 1137 (January): 121906. https://doi.org/10.1016/j.jchromb.2019.121906.

29 Vanhoenacker, G., Steenbeke, M., Sandra, K., and Sandra, P. (2018). Profiling nonionic surfactants applied in pharmaceutical formulations by using comprehensive two-dimensional LC with ELSD and MS detection. *LCGC North America* 36 (6): 385–393.

30 Stoll, D., Cohen, J., and Carr, P. (2006). Fast, comprehensive online two-dimensional high performance liquid chromatography through the use of high temperature ultra-fast gradient elution reversed-phase liquid

chromatography. *Journal of Chromatography A* 1122 (1–2): 123–137. https://doi.org/10.1016/j.chroma.2006.04.058.

31 Guiochon, G., Gonnord, M.F., Zakaria, M. et al. (1983). Chromatography with a two-dimensional column. *Chromatographia* 17 (3): 121–124. https://doi.org/10.1007/BF02271033.

32 Davydova, E., Schoenmakers, P.J., and Vivó-Truyols, G. (2013). Study on the performance of different types of three-dimensional chromatographic systems. *Journal of Chromatography A* 1271 (1): 137–143. https://doi.org/10.1016/j.chroma.2012.11.043.

1.2

Do you HILIC? With Mass Spectrometry? Then do it Systematically

Thomas Letzel

Analytisches Forschungsinstitut für Non-Target Screening GmbH (AFIN-TS GmbH), Am Mittleren Moos 48, 86167, Augsburg, Deutschland

Hydrophilic interaction liquid chromatography (HILIC) was first mentioned by name in 1990 by Andrew Alpert [1]. Since then, there has been a regular up and down in the use of this technique. After years of stagnation, the use of HILIC is now, and has been for some time, increasing rapidly and hopefully remains like that.

Ultimately, the "downs" in the use of HILIC resulted from a sometimes-misinterpreted retention mechanism or the often-nonsystematic approach to method optimization. This chapter attempts to outline a possible systematic approach for method optimization. Separating (very) polar molecules, i.e. molecules that have a negative logD (logP) value and are thus, by definition, (very) water-soluble, separation with HILIC is an obvious choice.

An (incomplete) overview of the available separation techniques based on their analyte logD values is shown in Figure 1.2.1 for classical reversed-phase liquid chromatography (RPLC), polarity-extended RPLC, HILIC, and ion chromatography (IC). It is obvious that molecules in the physicochemical transition from "better octanol-soluble" to "better water-soluble" (i.e. in the LogD value range around 0) as well as "charged" and "uncharged" molecules can be separated with several techniques. Other separation techniques such as capillary electrophoresis (CE), gas chromatography (GC), and supercritical fluid chromatography (SFC) are not considered (among other things because they would also require different device technology) but are still available.

Looking at the substances to be separated, i.e. at their polarity, logD value, or their pK_a value, there are often strong reasons to retard or separate them with HILIC. This is especially true, if the logD value is in the (very) negative range and a charged molecule is present. Figure 1.2.2 presents toluene (log P-value of 2.5) and 4-hydroxybenzoic acid (log D (pH 7) value of −1.24 and a simultaneous pK_a value of 4.54) as a prominent example for comparing a separation with RPLC and with HILIC [2]. As can be seen, the nonpolar molecule toluene is generally better retarded on an RPLC column at a pH value of 7.0, and the polar molecule 4-hydroxybenzoic acid on a HILIC column.

Optimization in HPLC: Concepts and Strategies, First Edition. Edited by Stavros Kromidas.
© 2021 WILEY-VCH GmbH. Published 2021 by WILEY-VCH GmbH.

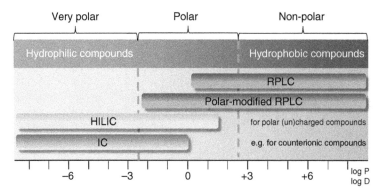

Figure 1.2.1 Scheme of the polarity ranges of separation techniques. Reversed-phase liquid chromatography (RPLC) is applicable for the separation of polar and nonpolar soluble analytes. Polar-modified RPLC allows extended separation deeper into the polar region. For the separation of very polar molecules, hydrophilic interaction liquid chromatography (HILIC) is typically used as well as ion chromatography for charged analytes.

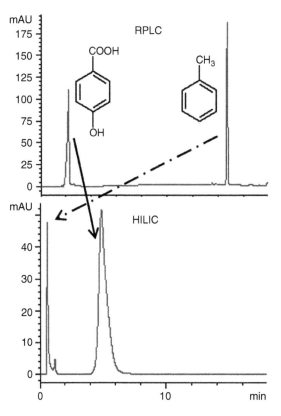

Figure 1.2.2 Toluene and 4-hydroxybenzoic acid as a prominent example for the comparison of separation by RPLC and by HILIC. *Source:* Bieber et al. [2].

For example, 4-hydroxybenzoic acid may have a logD value of 1.31 at pH 3, −1.24 at pH 7, or −4.14 at pH 12 (calculated with the logD Predictor [3] and a pK_a value of 4.54 for the acidic group [4]. Thus, it is crucial which pH conditions in the mobile phase are chosen and used in the HILIC (see also further down in the Section 1.2.2.3), because these conditions directly influence the polarity and the charge state of the molecule to be separated. In this case, RPLC could be used for the analysis of the molecule at a pH value of 3; HILIC is recommended at higher pH values. It is obvious that the pH value for the corresponding charge state of the molecules must be kept stable or – as described in the book chapter on "Gradient elution in HILIC" [5] – can be used to change the charge of molecules in pH gradients.

In this book chapter presented here, I do not describe the current state of knowledge about the separation mechanism of molecules taking into consideration the water layer on the particle surface (and thus the distribution equilibrium of the molecules between the mobile phase and the aqueous phase on the surface), the adsorption properties of the stationary phase, nor the possible electrostatic interaction sites of the stationary phase. Instead, the reader is referred to my earlier work [2, 6] or to the excellent work of colleagues [7–9]. Even if it is omitted here, knowledge of the mechanistic basics is essential to systematically develop a method. So please take the time to acquire this knowledge.

Assuming HILIC is short listed as a separation technique because of the appropriate polarity of the substances to be separated, have you made the decision to use it, and have you familiarized yourself (in the literature) with the separation mechanism? If yes, then there is nothing to prevent a successful implementation. In this chapter, the focus is also on mass spectrometric compatibility of the HILIC conditions, which ultimately simplifies method optimization a bit (since some parameters are predetermined by mass spectrometric detection; see below and under Section 1.2.3).

To ensure successful optimization, the steps or parameters should always be performed in the following order:

(1) Stationary phase
(2) Mobile phase with (a) Organic solvent (b) Salts (c) pH value
(3) Other settings or conditions specific to mass spectrometric detection

1.2.1 Initial Situation and Optimal Use of Stationary HILIC Phases

The variety of stationary HILIC phases available on the market has been increasing steadily for years, resulting in a wide range of materials (including ones specially developed for HILIC). However, without further knowledge of the properties of the substances to be separated (see the previous section), it is not easy for the user to make the right choice for the respective phase. Ultimately, the properties of the analytes determine whether charged or uncharged stationary phases should be used. However, before we deal with the functional surfaces of the HILIC columns in detail, let us first give a basic definition of HILIC stationary phases. Our definition of stationary HILIC phases states that typical HILIC phases with their "hydrophilic character should be able to accumulate water on the surface" [2]. However, not all phases are equally well suited for this purpose. First and foremost, the different hydrophilic properties of the materials influence the thickness of the water layer into which the analytes from the mobile phase can enter by a process called distribution. This, of course, has a direct influence on the retention of these analytes. Thus, the presence of charged and/or polar functional groups on the material surface leads to further stabilization of the water layer on the particle. On the other hand, these functional groups also have direct interactions, such as electrostatic interactions and/or hydrogen bonds, with corresponding analytes. It is therefore of great importance to choose the stationary phase based on the chemical properties of the analytes to be investigated.

HILIC phases are mostly based on classical silica particles and more recently also particles with polymer surfaces. In principle, two groups can be distinguished for the former (which will be dealt with in the following): silica particles with chemically bound polar functional groups, and silica particles with unchanged exposed silanol groups. A basic scheme of this classification is shown in Figure 1.2.3.

The first applications were originally carried out mainly with unmodified free silica particles. Early on, however, work was also carried out with semipolar phases as known from RPLC, such as cyano, diol, or amide phases (see also Table 1.2.1). The free silica phase is still one of the most popular materials today, but unfortunately it reacts very sensitively to small changes in the composition of the mobile phase.

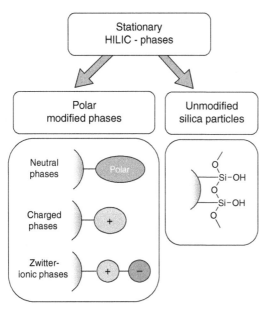

Figure 1.2.3 Classification of stationary HILIC phases (analogous to [2]). *Source:* Bieber et al. [2].

Table 1.2.1 Classification of functional HILIC phases and interaction possibilities with corresponding analytes.

Material and conditions	Electrostatic interaction	Dipole–dipole	Hydrogen bond
Silica, pH < 4.5	No	Yes	Donor + acceptor
Silica, pH > 4.5	Anionic → interacts with cations	Yes	Acceptor
Cyano	No	Yes	Acceptor
Diol	No	Yes	Donor + acceptor
Amide	Cationic → interacts with anions	Yes	Donor
Sufobetaine	Zwitterionic → weakly cationic and anionic	Yes	Acceptor
Phosphoryl choline			

These silica phases have free silanol groups on the surface, which are neutral at a pH below pH 4–5 and allow polar interactions such as dipole–dipole and hydrogen bonds with the analytes. If such a silica phase is operated at a pH above 4–5, it is deprotonated and can then also act as a cation exchanger, so that positively charged basic analytes are strongly retained.

In addition to original silica material, almost all column manufacturers offer semipolar phases from the original RPLC stock mentioned above or specially developed HILIC phases as part of their product range. A basic overview with detailed assessment of the polar interactions for the individual-phase materials can be found in Table 1.2.1.

Phases with polar functional groups are prepared by derivatization of the silanol groups on the surface. The derivatized phases are conventionally divided into neutral, charged, or zwitterionic phases, based on the charge state of the functional groups (see also Figure 1.2.3).

Neutral stationary phases contain polar functional groups, which are present in their neutral form in the typical HILIC range. Here, the retention of the analytes is mainly based on the above-mentioned accumulation in the aqueous surface and the hydrophilic interactions with the functional groups. Many of the stationary HILIC phases belong to this category.

The positively charged amino phase is also a stationary phase often used in HILIC. The functional group usually consists of an aminopropyl group with a primary amino group – which is positively charged - and the region of the so-called hydrocarbon linker ("spacer"). This type of stationary phase shows high affinity for anionic and acidic analytes. Due to the resulting electrostatic interaction, anions are strongly bound to the phase and separated via the anion exchange mechanism. However, amino phases can also be used very successfully for the separation of neutral polar molecules, which are strongly retained on the column due to the high hydrophilicity of these phases.

Zwitterionic stationary HILIC phases are already available in different versions and have the most universal use among all HILIC phases. Zwitterionic residues contain both a permanent positive and a permanent negative charge. These phases are very hydrophilic and at the same time contain moderate ion exchange properties. For this reason, these phases can be used for the separation of neutral, acidic, and basic organic molecules as well as for inorganic ions.

A good rule to select the most suitable HILIC phase is to consider that neutral analytes are usually less hydrophilic than the charged ones. In consequence, for neutral molecules, neutral, charged, or zwitterionic phases are well suited for the retention of the analyte. On the other hand, charged molecules are often too strongly retarded by their electrostatic attraction with the oppositely charged phase of HILIC materials that neutral and zwitterionic phases often give the better manageable results.

It is also worth mentioning here that mass spectrometric detection likewise depends on ions, which makes it therefore easily compatible with ionic analytes present in HILIC.

After selecting the correct stationary phase, the next step is to select and optimize the mobile phase.

1.2.2 Initial Situation and Optimal Use of the "Mobile" HILIC Phase

Depending on the stationary phase selected, the main characteristics of separation by HILIC are the properties of distribution, adsorption, and possibly electrostatic interaction. A common feature of all stationary phases, as described above, is the ability to form a water layer on the surface of the particles by water adsorption. Hydrophilic components are retained on the HILIC phases mainly because they preferentially accumulate in the formed water layer, whereas more hydrophobic components do not, or do so less well, and therefore elute earlier.

It is important to know what influence the following three conditions have on the properties of the mobile phase, which is responsible for the transport of the molecules. On the one hand, these properties influence the solubility (or distribution) of the analytes in the mobile phase (or in contact with the water layer), but they also have a direct effect on the water layer on the particle surface itself. In the following, the three parameters organic solvent, salt, and pH value are discussed individually.

1.2.2.1 Organic Solvent

Based on the foregoing, it can be stated that the first (but also most important) parameter for method development is the type and content of organic solvent in the mobile phase.

The type of solvent in HILIC is usually determined rather quickly (especially when using mass spectrometric detection). The polarity and thus the elution strength of the mobile phase can be controlled by using different organic solvents. Various water-miscible solvents that can be used contain alcohols or are cyclic ethers. In principle, the elution strength of the solvent in the mobile phase in HILIC separations decreases in the following sequence: methanol > ethanol > isopropanol > tetrahydrofuran > acetonitrile. The stronger the elution strength, the more difficult it is to dose the solvent and the less suitable it will be for the formation of the water layer, due to the protic properties of the alcohols. With the exception of methanol, these solvents are also of limited or no use in mass spectrometry. Due to its weak elution strength, nonprotic properties, and mass spectrometric compatibility, acetonitrile is the most commonly used solvent for HILIC separations and represents the best choice for beginning a method optimization. In other words, the organic solvent is already decided: acetonitrile.

What is more, the solvent with a strong elution strength (due to the necessity of a water layer on the HILIC surface) is also clear: water.

For separations in HILIC (as in RPLC), there is now the possibility to pump the mobile phase isocratically or in a gradient. In method optimization it is often useful to use a solvent gradient; on the one hand to determine the solvent composition for optimal elution of the molecules of interest and on the other hand to support peak sharpening or forcing. Thus, it is essential to first consider the distribution of the analytes between the mobile phase and the water layer at the surface of the stationary phase. In HILIC, gradient conditions start with a very high content of weak eluent, i.e. typically 95% acetonitrile (ranging from 90% to a maximum initial content of 98%). An example is shown in Figure 1.2.4. The subsequent elution of retarded molecules is achieved by increasing the strong eluent (water), typically to a value of up to 40%. At this water content in the running medium, the water layer on the particle surface can no longer be maintained, and most neutral, polar analytes will elute immediately. In this situation, the parameter "distribution between solvents" has no influence on the retention of the molecules anymore. Functional groups on the surface of the stationary phase will then only form adsorptive interactions with the analytes and, in the case of charged columns, also electrostatic interactions. In consequence, at this stage, coelution of many polar molecules and therefore often a poor chromatographic separation is seen. This fact must always be taken into consideration when very fast and steep solvent gradients are to be used. In RPLC applications this often works well, while in HILIC it does not, for the reasons mentioned above. However, due to the unfavorable conditions of the Van Deemter equation, the gradient should not be too flat either, because then the peaks become broadly (co)eluting and lose their shape. A generic elution gradient consisting of a water/acetonitrile mixture is shown in Figure 1.2.4 and can be used as a first step in detailed method development. In any case, this gradient will give a first impression of the elution properties of the substances to be separated and their polar properties.

As already mentioned, during operation, HILIC reacts much more sensitively to solvent changes than RPLC. For this reason, additional information and precautionary measures are given in the following that can help to optimize your method and which you should also take into account. First, the solvent of the sample to be injected is always a critical factor. Since high local water concentrations, as present

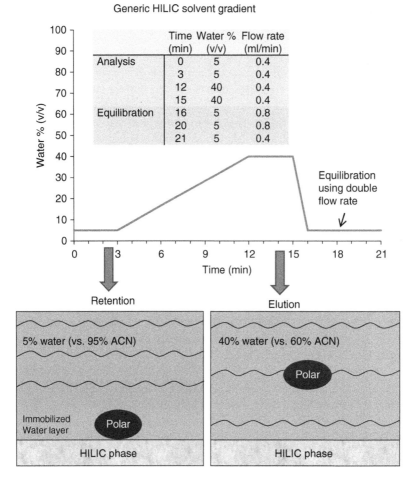

Figure 1.2.4 Typical HILIC gradient with course and composition of the mobile phase (above) and the corresponding interaction scheme of an analyte with the surface of the stationary phase. *Source:* Letzel [5]. © 2019, John Wiley & Sons.

in a water sample, destroy the fragile aqueous layer on the surface of the stationary phase, these samples used to be problematic. The solution is to keep the acetonitrile content as high as possible, depending on the analyte. Methanol addition up to 20% can also support the solubility of the analyte. In any case, the water content in the sample to be injected should be kept below 5% or the water sample should be introduced slowly. Slow injections of small volumes, for example, allow aqueous samples to be added under suitable conditions by mixing them with the mobile phase before they come in contact with the stationary phase and thus do not cause any change in the column surface. Alternative options for injecting water can be found at the end of the chapter. Second, wash solutions of the injection needle are also critical. Always use acetonitrile with 5–10% water without addition of acids or salts, because anything else may briefly change the composition of the solvent during injection, and this often leads to irreproducible changes in the retention time. Premixed solvents are also a problem, especially if different users employ different methods. Keep in mind that the retention with HILIC can change drastically even with small changes in the water (and salt) content. It is therefore imperative to prepare very detailed instructions for the laboratory, which show exactly how the solvent should be prepared. The use of industrially premixed acetonitrile/water solutions or solutions that are always prepared by the same protocol helps to increase the reproducibility of the analyses.

As shown in Figure 1.2.4, you should use a gradient to increase the flow rate when reconditioning the column, without exceeding the maximum pressure of the pump. In this way, you will avoid the perceived disadvantage of solvent gradients in HILIC due to the known "long" reconditioning time of the stationary phase. In contrast to RPLC, twenty to thirty instead of ten column volumes are required for sufficient reconditioning. By increasing the flow rate, the column can be reconditioned sufficiently quickly. Alternatively, it is possible to equilibrate for a shorter time at a constant flow rate. This leads to a so-called dynamic equilibrium and can also lead to reproducible results. Due to the incomplete equilibration, it must be ensured that the equilibration time and the flow rate between the runs always remain identical. However, for users that do not have extensive experience with the technique, this method is not advisable, since even small variations can have major effects on HILIC separation. Speaking of small changes and large effects, Figure 1.2.5 shows the two molecules 1-hydroxy- and 4-hydroxybenzoic acid (molecules 1 and 2, respectively) in their retention and performance on three stationary phases in an isocratic run at 90% acetonitrile (upper chromatograms) and at 80% acetonitrile (lower chromatograms). As can be clearly seen with 4-hydroxybenzoic acid, the retention time on all three phases is very strongly reduced, whereas only the cationically charged surface retains sufficient retention at 80% acetonitrile.

1.2.2.2 Salts

Salts are typically added to the mobile phase to control the electrostatic interactions between charged analytes and the stationary phase. Normally, this also leads to an improvement of the peak shape of the eluting analytes due to less pronounced peak tailing. Normally, only salts such as ammonium acetate and ammonium formate are used for mass spectrometric detection in HILIC, sometimes also ammonium bicarbonate. These salts dissolve very well in solvents with high acetonitrile contents and are volatile enough to be removed in the ionization process of the mass spectrometer. The use of less well-soluble salts, such as phosphates, or ion pair reagents, such as trifluoro acetic acid (TFA), should be avoided if possible.

The two salts ammonium acetate and ammonium formate cannot be used interchangeably, because ammonium formate produces a more acidic pH value in the mobile phase than ammonium acetate. However, this allows the use of a mixture of both salts in the separation at different pH values. Solutions containing

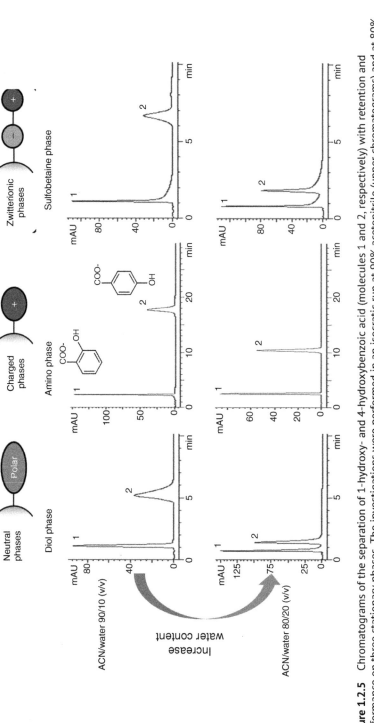

Figure 1.2.5 Chromatograms of the separation of 1-hydroxy- and 4-hydroxybenzoic acid (molecules 1 and 2, respectively) with retention and performance on three stationary phases. The investigations were performed in an isocratic run at 90% acetonitrile (upper chromatograms) and at 80% acetonitrile (lower chromatograms). *Source:* Modified from Greco and Letzel [6].

ammonium bicarbonate are not recommended because they are very unstable. Evaporation of carbon dioxide changes both the salt content and the pH value (up to alkaline).

In contrast to RPLC, the salt content in the HILIC solvent has a direct influence on the retention time of analytes (see also the chromatograms in Figure 1.2.6, where there is only a difference of 5 mM ammonium acetate in the solvent between the upper and lower values). The principal effect of increasing the salt content in the mobile phase is to reduce the electrostatic interactions in the case of charged molecules with charged or zwitterionic column materials (see also [2]). In the case of electrostatic attraction between analyte and stationary phase this leads to reduced retention, whereas in the case of electrostatic repulsion it leads to increased retention of the corresponding analyte.

The second effect of an increased salt content in the mobile phase is reflected in the layer thickness of the water layer on the surface of HILIC materials. The high organic solvent content forces the salt to preferentially accumulate in the water layer. An increasing salt concentration thus leads to an increasing thickness of the water layer, which is accompanied by a stronger retention of all polar analytes. For this reason, a significantly increased salt content for polar noncharged analytes causes increased retention even without the electrostatic interaction.

Salts used in HILIC separation also have additional important properties. One effect recently described by A. Alpert is the so-called salting out of cations with a very strong increase in the cosmotropic salinity and overlays the prolonged retention in the increased water layer [10]. According to the descriptions by Alpert and Jandera [11], higher overall salinity can lead to higher retention times as well as to lower retention times, which ultimately makes the simple interpretation and use of salt application more difficult. However, as these studies have mainly been conducted with nonvolatile salts, this effect is only briefly mentioned here.

However, it remains important to note that the salt content in the mobile phase should always be kept very stable and reproducible to avoid changing the properties of the water layer or the adsorptive interactions during a run, or at least to do so in a reproducible manner.

1.2.2.3 pH Value

As mentioned at the beginning of this chapter, the pH value of the mobile phase should always be adjusted according to the hydrophilicity of the analytes to be separated. In contrast to RPLC, the pH value of the mobile phase in HILIC should be deliberately chosen so that the analytes change into their ionic form. Since charged species are generally more polar than the neutral species, they can be retained longer, i.e. retarded better during separation with HILIC. At the same time, charged species can be easily detected by mass spectrometry.

It is important to note (also with regard to the pH effect of the silica phases) that the pH value in the solvent should be kept very stable and reproducible, so that the properties do not change during a run or do so in a reproducible manner.

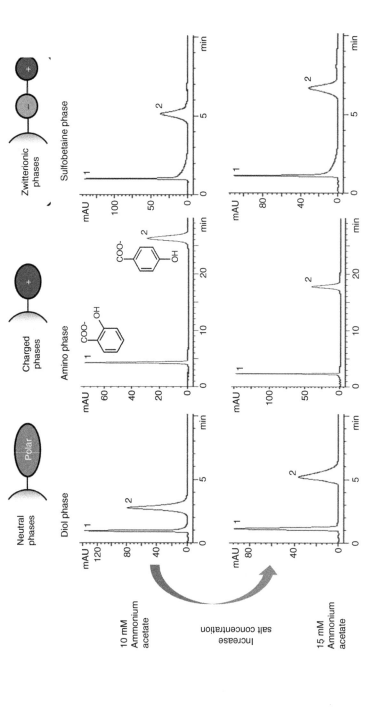

Figure 1.2.6 Chromatograms of the separation of 1-hydroxy- and 4-hydroxybenzoic acid (molecules 1 and 2, respectively) with retention and performance on three stationary phases. The investigations were carried out in an isocratic run with 10 mM ammonium acetate in the aqueous running medium (upper chromatograms) and 15 mM ammonium acetate (lower chromatograms). Source: Modified from Greco and Letzel [6].

1.2.3 Further Settings and Conditions Specific to Mass Spectrometric Detection

From the high acetonitrile content at the beginning of HILIC separations (and also during the course of the separation), it follows that mass spectrometry may very well be used for detection. This comes from the fact that – due to their low surface tension – the spray droplets produced by HILIC during ionization are effectively dried in the source and the molecules can be efficiently transported into the gas phase and – if they are charged – into the mass spectrometer.

There are now a number of examples of the versatile use of HILIC with mass spectrometric detection, even for very challenging polar molecules, more recently also with a perspective toward quantification. A selection of these reports is cited below and may easily be expanded. For example, there are HILIC studies on the separation of amino acids [12], metabolites in biological systems [13] as well as urine [14], antibiotics [15], lipids quantitatively in blood plasma [16], impurities of pharmaceutical substances [17], environmental chemicals [18, 19], flavonoids in plant extracts [20], as well as n-linked intact glycopeptides [21], proteins ([22], and intact proteins [23].

Finally, one should not forget or neglect the fundamental problems in using HILIC as a so-called front-end system for mass spectrometers. Problems arise because the often-sensitive detection or detection limits of mass spectrometers can be strongly influenced and disturbed by so-called column bleeding. A detailed study with 17 different columns (and their influence on 55 components) is highly recommended reading in this regard [24]. This study recommends consistent method optimization and determination of signal suppression effects for all substances to be analyzed. It is also essential to determine matrix effects of coeluting substances [18, 25].

Another noteworthy publication is also recommended when it comes to the effects of detection and peak shapes in HILIC-MS [26]. It shows that even small differences or fluctuations in the composition of the mobile phase and the injection solvents can have a major impact. This work again impressively demonstrates the necessity of accurate work in the preparation of mobile phase and injection solvent mentioned throughout this chapter.

Such studies are already bearing fruit, as solutions to the HILIC problems have recently been actively worked on. New developments such as FEED injection (Focused, Extended, Extra-control, Delay-volume free; personal communication from Agilent) will help make aqueous injections more effective in the future. In this new concept, after injection of a sample, the injection flow is mixed with the mobile phase via a "T-piece." This means that the analytes are not injected directly onto the column as an unmixed sample – in the worst case in the "wrong" solvent – but always diluted in the mobile phase. The lack of mixing in previous injection systems has led to the above-mentioned poor injectability of aqueous samples onto a HILIC phase. The final recommendation at this point is to inject the aqueous samples in only a small volume or to mix them with mobile phase (e.g. by means of a FEED injector) before they hit the column, or to apply them to the RPLC column in a serial RPLC–HILIC coupling [18].

1.2.4 Short Summary on Method Optimization in HILIC

If the physicochemical properties of the molecules to be analyzed allow separation by means of HILIC, the stationary phase should be selected with care, and then the mobile phase should be optimized step by step. The electrostatic interaction possibilities of some stationary phases make the separation appear more complicated at first, but even these can be assessed and put to use for the substances to be separated.

At this point, it is important to mention again that the composition of the mobile phase has a greater overall influence on the retention time than the other parameters. Therefore, the sequence for the HILIC method optimization (in combination with mass spectrometric detection) remains as follows:

(1) Stationary phase
(2) Mobile phase with (a) organic solvent (b) salts (c) pH value
(3) Other settings or conditions specific to mass spectrometric detection.

References

1 Alpert, A.J. (1990). Hydrophilic-interaction chromatography for the separation of peptides nucleic acids, other polar compounds. *Journal of Chromatography* 499: 177–196.
2 Bieber, S., Döteberg, H.-G., Greco, G. et al. (2016). Gradient, HILIC, SFC und trends in der HPLC. In: *HPLC-Tipps Band 3* (ed. S. Kromidas). Pirrot-Verlag 978-3-937436-58-6.
3 ChemAxon (2020); logD predictor. https://disco.chemaxon.com/calculators/demo/plugins/logd/ (accessed 5 May 2020).
4 PubChem. (2020). https://pubchem.ncbi.nlm.nih.gov/compound/135#section=Decomposition (accessed 20 May 2020).
5 Kromidas, S. (2019). *Gradient HPLC for Practitioners*. WILEY-VCH; ISBN: 9783527344086.
6 Greco, G. and Letzel, T. (2013). Main interactions and influences of the chromatographic parameters in HILIC separations. *Journal of Chromatographic Science* 51 (7): 684–693.
7 Buszewski, B. and Noga, S. (2012). Hydrophilic interaction liquid chromatography (HILIC) – a powerful separation technique. *Analytical and Bioanalytical Chemistry* 402 (1): 231–247.
8 Nováková, L., Havlíková, L., and Vlčková, H. (2014). Hydrophilic interaction chromatography of polar and ionizable compounds by UHPLC. *Trends in Analytical Chemistry* 63: 55–64.
9 Guo, Y. (2015). Recent progress in the fundamental understanding of hydrophilic interaction chromatography (HILIC). *Analyst* 140 (19): 6452–6466.

10 Alpert, A.J. (2018). Effect of salts on retention in hydrophilic interaction chromatography. *Journal of Chromatography. A* 1538: 45–53.

11 Jandera, P. (2011). Stationary and mobile phases in hydrophilic interaction chromatography: a review. *Analytica Chimica Acta* 692: 1–25.

12 Guerrasio, R., Haberhauer-Troyer, C., Mattanovich, D. et al. (2014). Metabolic profiling of amino acids in cellular samples via zwitterionic sub-2 μm particle size HILIC-MS/MS and a uniformly 13C labeled internal standard. *Analytical and Bioanalytical Chemistry* 406: 915–922.

13 Tang, D.-Q., Li, Z., Xiao-Xing, Y., and Choon, N.O. (2016). HILIC-MS for metabolomics: an attractive and complementary approach to RPLC-MS (review). *Mass Spectrometry Reviews* 35: 574–600.

14 King, A.M., Mullin, L.G., Wilson, I.D. et al. (2019). Development of a rapid profiling method for the analysis of polar analytes in urine using HILIC–MS and ion mobility enabled HILIC–MS. *Metabolomics* 15: 17.

15 Kahsay, G., Song, H., Van Schepdael, A. et al. (2014). Hydrophilic interaction chromatography (HILIC) in the analysis of antibiotics. *Journal of Pharmaceutical and Biomedical Analysis* 87: 142–154.

16 Lange, M. and Fedorova, M. (2020). Evaluation of lipid quantification accuracy using HILIC and RPLC MS on the example of NIST® SRM® 1950 metabolites in human plasma. *Analytical and Bioanalytical Chemistry* 412: 3573–3584.

17 Machairas, G., Panderi, I., and Geballa-Koukoula, A. (2018). Development and validation of a hydrophilic interaction liquid chromatography method for the quantitation of impurities in fixed-dose combination tablets containing rosuvastatin and metformin. *Talanta* 183: 131–141.

18 Bieber, S., Greco, G., Grosse, S., and Letzel, T. (2017). RPLC-HILIC and SFC with mass spectrometry: polarity-extended organic molecule screening in environmental (water) samples. *Analytical Chemistry* 89: 7907–7914.

19 Salas, D., Borrull, F., Fontanals, N., and Marcé, R.M. (2017). Hydrophilic interaction liquid chromatography coupled to mass spectrometry-based detection to determine emerging organic contaminants in environmental samples. *Trends in Analytical Chemistry* 94: 141–149.

20 Sentkowska, A., Biesaga, M., and Pyrzynska, K. (2016). Application of hydrophilic interaction liquid chromatography for the quantification of flavonoids in *Genista tinctoria* extract. *Journal of Analytical Methods in Chemistry*: 3789348. http://dx.doi.org/10.1155/2016/3789348.

21 Shu, Q., Li, M., Shu, L. et al. (2020). Large-scale identification of N-linked intact glycopeptides in human serum using HILIC enrichment and spectral library search. *Molecular & Cellular Proteomics* 19 (4): 672–689.

22 Fekete, S. and Guillarme, D. (2014). Ultra-high-performance liquid chromatography for the characterization of therapeutic proteins. *Trends in Analytical Chemistry* 63: 76–84.

23 Camperi, J., Combès, A., Fournier, T. et al. (2020). Analysis of the human chorionic gonadotropin protein at the intact level by HILIC-MS and comparison with

RPLC-MS. *Analytical and Bioanalytical Chemistry* 412: 4423–4432. https://doi.org/10.1007/s00216-020-02684-8.

24 Schulze, B., Bader, T., Seitz, W. et al. (2020). Column bleed in the analysis of highly polar substances: an overlooked aspect in HRMS. *Analytical and Bioanalytical Chemistry* https://doi.org/10.1007/s00216-020-02387-0 412: 4837–4847.

25 Mueller, K., Zahn, D., Froemel, T. et al. (2020). Matrix effects in the analysis of polar organic water contaminants with HILIC-ESI-MS. *Analytical and Bioanalytical Chemistry* 412: 4867–4879. https://doi.org/10.1007/s00216-020-02548-1.

26 Boulard, L., Dierkes, G., and Ternes, T. (2018). Utilization of large volume zwitterionic hydrophilic interaction liquid chromatography for the analysis of polar pharmaceuticals in aqueous environmental samples: benefits and limitations. *Journal of Chromatography. A* 1535: 27–43.

1.3

Optimization Strategies in LC–MS Method Development
Markus M. Martin

Chromatography and Mass Spectrometry Division, Thermo Fisher Scientific, HPLC Product Management, Dornierstr. 4, 82110, Germering, Germany

1.3.1 Introduction

With mass spectrometry-based detectors having left the innovation-savvy research laboratories and penetrating more and more routine analyses in quality control laboratories, the need for newly developed high-performance liquid chromatography–mass spectrometry (HPLC–MS) methods is also increasing. Today's mass spectrometers for routine applications have made considerable steps forward in usability and left the reputation of a diva for experts behind, which makes them much easier to use for both skilled and less experienced analytical chemists. However, there is still a whole universe of defined liquid chromatography (LC) methods in the various pharmacopoeias of the world that have been developed and validated decades ago and that still describe HPLC methods that are not compatible with the requirements of mass spectrometry. In the daily laboratory work, a user is typically confronted with two different types of tasks: it is either the development and validation of new HPLC–MS methods, or the challenge of making an established method compatible with mass spectrometry. Both aspects shall be discussed here with a focus on approaches and optimization potential. What goes beyond the scope of optimization considerations and thus is discussed in separate literature [1] are basic starting questions like which HPLC–MS systems and methods are best suited for which analytical goal and how to use them in the best possible way. This chapter is limited to electrospray ionization (ESI) and chemical ionization at atmospheric pressure (APCI) as the most common ionization methods of the HPLC–MS coupling, although the general ways of thinking apply to other ionization and coupling modes as well.

1.3.2 Developing New Methods for HPLC–MS Separations

The approach to the design of a new LC–MS separation method does not substantially differ from that for conventional LC methods. However, it is important to develop a UHPLC separation with a special focus on the dedicated requirements that are dictated by mass spectrometry. In parallel, the parameters for the mass spectrometer ion source, ion transfer optics, and the respective mass analyzer are optimized for the target analytes. Usually, these two steps are initially carried out separately from each other. Then, the UHPLC is coupled with the mass spectrometer to check if both sets work correctly together; after determining the influence of matrix effects, the final UHPLC–MS method is then subject to a suitability or validation.

This results in the following order of steps:

(1) Developing the LC separation under MS-compatible conditions (offline)
(2) Optimizing mass spectrometric source and instrument settings (offline)
(3) Verifying the mass spectrometric settings after coupling and determining matrix effects
(4) Finally checking the full method followed by method validation

The individual steps will be discussed in more detail in the following sections. How these detailed optimization steps can be merged and automated if necessary is briefly outlined in Section 1.3.2.5.

1.3.2.1 Optimizing the LC Separation

The optimization of an HPLC separation in general has already been discussed in various chapters of this book (see Chapters 1.9 and 3.1); therefore, here we will exclusively take a look at the aspects that are primarily important from an LC–MS analysis perspective.

1.3.2.1.1 Optimizing for Sensitivity and Limit of Detection – Which Column to Take?

A major difference in LC–MS analysis compared to conventional HPLC methods is the maximum flow rate that can be used for a separation. Responsible for this flow limitation is the ion source of the mass spectrometer; it is capable to only evaporate a limited amount of liquid mobile phase into the gas phase and to remove it quantitatively from the analyte molecules to keep the vapor away from the inner, high-vacuum sections of the mass spectrometer. ESI, which is used in over 80% of all published online LC–MS couplings (the remaining share is dominated by APCI with over 15%), can be supported by pneumatically nebulizing the LC effluent using an inert gas like nitrogen; for best possible ionization efficiency, this principle operates with flow rates of approximately 300 µl/min. All commercial ESI sources (except for nanospray sources) manage easily also higher flow rates of up to 1 ml/min and above. Numerous publications underline that the sensitivity of ESI-based detection methods only starts to suffer when it is no longer possible to effectively remove the amount of mobile phase; hence, there is not "the one" magical top limit for flow rates in LC–MS analyses. However, many LC–MS users still prefer a conservatively low LC flow rate of 300–500 µl/min instead of the optimal linear velocity demanded by the LC column particle size if the LC column was to be operated in the minimum of the van Deemter curve. One result out of this is that the inner diameter (ID) of the column for LC–MS applications does not exceed 2.1 mm to ensure that the LC separation runs as efficiently as possible. A UHPLC column of this diameter and an average particle size of the phase material dp of 2 μm has its optimal linear speed in the range of 5.5 mm/s, which translates into a flow rate of 1.1 ml/min. In practice, many LC–MS users reduce this flow rate in order not to overload the drying performance of the ion source, even if the LC separation then runs with a still-elevated impact of the longitudinal diffusion (B-term), leading to unwanted excessive band dispersion. Reducing the column inner diameter down to 1 mm is controversially discussed for several reasons though, even if many users expect a lower detection limit from the smaller inner diameter due to the sample molecules being diluted in a smaller peak volume. The pro and con of this approach is discussed in more detail in the literature [2]. In a nutshell, concentration-sensitive detectors (particularly ESI is described as a primarily concentration-sensitive ion formation process) provide the same detection limit on a 2.1-mm ID separation column as on a 1-mm ID column if following the scaling laws of chromatography; these teach that on a 2.1-mm ID column, 4.4 times the sample volume of a 1.0-mm ID column needs to be injected to run the separation at the same conditions of the adsorption isotherm. From a technical point of view, however, separations on 2.1-mm ID columns are more robust than those on 1-mm ID columns, which is due to both the column material used and the limited suitability of UHPLC systems for very low flow rates of less than 50 μl/min. This leaves us with only two reasons for using 1-mm ID separation columns: Either the sample amount is limited so that injection volumes of 4–5 μl per measurement consume too much sample, or the MS ion source restricts us actually to flow rates of about 100 μl/min or less. However, the latter rarely happens in practice. Commercial ESI sources with pneumatically assisted nebulization usually handle flow rates of up to 600–700 μl/min well without a notable loss of sensitivity. In principle, APCI is even more compatible with higher flow rates, as a minimum flow rate of approximately 200 μl/min is required to evaporate enough eluent for creating the ionizing reagent gas in the APCI source.

Take-home messages:

- If LC separations are correctly scaled over different column diameters, there will be no significant gain in sensitivity with concentration-sensitive detectors such as ESI-MS
- For most LC–MS separations, internal column diameters of 2.1 mm are sufficient; even UHPLC-phase materials with dp ≤ 2 μm can generally still be operated at or near their optimum efficiency
- Internal diameters of less than 2.1 mm are only without a true alternative in two cases – significantly limited sample volumes and/or the application of drastically reduced flow rates of less than 100 μl/min due to technical boundary conditions such as the design of the MS ion source

1.3.2.1.2 Optimizing Resolution vs. Sample Throughput

As with the development and optimization of a conventional HPLC method, the key initial question is about the goal of a separation method with MS detection.

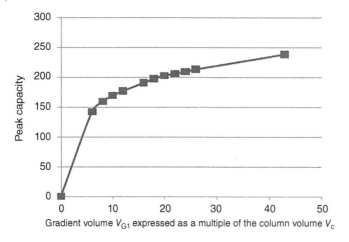

Figure 1.3.1 Peak capacity vs. gradient volume, expressed as a multiple of the LC column volume; measured for a 250 × 2.1 mm column with a particle size dp = 2.2 μm and a flow rate $F = 670\,\mu\text{l/min}$.

Each method has to find the right balance between as much analytical information as needed and the shortest possible analysis time, thus always leading to a trade-off between sample throughput and separation quality. Fast separations go hand in hand with a certain loss of analytical information, reflected by chromatographic resolution. This dilemma is nicely visualized when looking at how the peak capacity – in gradient separations defined as the quotient of the gradient time and the average peak baseline width – changes with the gradient volume [3], expressed as the product of gradient time and flow rate.

Figure 1.3.1 illustrates for a UHPLC application example how the peak capacity changes as a function of the gradient volume, here for better standardization expressed as a multiple of the column volume. The plot nicely shows that the peak capacity of a gradient separation initially rises steeply with increasing gradient volume before running into saturation for large gradient volumes. As described by the gradient volume concept, fast separations that are achieved by small gradient volumes on short, small-volume columns are therefore limited in separation performance and thus particularly suitable for screening experiments. The mass spectrometer offers an additional benefit here as it adds another separation dimension, which allows to detect and – with certain limitations – to quantify potentially coeluting substances individually. Depending on the mass spectrometer used, a complete baseline separation is therefore not mandatory for quantitation in MS detection, unlike in conventional HPLC. Highest possible peak capacities, however, require larger gradient volumes, longer columns, and thus longer run times.

The latter case can be managed well by mass spectrometry. However, in the case of rapid screening separations, users typically face two restrictions: First, too many coeluting substances may interfere with each other during the ionization in the ion source (*competitive ionization*) and thus negatively impact the quantitative result. Second, almost every mass spectrometer requires a certain period (*cycle time*) for each individual mass-to-charge ratio to be measured, which limits the data rate. Fast UHPLC separations can still overstretch the speed of many mass spectrometers: Depending on the device type, peak widths of five seconds and less can be too small to scan all desired mass signals quickly enough to generate the 25 data points per peak on average that are usually required for a statistically reliable quantitation. In most cases, we have to accept – and can still live with – around 10 data points per peak for quantitation. Nevertheless, while developing the method, it may be indicated to generate less efficient separations (i.e. broader peaks) and more resolution by selectivity than the chromatography would in principle enable, just to allow the mass spectrometry to keep pace with data generation. A common example is the quantitation in complex samples by triple quadrupole mass spectrometers, which provide the required detection selectivity by tandem MS in selected reaction monitoring (SRM) mode. The more SRM transitions the mass spectrometer has to process, the longer the cycle time and correspondingly lower the data rate are. Slowing down the chromatography can here measurably improve the quantitative result, depending on the MS type and device generation.

Take-home messages:

- High-resolution screening analyses with fast and/or steep gradients and low peak capacity might be too fast for the cycle time of a mass spectrometer
- If needed, sacrifice separation efficiency and artificially increase resolution to allow adequate MS detection
- Under the given detection conditions, the average cycle time of the mass spectrometer should allow at least 15, but not less than 8–10 data points per peak for quantitation

1.3.2.1.3 MS-Compatible Eluent Compositions and Additives

An MS-compatible mobile phase requires *all* components of the mobile phase to be highly volatile. Virtually all solvents fulfill this requirement in RP chromatography: water as the LC solvent with the highest evaporation enthalpy is very well compatible with ESI and APCI processes and is often inherently necessary for a properly working ionization process. Organic solvents in return enhance the drying of the eluate during the spray process due to their higher vapor pressure and by reducing the surface tension of the liquid droplets during the nebulization, thus promoting the evaporation of the solvent molecules. For a more in-depth assessment of this topic, it should be referred to the literature [4].

The demand for high volatility must also apply to all mobile-phase additives. Additives that form nonvolatile salts lead to increased suppression of ion formation in the ion source (*ion suppression*) and result in severe contamination of the MS source with salt deposits. This requires frequent and intensive cleaning of the ion source; despite various design improvements by MS manufacturers, it also leads to a rapidly noticeable decrease in detection sensitivity. For pH adjustment and buffering as well as to improve the chromatography, HPLC classics such as phosphates, borates, and generally salts containing alkali and earth alkaline ions should be avoided. Instead, volatile organic acids, bases, and their ammonium salts can be used. Acidic pH values are preferably generated using formic acid, acetic acid, or trifluoroacetic acid, typically applied in concentrations between 0.05% and 0.1% by volume. Aqueous ammonia solutions or alkylamines such as triethylamine are useful for a basic pH. Ammonium salts such as ammonium formate, ammonium acetate, or carbonate are recommended for buffering.

The use of oxidizing additives should also be avoided: chlorides as eluent additives lead, e.g. not only to ion suppression; under the influence of the electrospray, they can also form chlorine and thus chemically modify the analytes by chemical oxidation [5] and corrode components such as the spray capillary over time. The use of volatile ionic detergents is also not recommended, as they deposit on the surfaces of the ion optics and the mass analyzer when entering the mass spectrometer, causing device malfunctions in the medium term. It should be noted here that the additives commonly used in MS, due to their physicochemical properties differing sometimes significantly from conventional LC modifiers, often lead to selectivities that are different from what we might be used to from classical HPLC (cf. Section 1.3.3.1). This is particularly true if they impact secondary selectivity-controlling factors like the ionic strength or the tendency to form ion pairs with polar substances.

In addition to the type of eluent additive, its effective concentration also matters. LC methods usually benefit from a high additive concentration, as this can provide high buffer capacities or suppress electrostatic interactions between the analyte and the stationary phase more effectively. However, high additive concentrations induce a high conductivity, which leads to high currents in the ion source during the ionization process, often resulting in technical problems or spray instability. In practice, it is a good advice to add additives, if necessary at all, in a concentration not higher than 25–30 mmol/l and, if possible, to avoid multiply charged additive ions.

Take-home messages:

- Use an "MS-friendly" phase system and optimize the LC method by UV detection
- Use stationary phases with low bleeding
- Use volatile eluent systems, typically made of volatile organic acids or bases of less than 25–30 mmol/l in water/methanol or water/acetonitrile mixtures
- Avoid nonvolatile salts, reactive eluent additives, and all kinds of detergents
- Aim at a percentage of organic modifier at the point of elution of 10–20% or higher
- Ideal flow rate for ESI: 0.05–0.3 ml/min
- Ideal flow rate for APCI: 0.3–1 ml/min

1.3.2.2 Optimizing Ion Source Conditions

With the LC separation method being worked out, now the settings of the mass spectrometer must be optimized so that the detection of the target analytes succeeds as sensitively as possible. This so-called tuning of the mass spectrometer is individually different for each HPLC–MS method; it should not be mixed up with the mass spectrometer calibration, in which the accurately known molecular mass and, if needed, the fragmentation behavior of defined reference standards are used to calibrate the mass analyzer and the ion detector of the mass spectrometer. For MS tuning, it is a good advice to continuously infuse a solution of reference standards of the target analyte(s) into the ion source in a flow injection analysis (FIA) by a syringe pump and to optimize the MS source and ion transfer settings accordingly. As an alternative, a series of individual sample injections into the eluent stream by the LC autosampler under varying MS settings, leading to distinct peaks in the MS chromatogram, can also be used; optimization of MS settings is then done based on analyzing peak parameters like height, baseline noise and signal-to-noise ratio rather than comparing relative changes in a constant signal with the FIA approach. In the case of FIA, it is advisable to prepare the tuning solution of the reference standards with about a tenth of the concentration expected in the later analysis. The composition of the infusion solvent ideally reflects the mobile phase at the time of the chromatographic elution of the analyte zone. If this cannot be determined or approximated, the solvent composition averaged over the gradient program also does the job; a certain deviation of the detection result in the later separation from the tuning result may result then, though.

The flow rate applied equals to the one used in the LC separation. If the flow is so high that it cannot be meaningfully provided by a syringe pump, the UHPLC pump is a useful tool: it delivers the pure mobile phase, while a syringe pump in front of the ion source feeds a more concentrated tuning solution into the eluent flow via a T-piece. Few mass spectrometer control programs do offer automated tuning routines to then optimize ion source settings iteratively, similar to what is done for the calibration of the MS. In most software, however, the MS ion source parameters still need to be optimized manually; to do so, the ion source parameters are varied stepwise one after the other so that a uniform baseline with low noise and without jumps, dips, or spikes results – the perfect indication for a stable spray process. The gas flow and pressure settings with pneumatically assisted nebulization and the drying temperature in the MS ion source have the strongest impact on the spray stability. A higher signal intensity, in return, is achieved by an increased (selective) ion formation yield as well as an increased transfer of analyte ions from the source through the vacuum lock into the ion transfer area of the mass spectrometer. In ESI mode, this is preferably controlled via the high voltage for the electrospray and via the acceleration voltages along the ion transfer path in the high vacuum area; for APCI, the charge density on the corona discharge needle is an additional setting to work with.

Unfortunately, the ion source parameters are not independent of each other. For the sake of time, a multidimensional check of the optimal settings is often not possible in everyday laboratory work. With the sequence of optimization steps described here, however, we usually find quickly a stable and fully adequate working space for the settings. The sequence of the optimization parameters is based on the physical processes needed for the eluate removal and the charge transfer to the analytes: drying of the eluate (nebulizing and evaporation efficiency, determined by size, properties, and temperature of the formed aerosol) and charge transfer in the gas phase. Based on this, it is advisable to optimize the source settings in the following order for optimal signal yield and minimized noise. Please note that, depending on the device manufacturer, not all parameters are accessible on every mass spectrometer:

- Ion source temperature; with APCI: also APCI heater temperature
- Nebulizer gas pressure

- Drying gas flow
- Ionization voltage; with APCI: also corona needle current
- Transfer line voltage

Commercial spray units allow to adjust the protrusion length of the spray capillary from the spray unit; in the case of nano-LC sources and of some ion sources for analytical flow rates, it is also possible to optimize the spatial position of the spray unit relative to the vacuum inlet of the mass spectrometer. The protrusion of the spray capillary has a substantial impact on the signal intensity. With nano-ESI sources, the spatial position of the spray unit relative to the mass spectrometer vacuum inlet has a great influence on the signal quality and strength, especially on the ion yield in the mass spectrometer, while for ion sources designed for analytical flow rates, the factory setting of the ion source hardly offers any optimization potential. As the position of the spray capillary in the spray unit is very crucial for the ion yield, regular checks of the sprayer for correct adjustment and possible damage of the spray capillary tip are highly recommended.

It should be mentioned at this point that many mass spectrometers offer the possibility of periodically switching between negative and positive polarity back and forth to detect quasi-simultaneously both positively and negatively charged analyte molecules in one LC–MS run. For a screening run, for example with in-process control, this can be a useful tool that helps to avoid an additional LC–MS analysis. However, the capabilities of this approach are limited: First, the expected yield of oppositely charged analyte ions is usually fairly low. LC–MS runs are mostly done at the edges of the pH range, e.g. acidified by formic acid or in alkaline using ammonia. At pH values below 2.5 or above 9, only strongly acidic or basic compounds are still charged and thus can be detected as ions in the oppositely charged mode. Second, it is not only the polarity of the ion source but also that of the entire ion path in the mass spectrometer which has to be inverted in cycles. Depending on the type of MS, it may take seconds for the mass spectrometer to stabilize again, and during this time the mass spectrometer is blind to any analyte molecules and not acquiring any data. If the respective LC–MS system enables such polarity switches rapidly enough, then it is an attractive additional option for screening experiments. A solid quantitation, however, is advised to always be carried out in one fixed polarity mode due to greater robustness.

Take-home messages:

- Dissolve the sample standards in the LC solvent; the concentration should be approximately 1/10 of the expected concentration of the injected real sample
- Infuse the standard solution at the flow rate used later in the LC separation
- Adjust gas flows and drying temperature (with APCI also: APCI heating block temperature) for a stable spray (rarely needed in routine cases: adjustment of the ESI spray capillary position)
- Optimize the ionization voltage (ESI/APCI; with APCI also: corona needle current): a higher voltage leads to greater ion yield, but exceeding an optimum increases the risk of glow discharge and sample fragmentation

- Optimize the voltage of the ion transfer capillary: a higher voltage increases the number of ions in the mass analyzer, but exceeding an optimum increases the risk (or chance) of ion fragmentation due to excessive collision with residual gas molecules (in-source fragmentation)

1.3.2.3 Optimizing MS Detection

After the successful optimization of the ion source settings which aims at the highest yield of the target ion adducts, the next step is to adjust the settings of the mass selector and detector so that they offer the highest detection sensitivity. The exact name and type of setting parameters vary depending on the type of mass spectrometer and the manufacturer. What they all have in common is the task of the electrical optics in the medium- and high-vacuum section of the mass spectrometer. The ion optics in front of the actual mass analyzer pursues two goals: maximizing the removal of neutral molecules, solvent residues and residual gas, and focusing the beam of analyte ions. The ion optics usually consists of various multipoles (quadrupole, hexapole, octapole, etc.) – alternatively of an ion funnel as a staggered sequence of focusing lenses, various deflection devices (e.g. curved multipoles), and acceleration electrodes in the form of baffle plates with narrow holes (*skimmer*). Given the variety of design types, it is very difficult to describe a general sequence of optimization steps here, especially since the optimizations depend on the type of component and its position along the ion flight path. The same applies to the mass analyzer itself; multistage quadrupoles understandably offer settings different from those of ion traps, orbitraps, or flight tubes. Changing the settings of the mass analyzer affects the mass accuracy (which is optimized during a calibration routine) more than the sensitivity; the parameters are therefore usually optimized universally for a specific mass range, but not for each individual LC–MS analysis problem. For the best achievable ion beam focus and selective ion transfer into the mass analyzer, the manuals of the instrument manufacturers usually provide detailed information and helpful assistance. In practice, however, it is only worthwhile to optimize these intrinsic MS parameters in few cases, e.g. when the detection limit of an expert system must be pushed to the limit in a research environment. In routine industrial use, the time invested in a further optimization is often not worth the gain in sensitivity.

Finalizing the MS optimization, it is recommended to continuously record the infusion of the tuning solution with the optimized settings for 1–5 minutes and store it as a reference data set. The data is then inspected regarding the spray or signal stability. The LC–MS data traces are used to determine baseline noise; a representatively averaged set of MS spectra is evaluated with respect to the spectrum quality, including the shape of the mass signals and impurity signals, the signal-to-noise ratio for the characteristic isotope signals, the isotope intensity distribution, and the achieved mass resolution. With high-resolution/accurate mass (HRAM) measurements, the mass accuracy (in ppm) should also be determined as a measure of the deviation between the theoretical and experimentally measured mass.

Take-home messages:

- Sensitivity can be further increased by sequentially optimizing the ion optics settings within the medium- and high-vacuum range of the mass spectrometer
- Manufacturer information helps to identify the most effective settings
- Optimizing the mass analyzer (calibration) only makes sense for certain mass ranges, but not for each individual sample type
- Record a reference data set of 1–5 minutes length
- Evaluate the signal stability, the quality of the mass spectrum, signal to noise, and mass resolution

1.3.2.4 Verifying the Hyphenated Method

In the final step, LC separation and MS detection are combined to prove the functionality of the full method. If we are in a hurry, we could just install the required separation column, fluidically couple the UHPLC system with the MS ion source inlet, and start a separation of a (matrix-free) standard solution with the separately optimized conditions for UHPLC and MS. The chance that this will work right away is quite high. Often, however, we have to correct some MS source parameters because the actual composition of the eluent in the solvent gradient for an analyte differs significantly from the isocratic composition of the tuning experiment. The source parameters optimized for an isocratic eluent composition are thus not ideally suited for significantly different mixing ratios of water and the organic modifier. This is particularly observed in the water-rich, "wet" part of the elution program; here, droplet formation occurs more frequently, which is indicated by spikes in the signal baseline. In extreme cases, we would have to work with source settings individually adapted to different time segments of the chromatogram. The MS control software available on the market does not allow continuous changes of source parameters, like a nebulizer gas pressure gradient over time. However, it is possible to split the separation time window into different time segments of MS detection for which independent settings can be defined. In many cases, however, this is not necessary; with a little bit of experience, we can quickly get to a global set of MS settings with which we can work reliably across the entire elution window.

Those who prefer a more methodical proceeding can add an intermediate step before the first LC–MS separation, in which the UHPLC–MS setup, including the separation column, is run; however, instead of injecting the sample via the LC autosampler, again a syringe pump is connected between the column and the mass spectrometer via a T-piece as we did it for the MS tuning. The UHPLC system is then running a blank gradient without sample injection through the autosampler; again, an analyte solution is continuously fed in behind the column via the T-piece. With this setup, it can be quickly determined whether the MS detection parameters optimized during MS tuning allow sufficiently good results over the entire gradient elution range and how the analyte signal changes in the mass spectrometer depending on the solvent composition. Likewise, the blank gradients without post-column infusion can be used to determine to which extent the baseline noise changes across the gradient, e.g. caused by solvents of inappropriate purity, increased column bleeding or contamination from the instrument. With the same setup, matrix influences (ion suppression) can be determined in a similar way using a matrix solution as feed [1]; this is of paramount importance for crude real samples with little sample preparation. After such a final suitability check of the LC–MS method, the usual process steps of a method validation can be started.

Take-home messages:

- With the settings optimized in MS tuning, the baseline stability can suffer from the changing solvent composition of the gradient elution in some sections of the gradient
- *The time-saving suitability test*: set up the coupled method and test an injection, modify source settings where necessary
- *More detailed suitability test*: record a blank gradient run over the separation column with post-column infusion of the analyte solution in the mass spectrometer before the sample is finally analyzed in the LCMS separation
- Determine matrix effects by post-column infusion

1.3.2.5 Method Development Supported by Software-based Parameter Variation

As the Sections 1.3.2.1–1.3.2.4 showed, the development and optimization of an LC–MS method involves many individual steps, which can be automated to some extent to save a considerable amount of time and, due to a more comprehensive investigation of the parameter space, can also lead to more robust methods. The market offers a large number of software packages offering the simulation of liquid chromatograms, statistical test planning, quality-by-design concepts, and automated method development (see Chapters 2.1, 2.2, 4.5). However, it does not always have to be an additional expert program. Common chromatography data systems (CDS) also offer intelligent options with onboard tools to let machines automatically find the way to the best possible separation and detection conditions. One approach shall be mentioned in brief here. Some chromatography data systems such as Thermo Scientific™ Chromeleon™ offer the option of declaring own variables for instrument method files (so-called *custom variables*) and filling them with values from a sequence table when called up by the software (Figure 1.3.2). This eliminates the effort of creating separate method files for each variation of separation and detection setting, thus opening an elegant and quick way to optimize not only LC separations automatically, but also the MS detection conditions. The user declares the parameters of the supported ion sources as method variables; the software sequence table wizard then generates the sequence table with the incremental changes of the various source parameters for each injection. While running the sequence, the CDS then carries out each injection with the specifically defined source settings. Afterwards, the Chromeleon query tools allow to filter the generated sequence data set for the best results, for instance by ranking according to number of detected peaks, maximum signal-to-noise ratio, or other performance criteria. Easy-to-read visualization diagrams help to identify the injection with the best separation and detection

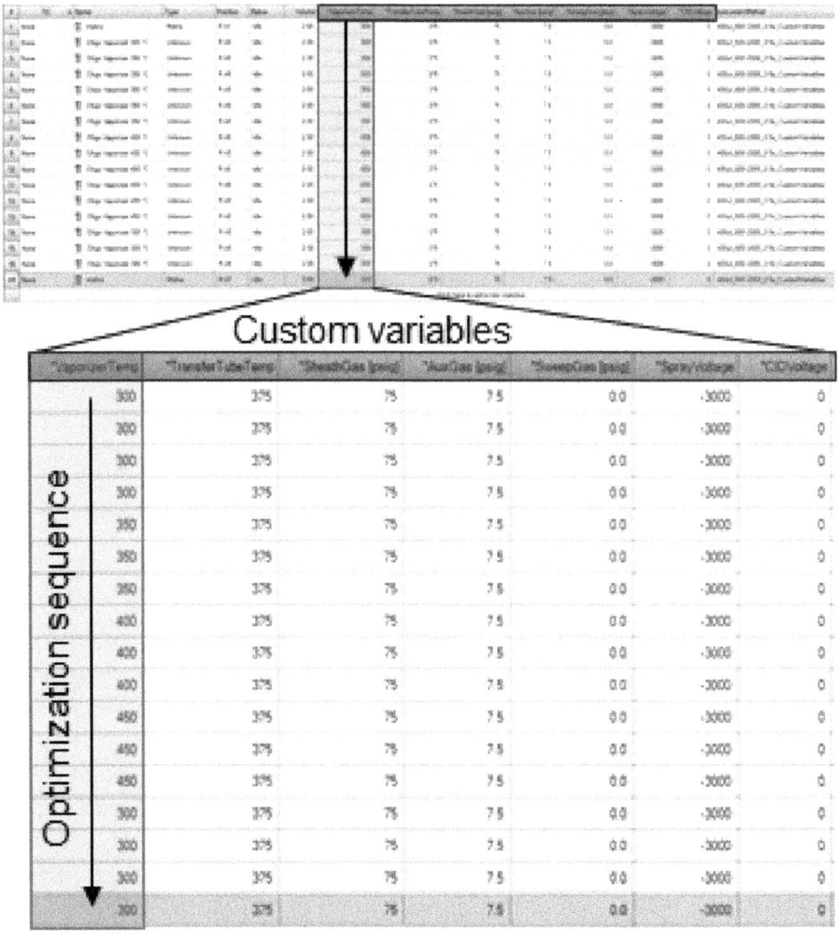

Figure 1.3.2 Automatically generated sequence table for the automated determination of optimal MS ion source settings for a single-quadrupole mass spectrometer using custom variables in Thermo Scientific Chromeleon 7. The method is used in the quality control in the manufacture of synthetic oligonucleotides. Source: Meding [6]. © 2020, Thermo Fisher Scientific Inc.

result, with all data being interactively linked to the report. This allows an automated measuring and data analysis for a wide set of parameters with high-granularity variations to determine the most suitable HPLC and MS settings for a given separation problem [6].

1.3.3 Transferring Established HPLC Methods to Mass spectrometry

As long as regulated environments still require numerous HPLC methods that were developed and validated at a time when LC–MS coupling was by no means part of routine analysis – the pharmacopoeias of the USA, Europe, and China are full of them -, LC analytical scientists will need to make established existing methods fully or partially MS-compatible as part of their daily business. The reasons for this can be manifold. For instance, stricter regulatory requirements, e.g. for the detection of genotoxic impurities, may require detection limits and a detection selectivity that were not yet achievable at the time the method was developed, but can now be realized by means of mass spectrometry. Also, new, unknown impurities detected during in-process control or quality control analysis typically require a mass spectrometric characterization. The latter case may happen rarely with long-established manufacturing processes in drug production, as the processes and synthesis paths have been known for years or decades, but even in the manufacturing of generics unpleasant surprises may occur every once in a while. Depending on the task and urgency, it is worth converting the entire LC separation into an LC–MS analysis. However, this usually comes with quite some development and validation effort and is therefore time consuming. If a quick analytical statement is needed as part of a failure analysis and troubleshooting, not always the entire LC method may have to be connected to a mass spectrometer; this would be overdone if only an examination of a specific (unknown) component in the chromatogram was requested. In such a case, selective online fractionation of the compound of interest by means of a heartcut approach can be very helpful, where the solvent of the originating method is exchanged for an MS-compatible mobile phase in a second separation dimension.

1.3.3.1 Transfer of an Entire HPLC Method to a Mass Spectrometer

If an existing LC method is to be repeated under MS-compatible conditions, the same limitations and requirements apply as for the development of new methods (see Section 1.3.2.1.1). Depending on the age of the original method being unsuitable for MS, we must be prepared for some unpleasant surprises. These include

- Type and generation of the stationary phase:
  Older HPLC phases come from a time when the stability of the applied stationary phase has not yet been optimized for MS applications. Particularly RP phases with embedded polar groups (depending on the manufacturer named *polar-embedded group* phases, *shield* phases, or similarly) of the first generation, developed in the late 1990s to the early 2000s, suffer from a fairly high level of column bleeding, which leads to an increased background noise in the mass spectrometer, typically increasing with rising organic content in the gradient program. True veterans of classic C18 phase chemistry, developed more than 20–25 years ago, also elute more metal ions than today's phase materials, which impairs the ion formation of the analytes. This may be an acceptable restriction for the identification of an unknown substance in "fire alert mode," provided the component of interest can still be detected. For an LC–MS analysis to be validated, however, this is not a viable option; an up-to-date LC phase must be used here. Experience has shown that most RP columns that were developed in the 2010s meet all the requirements for MS accessibility. HPLC columns, the development history of which goes back

to the last millennium and which are still used today in many traditional methods in the QC area, are not recommended per se for LC–MS analysis.
- Changed selectivities due to MS-compatible mobile phases:
  As described in Section 1.3.2.1.1, many conventional HPLC additives and buffers are not MS-compatible due to their low volatility. Replacing them by highly volatile alternatives is therefore imperative but may lead to unwanted changes in selectivity. The allround buffer of HPLC, the phosphate system with its three buffer points at (around) pH 2, 7, and 12, is very popular with classic LC methods, and an acid/base mixture compatible with mass spectrometry buffering at the same pH is not always easy to find. A popular example of this is ammonium acetate, which due to its two pKa values of 4.75 (acetic acid/acetate) and 9.2 (ammonium/ammonia) has *no* buffer capacity at pH 7 but reacts particularly sensitive to changes in pH there. In addition, most MS-compatible buffer systems consist of monovalent ions (such as triethylammonium, ammonium, formate, or acetate), while phosphate ions are bivalent to trivalent, depending on the pH value. With the same buffer concentration, this changes the ionic strength, which in the case of ionic analytes (acids, bases, permanent ions) can influence the interaction with polar surfaces and thus the retention. In both cases, a change in selectivity relative to the original method can be seen, which can lead to a partial loss of resolution or even a major change in elution pattern due to peak reversal. Unfortunately, it is usually more the rule than the exception that established HPLC separations result in a changed chromatogram after switching to MS-compatible mobile solvents.

These symptoms can be corrected with a few adjustments in the separation system like the gradient profile, depending on how big the deviations between the original LC separation and the MS-related separation system are. Some analytes are rather insensitive against a change in the mobile and stationary phase, so that a switch to a modern stationary phase and the translation of the recipe for the mobile phase into the MS world has little to no influence on the separation result. Minor modifications of the solvent mixture or the gradient program can then eliminate remaining flaws such as a not fully achieved baseline separation; alternatively, we generate the additional selectivity that the LC separation is no longer able to achieve by using the mass spectrometer, e.g. with tandem MS detection. Particularly for complex samples already having critical peak pairs with a resolution of $R_S < 2$ in the original method, the mere translation of a conventional LC method quickly becomes more complex than a complete new method development, which then also produces more consistent and state-of-the-art results, featuring increased speed and improved resolution (UHPLC methods).

1.3.3.2 Selected Analysis of an Unknown Impurity – Solvent Change by Single-/Multi-Heartcut Techniques

It may not always be required to adapt an existing HPLC method so that it becomes fully MS-compatible – it depends on the aim of the analytical laboratory and the technologies applied. Characterization laboratories, which mainly have to

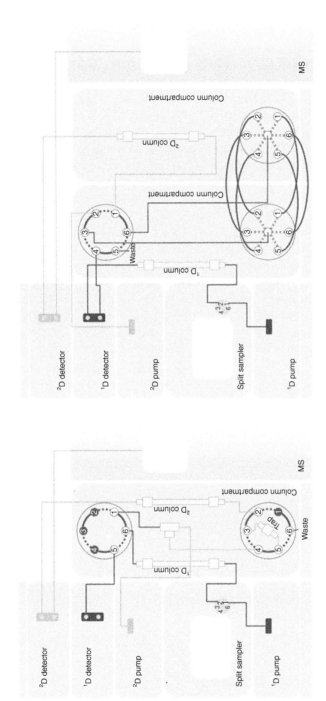

Figure 1.3.3 Left: single-heartcut setup for a dedicated isolation of an unknown peak in the first dimension (^{1}D column), "parking" of this compound on a trap column (installed in the bottom valve) while reducing the solvent strength, followed by elution via an (optional) second column (^{2}D column) into the mass spectrometer; pump ^{1}D (bottom) delivers the mobile phase for the conventional HPLC method, while pump ^{2}D (top) provides the MS-compatible mobile phase; Right: multi-heartcut setup for repetitive fractionation using storage loops (lower two multi-position valves) and transfer into a mass spectrometer. Source: Grübner and Greco [7]. © 2020, Thermo Fisher Scientific Inc.

investigate and determine unknown impurities, e.g. to support the development of chemical syntheses or production processes, be it in the in-process control or in release control analysis, often have to extract just one certain substance in the chromatogram; this compound may then be further subject to HRAM measurement and tandem-MS experiments to propose a structure for this new unknown in a chromatogram. In such a situation, speed usually matters most – the faster the result of the analysis is available, the sooner it can be decided whether, for example, a synthesis batch can yet be worked up or has to be discarded. A complex method translation to MS-compatible conditions usually then costs too much time. It would be much more useful to selectively cut out those peaks to be investigated from a separation with a non-MS-compatible phase system and then to change the eluent before transferring them into a mass spectrometer. This quite common application has recently been gaining more attention, as 2D-LC systems are more and more adopted; thanks to modern software support, multidimensional liquid chromatographs are meanwhile much easier to use than they were a decade or two ago. Figure 1.3.3 illustrates how comparably simple setups of a so-called single or multi-heartcut system may be realized.

If the component we are looking for elutes in the first, the non-MS-compatible separation dimension, the peak volume is directed by a switching valve to a trap column in the second dimension for the time of elution. While the remaining separation of the first dimension continues after switching back the valve, the trapped compound elutes free from MS-incompatible additives in the second dimension by a second LC pump; optionally, a solvent strength reduction supports the trapping and cleanup before the mass spectrometer allows a further in-depth analyte investigation. This setup can be further optimized as needed, enabling for instance not merely a solvent exchange, but a true 2D separation, which allows to verify the purity of the trapped peak fraction by an alternative/orthogonal selectivity in the second dimension and to further separate any coeluting components. Cutting multiple peak fractions (multi-heartcut), including quantitation in the second dimension, is also feasible with appropriate upgrades of this setup. Describing this in detail is beyond the scope of this chapter, which is why we refer to literature for further reading [7].

Abbreviations

2D	two-dimensional
APCI	atmospheric pressure chemical ionization
CDS	chromatography data system
ESI	electrospray ionization
FIA	flow injection analysis
HPLC	high-performance liquid chromatography
HR/AM	high resolution/accurate mass
ID	inner diameter
LC	liquid chromatography
MS	mass spectrometry

ppm   parts per million
QC   quality control
RP   reversed-phase
SRM   selected reaction monitoring
UHPLC   ultrahigh-performance liquid chromatography
UV   ultraviolet (radiation).

References

1 Martin, M.M. (2017). Technical aspects and pitfalls of LC/MS hyphenation. In: *The HPLC-MS Handbook for Practitioners* (ed. S. Kromidas), 19–70. Weinheim: Wiley-VCH.
2 Martin, M.M. (2019). *Aspects of gradient elution in LC-MS analysis*. In: *Gradient HPLC for Practitioners: RP, LC-MS, Ion Analytics, Biochromatography, SFC, HILIC* (ed. S. Kromidas), 189–211. Weinheim: Wiley-VCH.
3 Engelhardt, H. and Elgass, H. (1978). *Journal of Chromatography* 158: 249–259.
4 Dams, R., Benijts, T., Günther, W. et al. (2002). *Rapid Communications in Mass Spectrometry* 16: 1072–1077.
5 Hoffmann, T. and Martin, M.M. (2010). *Electrophoresis* 31 (7): 1248–1255.
6 Meding, S. (2020). Quality control of oligonucleotides with a single quadrupole mass spectrometer. Thermo Fisher Scientific Technical Note 73670 (Online). https://assets.thermofisher.com/TFS-Assets/CMD/Technical-Notes/tn-73670-lc-ms-oligonucleotides-single-quad-tn73670-en.pdf (accessed 16 April 2021).
7 Grübner, M. and Greco, G. (2019). Flexible HPLC instrument setups for double usage as one heart-cut-2D-LC system or two independent 1D-LC systems. Thermo Fisher Scientific Technical Note 73298 (Online). https://assets.thermofisher.com/TFS-Assets/CMD/Technical-Notes/tn-73298-hplc-1d-lc-or-heart-cut-2d-lc-tn73298-en.pdf (accessed 16 April 2021).

1.4

Chromatographic Strategies for the Successful Characterization of Protein Biopharmaceuticals

Szabolcs Fekete[1,2], Valentina D'Atri[1,2], and Davy Guillarme[1,2]

[1] School of Pharmaceutical Sciences, University of Geneva, CMU-Rue Michel Servet 1, 1211, Geneva 4, Switzerland
[2] University of Geneva, Institute of Pharmaceutical Sciences of Western Switzerland, CMU-Rue Michel Servet 1, 1211, Geneva 4, Switzerland

1.4.1 Introduction to Protein Biopharmaceuticals

Protein biopharmaceuticals have emerged as important therapeutic options for the treatment of various diseases, and, in particular, cancers. Given their obvious benefits in terms of efficacy and side effects, monoclonal antibodies (mAbs) and related products are considered today as the most successful therapeutic biopharmaceutical products. Until now, more than 80 mAbs have reached the market and more than 600 are in clinical phases [1]. To increase the potential success rates and improve the quality and safety of protein biopharmaceutical products as approved drugs, a comprehensive set of analytical and functional techniques have to be used at the very early stage of the R&D process. In particular, there are a number of chemical and enzymatic modifications that can occur during the production and storage of any mAb-related products that need to be evaluated [2].

Figure 1.4.1 summarizes some of the most important modifications, responsible for the microheterogeneity, which is an inherent property of therapeutic proteins. The analyst has to develop numerous methods to identify, characterize, and routinely monitor such modifications, using state-of-the-art analytical methods to ensure efficacy, safety, and overall quality of the biopharmaceutical drug. Among the available analytical strategies, liquid chromatography (LC) is often applied for the detailed evaluation of these new protein-based drugs [3]. As reported in Figure 1.4.1, various modes of chromatography can be applied to successfully characterize charge variants (ion exchange chromatography, IEX), hydrophobic variants (hydrophobic interaction chromatography, HIC, and reversed-phase liquid chromatography, RPLC), glyco-variants (hydrophilic interaction chromatography, HILIC), or size variants (size exclusion chromatography, SEC).

To simplify the complexity of protein biopharmaceuticals prior to their analysis, it is also important to keep in mind that they can be analyzed either: (i) at the intact level (150 kDa for mAbs), (ii) at subunits level, after digestion (specific cleavage in

Optimization in HPLC: Concepts and Strategies, First Edition. Edited by Stavros Kromidas.
© 2021 WILEY-VCH GmbH. Published 2021 by WILEY-VCH GmbH.

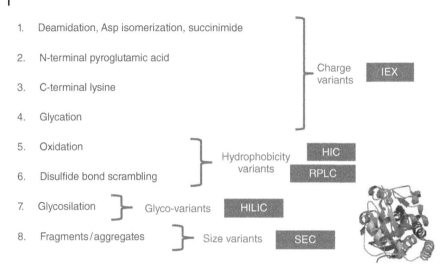

Figure 1.4.1 Common modifications observed in therapeutic proteins and chromatographic methods employed for their characterization.

the hinge region) with IdeS (or papain, pepsin) and/or reduction with dithiotreitol (DTT) or tris(2-carboxyethyl)phosphine (TCEP) (25–50–100 kDa for mAbs), or (iii) at the peptide level after digestion with trypsin (or chymotrypsin) (1–4 kDa) [4].

The chromatographic strategies above described have to be sufficiently fast, due to the large number of samples to be analyzed. More importantly, they should offer an adequate resolution (high selectivity and peak capacity) to separate closely related compounds (as example, during protein deamidation, only one amide functional group is replaced by one carboxylic acid group in the whole protein structure). Last but not least, it is also of prime importance that these chromatographic approaches should preferably be compatible with mass spectrometry (MS), due to the high level of information that can be attained with this detector [5, 6]. This last point is particularly difficult to reach with nondenaturing modes of chromatography (IEX, SEC, and HIC).

1.4.2 From Standard to High-Performance Chromatography of Protein Biopharmaceuticals

As previously mentioned, various chromatographic modes can be used for the characterization of protein biopharmaceuticals. From an historical point of view, IEX, SEC, and HIC have been largely used in the past, since they are considered as nondenaturing strategies. However, the performance achieved with these chromatographic modes was quite limited both in terms of resolution and throughput. For this reason, some improvements were brought to these historical methods in the last years, as reported below.

SEC is the gold standard technique to separate protein aggregates and fragments (see also Chapter 1.5). In SEC, there has been a clear evolution in terms of stationary-phase dimensions in the last few years. As example, it was demonstrated that columns of 150 × 4.6 mm packed with sub-2-μm particles can yield the same separation quality than conventional 300 × 7.8 mm SEC columns packed with 5-μm particles, but are 3–4 times faster and cheaper [7]. However, to properly use these new UHP-SEC columns (ultra-high performance, UHP), the extra-column volume of LC systems needs to be decreased as much as possible, and therefore such columns are only compatible with modern UHPLC instruments [8]. Besides the column dimensions, the chemistry has also evolved toward more inert stationary phases, to limit as much as possible the possible electrostatic and hydrophobic interactions that can take place in SEC with proteins having high isoelectric point (pI) or possessing hydrophobic character, respectively [9]. On the other hand, the mobile-phase conditions in SEC have not changed so much and it is still recommended to use large proportion of salts (mostly KCl or NaCl at 100–300 mM) as well as phosphate buffer having relatively neutral pH. More information on the recent trends in SEC can be found in [10]. In conclusion, the modern version of SEC consists in using columns of 150 × 4.6 mm packed with sub-3-μm particles, combined with a mobile phase containing at least 100 mM of phosphate buffer at physiological pH and KCl at 100–300 mM.

IEX can be considered as the reference technique for the evaluation of charge heterogeneity. Historically, cation exchange remains the most widely used mode for proteins characterization, and proteins are typically eluted from an IEX column using a linear salt gradient (e.g. NaCl), at a mobile-phase pH of 5.5–7.5. Besides salt gradients, pH- or salt-mediated pH-gradient (combination of salt and pH gradient) modes have also seen several applications in the last few years. To perform a suitable pH gradient, the mobile phase should contain a mixture of various components having a broad range of pK_a and evenly distributed along the investigated pH range [11, 12]. This allows to achieve slightly different selectivities compared to salt gradient, and also to have a generic mobile-phase recipe (often referred as "platform method"), as reported in Figure 1.4.2. In addition, some providers are now offering commercial pH-gradient mobile phases that drastically simplify the preparation of eluents. Besides these improvements in terms of mobile phase, the particulate stationary phases used in current IEX practice are still made of nonporous materials (to limit mass transfer resistance of large biomolecules); therefore, there is limited interest in decreasing particle size, and for this reason, columns packed with 5-μm particles remain the standards. However, the columns are shorter than they were in the past (50–100 mm vs. 150–250 mm) since the performance is not significantly influenced by the column length in IEX. Various state-of-the-art IEX columns were recently compared in [13], and more information on the current trends in IEX can be found in [14]. Based on the previous statements, modern IEX consists in using relatively short columns of 50–100 mm, packed with 5-μm nonporous particles, in combination with either salt or pH gradient.

HIC is an analytical approach that is used today almost exclusively for the characterization of cysteine-conjugated antibody–drug conjugates (ADCs), which have to be analyzed under their native form (no or only partial covalent bond between the light and heavy chains in these products) [15]. For all other applications, RPLC

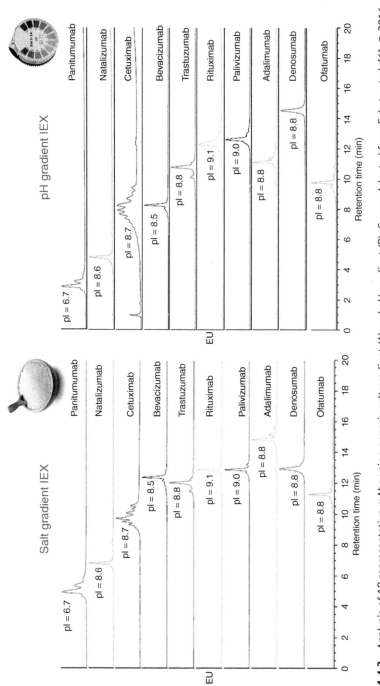

Figure 1.4.2 Analysis of 10 representative mAbs using generic salt gradient (A) and pH gradient (B). *Source:* Adapted from Fekete et al. [6]. © 2016, American Chemical Society.

is able to provide the same information, but with much less salts in the mobile phase and higher kinetic performance. So, the use of HIC tends to decrease today for the analytical characterization of biopharmaceuticals. More information on the recent trends in HIC (i.e. use of columns packed with sub-3-µm particles, addition of 10–20% organic modifier in the mobile phase) can be found in [6].

Besides the use of nondenaturing chromatographic techniques, there has been a growing interest for denaturing techniques, such as RPLC and HILIC, due to their inherent compatibility with MS. However, it is not easy to analyze such very large molecules with these two modes and, for this reason, there have been a lot of improvements brought to these LC modes over the years.

Because of its versatility, flexibility, robustness, and compatibility with MS, RPLC is very popular for the analysis of peptides and proteins. However, to limit peak distortion (which occurs due to secondary ionic interactions between proteins and stationary phase), it is highly recommended to add trifluoroacetic acid (TFA) in the mobile phase to (i) form ion pairs with the positive charges at the surface of the proteins and also to (ii) decrease mobile-phase pH, thus limiting the amount of charged residual silanols at the surface of the RPLC stationary phase [16]. Secondly, to avoid band broadening, due to the slow mass transfer of large molecules having low-diffusion coefficients, it is necessary to use modern stationary phases packed with wide-pore sub-2-µm fully porous or sub-3-µm superficially porous particles, in combination with elevated mobile-phase temperature (70–90 °C) [17]. However, the analysis time should remain reasonable (10 minutes maximum) at such temperature and pH to avoid protein degradation during its analysis. Another issue related to the use of RPLC for proteins is the possible adsorption (either reversible or irreversible) at the surface of the stationary phase, which can be limited by using less hydrophobic materials (e.g. C4 vs. C18), together with elevated temperature [18]. To limit the adsorption within the LC system, so-called bio-inert (or metal-free, stainless-steel free) LC instruments are now available from most of the providers. Recently, some innovative column chemistries (polyphenyl) have been developed and released, which allow to analyze proteins at moderate conditions (more reasonable temperature (50–60 °C), and with lower amount of TFA (0.02–0.05%) [19]. Such conditions limit the possible denaturation of proteins and also improve MS sensitivity. More information on the recent trends in RPLC can be found in [6]. In modern RPLC of proteins, it is recommended to use columns of 100–150 × 2.1 mm packed with sub-2 µm fully porous or sub-3 µm superficially porous particles. There is currently a trend, which consists in using less TFA and lower mobile-phase temperature, to reach milder conditions, thus limiting proteins denaturation.

Finally, one of the most recent chromatographic approaches suggested for the characterization of proteins was HILIC, which separates protein variants based on their differences in hydrophilic properties (i.e. glycosilation can be easily characterized). This strategy becomes viable thanks to the commercialization of wide-pore, amide-bonded stationary phase packed with sub-2-µm particles in 2016 [20, 21]. Similar to what is done in RPLC, TFA is required in the mobile phase and elevated temperature also has to be used. As illustrated in Figure 1.4.3, it is possible to profile the glycans using generic conditions, even for protein subunits of 25 kDa.

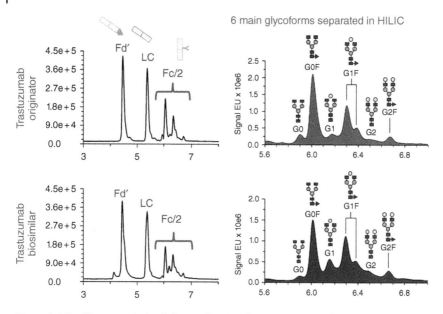

Figure 1.4.3 Glycan analysis of digested/reduced trastuzumab subunits under HILIC conditions. Comparison of originator and biosimilar.

Some additional information on the use of HILIC for proteins can be found in [22]. In conclusion, amide column packed with sub-2-μm fully porous particles should be preferentially used for successful HILIC operation of large protein fragments, in combination with elevated temperature and 0.1% TFA in the mobile phase.

1.4.3 Online Coupling of Nondenaturing LC Modes with MS

Besides the chromatographic information, it is also important today to have MS data on the different peaks eluted from the chromatographic column, for further identification. Obviously, RPLC and HILIC modes are inherently compatible with MS (see also Chapters 1.2 and 1.3). The only constraint is the use of TFA to achieve suitable peak shape. TFA can indeed hamper MS sensitivity (sensitivity is on average decreased by a factor 10 when using a mobile phase containing 0.1% TFA vs. 0.1% FA).

In IEX, traditional salt- or pH-based eluents are often unsuitable for direct coupling to MS, due to the nonvolatility or high ionic strength of salts. In the last few years, a lot of attempts have made to make IEX compatible with MS. For this purpose, various combinations of volatile salts/buffers (i.e. formic acid, acetic acid, ammonium formate, ammonium acetate, ammonium carbonate, ammonium bicarbonate, etc.) at different ionic strengths and mobile-phase pH were tested [23–26]. Among the conditions that were reported in the literature until now, pH gradient was preferred, in combination with an ionic strength gradient (salt-mediated pH gradient),

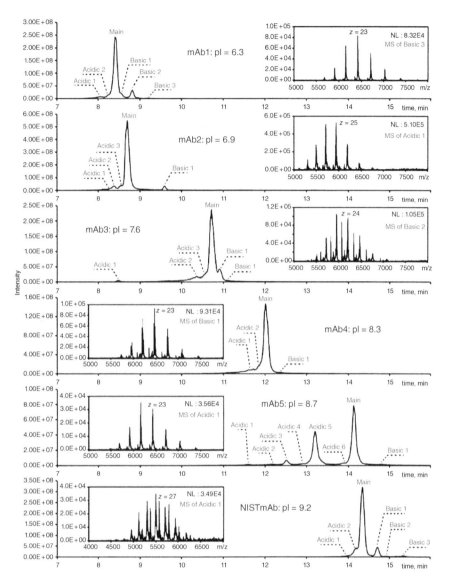

Figure 1.4.4 IEX-MS analysis of different mAb samples with pI values ranging from 6.3 to 9.2. The raw mass spectrum of an exemplary minor charge variant peak from each sample is displayed in the corresponding inset. *Source:* Adapted from Yan et al. [26]. © 2018, American Chemical Society.

to obtain sharp peaks (peak compression effect). As example, the recipe suggested by Yan et al. [26] was found to be quite generic and based on the use of 20 mM ammonium acetate (pH 5.6) in A and 140 mM ammonium acetate +10 mM ammonium bicarbonate (pH 7.4) in B. Figure 1.4.4 shows the chromatograms obtained using these conditions for various mAbs having pI between 6.3 and 9.2, including the MS identification of the charge variants. It is also important to notice that Waters has recently launched a commercial IEX-MS mobile phase, called IonHance CX-MS buffer. It consists of 50 mM ammonium acetate +2% AcN, pH 5.0 (acetic acid) in A, and 160 mM ammonium acetate +2% AcN, pH 8.5 (ammonium hydroxide) in B. This buffer was recently applied in [27], for the analysis of three mAbs having high pI under their native and proteolyzed forms. In conclusion, IEX-MS should ideally be performed using salt-mediated pH gradient, in absence of carbonate/bicarbonate to achieve highest possible MS sensitivity.

Similar to what was previously described in IEX, it was recently demonstrated that SEC with 100 mM ammonium acetate as main mobile-phase component can also be coupled online to MS. However, reliable results were only obtained for acidic mAbs, having limited ionic interactions with the stationary phase [28, 29]. When analyzing basic mAbs (which corresponds to a large majority of commercial mAbs), significant peak broadening and tailing was observed, as well as adsorption of the aggregates. The amount of salts in the mobile phase was therefore not sufficient to limit the ionic interactions that can take place. So, there is a need for more inert SEC stationary phases. Alternatively, some other research groups have also reported use of so-called salt-free eluents (containing, for example, acetonitrile, TFA, and formic acid in various ratios) [30]. However, under these conditions, mAb aggregates cannot be evaluated, since they are often unstable, and these conditions are considered as denaturing.

For additional information on the topic of combining nondenaturing chromatographic modes with MS, the readers can refer to [31].

1.4.4 Multidimensional LC Approaches for Protein Biopharmaceuticals

Due to the inherent complexity of protein biopharmaceutical samples, and the difficult hyphenation of some modes of chromatography with MS, it could be of interest to use multidimensional LC. As reported in Figure 1.4.5, there are various possible combinations of chromatographic modes to perform two-dimensional separations. In the first dimension, any mode of chromatography can be used. However, in the second dimension, there is a need to have an orthogonal separation mechanism to obtain complementary information, and to preferentially use a chromatographic mode inherently compatible with MS. Based on these recommendations, RPLC is the most widely used mode until now in the second dimension even if it is considered as denaturing mode. To avoid possible sample denaturation, it could be of interest to use SEC with ammonium acetate buffer in the second dimension as an alternative, to remove nonvolatile salts from the first dimension.

The first attempts to characterize protein biopharmaceuticals using 2D-LC–MS were performed at the peptide level [32] using RPLC × RPLC, IEX × RPLC, and HILIC × RPLC all coupled to MS, for the analysis of trastuzumab sample. The applicability of 2D-LC in the assessment of identity, purity, and comparability was successfully demonstrated by the analysis of different innovator production batches, a biosimilar in development and stressed samples.

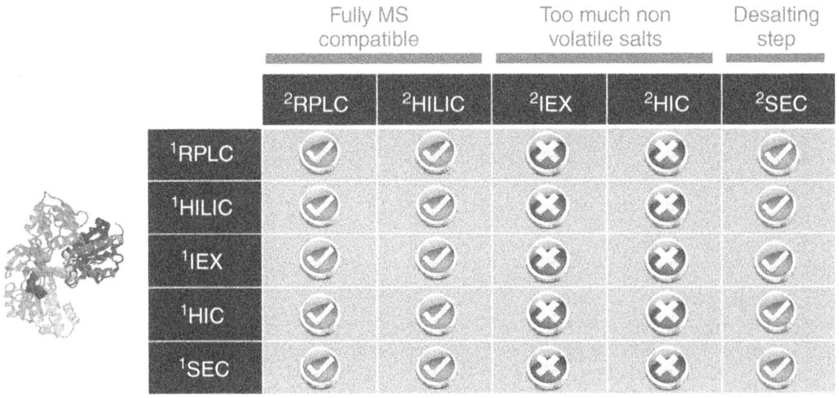

1.4.4 Multidimensional LC Approaches for Protein Biopharmaceuticals

	²RPLC	²HILIC	²IEX	²HIC	²SEC
	Fully MS compatible		Too much non volatile salts		Desalting step
¹RPLC	✓	✓	✗	✗	✓
¹HILIC	✓	✓	✗	✗	✓
¹IEX	✓	✓	✗	✗	✓
¹HIC	✓	✓	✗	✗	✓
¹SEC	✓	✓	✗	✗	✓

Figure 1.4.5 Possible combinations of chromatographic approaches in the first and second dimensions, for the analytical characterization of protein biopharmaceuticals.

Next to this pioneering work, the multidimensional LC approach was applied at the intact or subunits levels of protein, for the characterization of mAb and ADC samples. IEX × RPLC–MS was applied to obtain direct MS identification of IEX peaks of rituximab after RPLC desalting [33]. However, this strategy offers a limited interest today, since IEX can be directly coupled to MS using volatile salts (see Section 1.4.3). On the other hand, SEC × RPLC–MS and SEC × SEC–MS were also used to directly assess the identity of peaks eluted from the SEC [10, 34, 35]. In both cases, phosphate buffer was used in the first dimension to have suitable SEC performance. Since aggregates are often noncovalent, it is hazardous to use RPLC in the second dimension (i.e. use of high temperature, presence of organic solvent, and acidic additives in the mobile phase) and for this reason, the use of SEC with ammonium acetate could be beneficial, to eliminate phosphate salts before MS. HIC × RPLC–MS and HIC × SEC–MS were also used for the characterization of cysteine-conjugated ADC [36, 37]. Again, SEC was beneficial in the second dimension to maintain the noncovalent native structure of ADC sample. Finally, HILIC × RPLC–MS was also applied to further increase the amount of information that can be obtained from a unidimensional separation [38]. As shown in Figure 1.4.6, the two separation dimensions were highly complementary and huge amount of information could be obtained on the glycans, thanks to the HILIC dimension. However, to ensure the proper compatibility between the two dimensions, active solvent modulation has to be used to limit the negative impact of acetonitrile-rich mobile phase on the RPLC separation.

For more information on the use of multidimensional LC techniques for protein biopharmaceuticals, the readers can refer to [39]. In conclusion, to successfully develop 2D-LC–MS methods for biopharmaceuticals, it is highly recommended to use RPLC in the second dimension. As an alternative, SEC can also be useful if the proteins can be denatured. It is also important to perform very fast analysis in the second dimension, ideally by combining short columns and high flow rates.

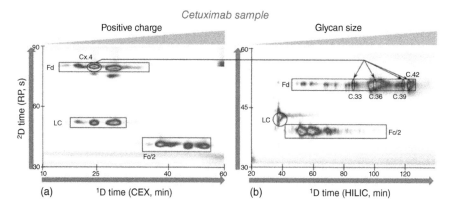

Figure 1.4.6 CEX × RPLC–MS and HILIC × RPLC–MS profiles of cetuximab after reduction and digestion (analysis of 25-kDa subunits). *Source:* Adapted from Stoll et al. [39]. © 2018, American Chemical Society.

1.4.5 Conclusion and Future Trends in Protein Biopharmaceuticals Analysis

Due to the high demand from the pharmaceutical industry and strong investments in the biopharmaceutical market, providers of chromatographic instrumentation and consumables have largely invested to develop and commercialize innovative solutions for the analysis of protein biopharmaceuticals. Among the most interesting innovations, we can cite the development of bio-inert systems, RPLC and HILIC columns with improved thermal stability (up to 90 °C), commercial mobile-phase buffers for IEX and IEX-MS operation, columns packed with sub-2-μm fully porous or sub-3-μm superficially porous particles, and multidimensional LC. Thanks to these advances, it is possible today to achieve superb performance in all chromatographic modes (i.e. IEX, SEC, HIC, RPLC, and HILIC), whatever be the level of analysis for biopharmaceutical products (i.e. intact, subunits, tryptic digest).

Obviously, for the coming years, there will be a continuous requirement for MS hyphenation of all chromatographic modes (in particular the nondenaturing ones), which can be achieved through the development of innovative volatile mobile phases and more inert stationary phases. There is also a need for improving throughput, selectivity, robustness, and sensitivity of the current method. As example, a promising solution was recently suggested (use of multi-isocratic elution mode) to improve selectivity for the analysis of biopharmaceutical products [40], while it is also possible to use difluoroacetic acid (DFA) rather than TFA to improve sensitivity without scarifying peak shapes. In terms of multidimensional LC, the possibilities are almost infinite and various additional combinations of chromatographic modes will be certainly explored in the coming years. Remarkably, it is also possible to include some online reduction or digestion steps directly within the multidimensional LC setup. The separation of charge variants in the first dimension, followed by identification and localization of the charge modification,

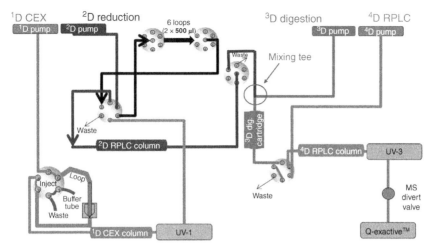

Figure 1.4.7 Multidimensional LC–MS/MS setup consisting in IEX in the first dimension, followed by reduction, tryptic digestion, and RPLC–MS/MS, for the identification and localization of charge modification on the protein structure. *Source:* Adapted from Goyon et al. [41]. © 2019, Elsevier.

using online reduction, tryptic digestion, and RPLC–MS/MS of peptides have been demonstrated in [41] and illustrated in Figure 1.4.7.

References

1. Strohl, W.R. (2018). Current progress in innovative engineered antibodies. *Protein & Cell* 9 (1): 86–120. https://doi.org/10.1007/s13238-017-0457-8.
2. Beck, A., Wagner-Rousset, E., Ayoub, D. et al. (2013). Characterization of therapeutic antibodies and related products. *Analytical Chemistry* 85 (2): 715–736. https://doi.org/10.1021/ac3032355.
3. Duivelshof, B.L., Fekete, S., Guillarme, D., and D'Atri, V. (2019). A generic workflow for the characterization of therapeutic monoclonal antibodies—application to daratumumab. *Analytical and Bioanalytical Chemistry* 411 (19): 4615–4627. https://doi.org/10.1007/s00216-018-1561-1.
4. Bobaly, B., D'Atri, V., Goyon, A. et al. (2017). Protocols for the analytical characterization of therapeutic monoclonal antibodies. II – Enzymatic and chemical sample preparation. *Journal of Chromatography B* 1060 (August): 325–335. https://doi.org/10.1016/j.jchromb.2017.06.036.
5. Beck, A., Sanglier-Cianférani, S., and Van Dorsselaer, A. (2012). Biosimilar, biobetter, and next generation antibody characterization by mass spectrometry. *Analytical Chemistry* 84 (11): 4637–4646. https://doi.org/10.1021/ac3002885.
6. Fekete, S., Guillarme, D., Sandra, P., and Sandra, K. (2016). Chromatographic, electrophoretic, and mass spectrometric methods for the analytical characterization of protein biopharmaceuticals. *Analytical Chemistry* 88 (1): 480–507. https://doi.org/10.1021/acs.analchem.5b04561.

7 Goyon, A., Beck, A., Colas, O. et al. (2017). Evaluation of size exclusion chromatography columns packed with sub-3 μm particles for the analysis of biopharmaceutical proteins. *Journal of Chromatography A* 1498: 80–89. https://doi.org/10.1016/j.chroma.2016.11.056.

8 Fekete, S. and Guillarme, D. (2018). Influence of connection tubing in modern size exclusion chromatography and its impact on the characterization of mAbs. *Journal of Pharmaceutical and Biomedical Analysis* 149: 22–32. https://doi.org/10.1016/j.jpba.2017.10.019.

9 Goyon, A., Beck, A., Veuthey, J.-L. et al. (2017). Comprehensive study on the effects of sodium and potassium additives in size exclusion chromatographic separations of protein biopharmaceuticals. *Journal of Pharmaceutical and Biomedical Analysis* 144: 242–251. https://doi.org/10.1016/j.jpba.2016.09.031.

10 Goyon, A., Fekete, S., Beck, A. et al. (2018). Unraveling the mysteries of modern size exclusion chromatography – the way to achieve confident characterization of therapeutic proteins. *Journal of Chromatography B* 1092: 368–378. https://doi.org/10.1016/j.jchromb.2018.06.029.

11 Fekete, S., Beck, A., Fekete, J., and Guillarme, D. (2015a). Method development for the separation of monoclonal antibody charge variants in cation exchange chromatography, Part I: Salt gradient approach. *Journal of Pharmaceutical and Biomedical Analysis* 102: 33–44. https://doi.org/10.1016/j.jpba.2014.08.035.

12 Fekete, S., Beck, A., Fekete, J., and Guillarme, D. (2015b). Method development for the separation of monoclonal antibody charge variants in cation exchange chromatography, Part II: pH gradient approach. *Journal of Pharmaceutical and Biomedical Analysis* 102: 282–289. https://doi.org/10.1016/j.jpba.2014.09.032.

13 Murisier, A., Farsang, E., Horváth, K. et al. (2019). Tuning selectivity in cation-exchange chromatography applied for monoclonal antibody separations, part 2: evaluation of recent stationary phases. *Journal of Pharmaceutical and Biomedical Analysis* 172: 320–328. https://doi.org/10.1016/j.jpba.2019.05.011.

14 Fekete, S., Beck, A., Veuthey, J.-L., and Guillarme, D. (2015). Ion-exchange chromatography for the characterization of biopharmaceuticals. *Journal of Pharmaceutical and Biomedical Analysis* 113: 43–55. https://doi.org/10.1016/j.jpba.2015.02.037.

15 Beck, A., D'Atri, V., Ehkirch, A. et al. (2019). Cutting-edge multi-level analytical and structural characterization of antibody-drug conjugates: present and future. *Expert Review of Proteomics* 16 (4): 337–362. https://doi.org/10.1080/14789450.2019.1578215.

16 Fekete, S., Veuthey, J.-L., and Guillarme, D. (2012). New trends in reversed-phase liquid chromatographic separations of therapeutic peptides and proteins: Theory and applications. *Journal of Pharmaceutical and Biomedical Analysis* 69: 9–27. https://doi.org/10.1016/j.jpba.2012.03.024.

17 Fekete, S., Beck, A., Wagner, E. et al. (2015). Adsorption and recovery issues of recombinant monoclonal antibodies in reversed-phase liquid chromatography. *Journal of Separation Science* 38 (1): 1–8. https://doi.org/10.1002/jssc.201400996.

18 Fekete, S., Rudaz, S., Veuthey, J.-L., and Guillarme, D. (2012). Impact of mobile phase temperature on recovery and stability of monoclonal antibodies using recent reversed-phase stationary phases. *Journal of Separation Science* 35 (22): 3113–3123. https://doi.org/10.1002/jssc.201200297.

19 Bobály, B., Lauber, M., Beck, A. et al. (2018). Utility of a high coverage phenyl-bonding and wide-pore superficially porous particle for the analysis of monoclonal antibodies and related products. *Journal of Chromatography A* 1549: 63–76. https://doi.org/10.1016/j.chroma.2018.03.043.

20 Fekete, S., Veuthey, J.-L., Beck, A., and Guillarme, D. (2016). Hydrophobic interaction chromatography for the characterization of monoclonal antibodies and related products. *Journal of Pharmaceutical and Biomedical Analysis* 130: 3–18. https://doi.org/10.1016/j.jpba.2016.04.004.

21 Periat, A., Fekete, S., Cusumano, A. et al. (2016). Potential of hydrophilic interaction chromatography for the analytical characterization of protein biopharmaceuticals. *Journal of Chromatography A* 1448: 81–92. https://doi.org/10.1016/j.chroma.2016.04.056.

22 Periat, A., Krull, I.S., and Guillarme, D. (2015). Applications of hydrophilic interaction chromatography to amino acids, peptides, and proteins. *Journal of Separation Science* 38 (3): 357–367. https://doi.org/10.1002/jssc.201400969.

23 Bailey, A.O., Han, G., Phung, W. et al. (2018). Charge variant native mass spectrometry benefits mass precision and dynamic range of monoclonal antibody intact mass analysis. *MAbs*: 1–12. https://doi.org/10.1080/19420862.2018.1521131.

24 Füssl, F., Cook, K., Scheffler, K. et al. (2018). Charge variant analysis of monoclonal antibodies using direct coupled pH gradient cation exchange chromatography to high-resolution native mass spectrometry. *Analytical Chemistry* 90 (7): 4669–4676. https://doi.org/10.1021/acs.analchem.7b05241.

25 Leblanc, Y., Ramon, C., Bihoreau, N., and Chevreux, G. (2017). Charge variants characterization of a monoclonal antibody by ion exchange chromatography coupled on-line to native mass spectrometry: case study after a long-term storage at +5 °C. *Journal of Chromatography B: Analytical Technologies in the Biomedical and Life Sciences* 1048: 130–139. https://doi.org/10.1016/j.jchromb.2017.02.017.

26 Yan, Y., Liu, A.P., Wang, S. et al. (2018). Ultrasensitive characterization of charge heterogeneity of therapeutic monoclonal antibodies using strong cation exchange chromatography coupled to native mass spectrometry. *Analytical Chemistry* 90 (21): 13013–13020. https://doi.org/10.1021/acs.analchem.8b03773.

27 Leblanc, Y., Faid, V., Lauber, M.A. et al. (2019). A generic method for intact and subunit level characterization of mAb charge variants by native mass spectrometry. *Journal of Chromatography B: Analytical Technologies in the Biomedical and Life Sciences* 1133 (October): 121814. https://doi.org/10.1016/j.jchromb.2019.121814.

28 Ehkirch, A., Hernandez-Alba, O., Colas, O. et al. (2018). Hyphenation of size exclusion chromatography to native ion mobility mass spectrometry for the analytical characterization of therapeutic antibodies and related products. *Journal of Chromatography B: Analytical Technologies in the Biomedical and Life Sciences* 1086 (March): 176–183. https://doi.org/10.1016/j.jchromb.2018.04.010.

29 Goyon, A., D'Atri, V., Colas, O. et al. (2017). Characterization of 30 therapeutic antibodies and related products by size exclusion chromatography: Feasibility assessment for future mass spectrometry hyphenation. *Journal of Chromatography B* 1065–1066: 35–43. https://doi.org/10.1016/j.jchromb.2017.09.027.

30 Lazar, A.C., Wang, L., Blättler, W.A. et al. (2005). Analysis of the composition of immunoconjugates using size-exclusion chromatography coupled to mass spectrometry. *Rapid Communications in Mass Spectrometry* 19 (13): 1806–1814. https://doi.org/10.1002/rcm.1987.

31 Farsang, E., Guillarme, D., Veuthey, J.L. et al. (2020). Coupling non-denaturing chromatography to mass spectrometry for the characterization of mAbs and related products. *Journal of Pharmaceutical and Biomedical Analysis* 185: 113207.

32 Vanhoenacker, G., Vandenheede, I., David, F. et al. (2015). Comprehensive two-dimensional liquid chromatography of therapeutic monoclonal antibody digests. *Analytical and Bioanalytical Chemistry* 407 (1): 355–366. https://doi.org/10.1007/s00216-014-8299-1.

33 Stoll, D.R., Harmes, D.C., Danforth, J. et al. (2015). Direct identification of rituximab main isoforms and subunit analysis by online selective comprehensive two-dimensional liquid chromatography-mass spectrometry. *Analytical Chemistry* 87 (16): 8307–8315. https://doi.org/10.1021/acs.analchem.5b01578.

34 Ehkirch, A., Goyon, A., Hernandez-Alba, O. et al. (2018). A novel online four-dimensional SEC × SEC-IM × MS methodology for characterization of monoclonal antibody size variants. *Analytical Chemistry* 90 (23): 13929–13937. https://doi.org/10.1021/acs.analchem.8b03333.

35 He, Y., Friese, O.V., Schlittler, M.R. et al. (2012). On-line coupling of size exclusion chromatography with mixed-mode liquid chromatography for comprehensive profiling of biopharmaceutical drug product. *Journal of Chromatography A* 1262: 122–129. https://doi.org/10.1016/j.chroma.2012.09.012.

36 Ehkirch, A., D'Atri, V., Rouviere, F. et al. (2018). An online four-dimensional HIC × SEC-IM × MS methodology for proof-of-concept characterization of antibody drug conjugates. *Analytical Chemistry* 90 (3): 1578–1586. https://doi.org/10.1021/acs.analchem.7b02110.

37 Sarrut, M., Fekete, S., Janin-Bussat, M.-C. et al. (2016). Analysis of antibody-drug conjugates by comprehensive on-line two-dimensional hydrophobic interaction chromatography x reversed phase liquid chromatography hyphenated to high resolution mass spectrometry. *Journal of Chromatography B* 1032: 91–102. https://doi.org/10.1016/j.jchromb.2016.06.049.

38 Stoll, D.R., Harmes, D.C., Staples, G.O. et al. (2018). Development of comprehensive online two-dimensional liquid chromatography/mass spectrometry using hydrophilic interaction and reversed-phase separations for rapid and deep profiling of therapeutic antibodies. *Analytical Chemistry* 90 (9): 5923–5929. https://doi.org/10.1021/acs.analchem.8b00776.

39 Stoll, D., Danforth, J., Zhang, K., and Beck, A. (2016). Characterization of therapeutic antibodies and related products by two-dimensional liquid chromatography coupled with UV absorbance and mass spectrometric detection. *Journal* *of Chromatography B* 1032: 51–60. https://doi.org/10.1016/j.jchromb.2016.05.029.

40 Fekete, S., Beck, A., Veuthey, J.-L., and Guillarme, D. (2019). Proof of concept to achieve infinite selectivity for the chromatographic separation of therapeutic proteins. *Analytical Chemistry* 91 (20): 12954–12961. https://doi.org/10.1021/acs.analchem.9b03005.

41 Goyon, A., Dai, L., Chen, T. et al. (2019). From proof of concept to the routine use of an automated and robust multi-dimensional liquid chromatography mass spectrometry workflow applied for the charge variant characterization of therapeutic antibodies. *Journal of Chromatography* . *Journal of Chromatography A*: 460740. https://doi.org/10.1016/j.chroma.2019.460740.

1.5

Optimization Strategies in HPLC for the Separation of Biomolecules

Lisa Strasser, Florian Füssl, and Jonathan Bones

Characterisation and Comparability Lab, NIBRT LTD., Foster's Avenue, Mount Merrion, Blackrock, Co. Dublin, A94 X099, Dublin, Ireland

High performance liquid chromatography (HPLC) finds wide application for the separation of biomolecules, offering different operation modes, which separate by physiochemical properties like size, charge, or hydrophobicity. Each operation mode is versatile and the analyst can influence a number of parameters to tailor a method to the respective experimental needs. However, optimization of a method often entails trade-offs, for example if high selectivity is the ultimate goal, the method will likely not be as fast as possible. Or when online mass spectrometric (MS) detection is desired, MS-compatible buffers have to be chosen, which can cause poorer separation or peak shape than conventional buffers. In any case, before approaching method optimization, it needs to be determined whether the main goal is improving the separation, sensitivity, or speed and how to prioritize these metrics.

1.5.1 Optimizing a Chromatographic Separation

Many chromatographic run parameters have profound impact on the quality of the separation and can be tuned to get the best possible outcome. The separation can be characterized by metrics like selectivity, resolution, or efficiency. A separation can be improved in more than one way, for example by increasing the space between peaks or by forcing a sharp peak shape. A potent way to obtain better separations is the use of more efficient columns. To improve a chromatographic separation simply the use of a longer column is often sufficient as can be seen in Figure 1.5.1.

Optionally, a column of same dimensions can be used but with smaller particle size. It should be mentioned however that a decrease in particle size is increasing the system backpressure and can subject biomolecules to shear stress, potentially causing undesired phenomena like fragmentation or aggregation. This can be counteracted by lowering the sample amount or by adjustments of the flow rate. Also, the use of superficially porous particles instead of porous particles is an effective way to

Optimization in HPLC: Concepts and Strategies, First Edition. Edited by Stavros Kromidas.
© 2021 WILEY-VCH GmbH. Published 2021 by WILEY-VCH GmbH.

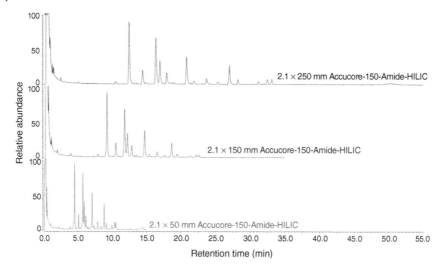

Figure 1.5.1 Separation of released and labeled N-glycans from a monoclonal antibody by HILIC chromatography with columns of three different lengths. Mobile phase A was 50 mM ammonium formate, pH 4.4, and mobile phase B was acetonitrile. The gradients applied were 80–60%B in 48.1, 31.05, and 13.45 minutes for columns of 250, 150, and 50 mm length, respectively. The flow rates were 1.4, 1.3, and 1 ml/min and injection volumes 33, 20, and 11.5 μl, respectively. Column temperature was 60 °C and fluorescence detection was performed.

increase the separation efficiency. The lower porosity of core–shell particles allows for good mass transfer what, in consequence, prevents peak broadening [1].

Increasing the temperature usually leads to a faster separation with earlier retention but can also increase the efficiency of the separation. Figure 1.5.2 shows the separation of monoclonal antibody (mAb) charge variants via pH-gradient-driven cation exchange chromatography (CEX). An increase from 20 to 60 °C leads to an improved selectivity, which can especially be seen when comparing the separation of the main peak with the most prominent acidic variant eluting shortly before.

Unsuitable sample diluents can be a reason for insufficient column binding and poor chromatographic performance. Protein formulations are often solutions of high ionic strength. If such a solution is injected for ion exchange chromatography (IEC) with a salt gradient, the formulation could cause insufficient protein binding to the column. The opposite is true in case of hydrophobic interaction chromatography (HIC) where the gradient starts with a mobile phase of high ionic strength. In cases where IEC with a pH-gradient is performed, the pH of the sample diluent should be adjusted in a way to prevent insufficient column binding. The same principle is crucial for reversed-phase liquid chromatography (RP-LC) where the sample should not be dissolved in organic solvents, while the opposite is true for modes like normal-phase liquid chromatography (NP-LC) or hydrophilic interaction liquid chromatography (HILIC).

Some chromatographic modes, which operate via gradient separation, can greatly benefit from the optimization of the gradient. Figure 1.5.3 shows a pH-gradient-

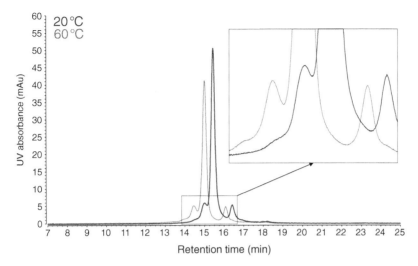

Figure 1.5.2 Separation of intact NIST antibody via pH-gradient-based CEX chromatography at different temperatures to show differences in selectivity. Separation was performed on a Thermo Fisher Scientific MabPac SCX-10 column (4 × 250 mm, 10 μm particle size), with a flow rate of 1 ml/min and UV detection at 280 nm. For gradient elution, Thermo Fisher Scientific CX-1 buffers were used with a gradient of 30–80% B in 30 minutes. Injection volume was 5 μl.

Figure 1.5.3 Separation of intact infliximab via pH-gradient based CEX chromatography at different temperatures to show differences in selectivity. Separation was performed on a Thermo Fisher Scientific MabPac SCX-10 column (2.1 × 50 mm, 5 μm particle size), with a flow rate of 0.4 ml/min and UV detection at 280 nm. Mobile phase A was 25 mM ammonium bicarbonate and 30 mM acetic acid, mobile phase B was 10 mM ammonium hydroxide and 2 mM acetic acid. Injection volume was 5 μl.

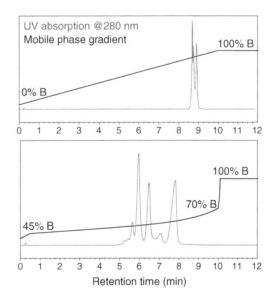

based CEX separation of intact infliximab charge variants with different gradients. Whereas for a gradient from 0% to 100% B in 10 minutes only 2–3 peaks can partially be distinguished, a refinement of the gradient slope yields the separation of at least 7 variants of considerably different abundance.

Another way to optimize the separation is to minimize undesired analyte-column interactions. This can be performed by modification of the stationary phase, by e.g. end capping of silanol groups, or by modifications of the mobile phase. In size exclusion chromatography (SEC), for example, the use of inert and uncharged column material can prevent undesired analyte-column interactions, which can cause peak broadening and a loss of efficiency (see also Chapter 1.4). Another way to avoid such interactions is the use of a sufficient amount of ions in the mobile phase to saturate charged groups on the stationary phase, thereby making them less attractive to analyte interaction. In case of the separation of proteins, the pH of the mobile phase can also be adjusted in a way to be close to the isoelectric point of the protein under investigation. Choosing such conditions will lead to a close to neutral net charge of the protein and lower the likelihood for interactions. Albeit seeming trivial, dead volume can play a significant role for the quality of a separation, especially in separation modes like SEC where no analyte focusing on the column occurs. Hence, capillary diameter and length, injection loop size, and injection volumes should be kept low when aiming for maximum efficiency.

A high degree of analyte-column interaction is not only a problem in SEC but also in other separation modes. HIC chromatography, for example, is frequently used for the characterization of cysteine-conjugated antibody–drug conjugates (ADCs), which can be highly hydrophobic and labile molecules. HIC columns are available with stationary phases of varying hydrophobicity, but not all of them are equally suited for the analysis of hydrophobic compounds like ADCs. When using a very nonpolar stationary phase, highly conjugated ADC species might not or only partially elute from the column by an inverse salt gradient, which can compromise separation or analysis in general. This can be counteracted by the addition or increase of the amount of organic solvent in the mobile phase B. However, too much organic solvent can, on the other hand, possibly lead to denaturation of the molecule or the dissociation of heavy and light chains of the antibody. Another option for improving a HIC separation is the use of a different salt or ionic strength. The best mobile phase conditions are always dependent on the stationary phase used and the particular molecule(s) under investigation. A guideline for the use of the ideal salt is represented by the Hofmeister series, which lists salts together with their ability to salt-out proteins [2].

Some chromatographic modes benefit from the addition of mobile phase modifiers. Mixtures of proteins and peptides, for example, are often separated by reversed-phase chromatography with mobile phases containing ion pairing reagents. The most commonly used modifiers are formic acid (FA), trifluoroacetic acid (TFA), and acetic acid in a range of 0.01–0.5% (v/v). The type of modifier used is dependent on the main experimental aim. TFA, for example, is known to be one of the best choices when the ultimate goal is the best possible separation and peak shape. Modern HPLC of proteins and peptides is often performed in conjunction with mass spectrometry. In such cases, TFA may not be the right choice as it usually causes ion suppression, which is going along with signal loss in the MS instrument. Alternatively, the concentration of TFA can be lowered to values of 0.01%, or other modifiers like formic acid can be used. Figure 1.5.4 shows RP-LC separations of trastuzumab and infliximab subunits after IdeS digestion and reduction, with TFA and FA as mobile phase additives. TFA is better

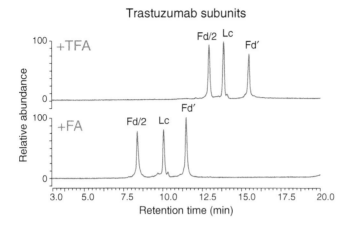

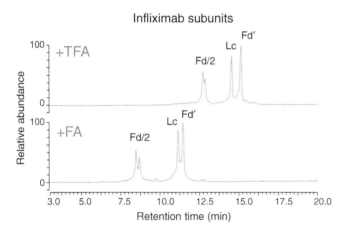

Figure 1.5.4 Ion pair (IP)-RP-HPLC separation of trastuzumab and infliximab subunits after IdeS digestion and reduction. Separation was performed on a Thermo Fisher Scientific MabPac RP column (2.1 × 100 mm), with a flow rate of 0.3 ml/min and UV detection at 280 nm. Mobile phase A was water with 0.1% of the respective modifier, mobile phase B was 90% acetonitrile with 0.1% of the respective modifier. A gradient from 25–45% B in 16 minutes was applied, 2 µl of drug product were injected on column.

in promoting separation and sharpness of peaks, while formic acid resulted in superior MS data.

1.5.2 Optimizing the Speed of an HPLC Method

Speed can be essential for various applications, for example, for high-throughput screening of attributes of therapeutic proteins early in the biopharmaceutical production pipeline. One way to increase the analysis speed while maintaining column efficiency is, for example, to use a small inner-diameter column with smaller-size particles at a higher flow rate, which will however, come at the cost of an increased system backpressure. Thus, high-speed methods require UHPLC systems capable of

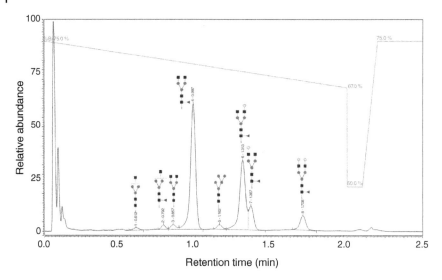

Figure 1.5.5 HILIC separation of released and labeled N-glycans of a monoclonal antibody. The separation column was a Thermo Fisher Scientific Accucore 150-Amide HILIC column with dimensions of 2.1 × 50 mm and 2.6 μm particle size. Mobile phase A was 50 mM ammonium formate, pH 4.5, mobile phase B was acetonitrile. The gradient applied was 75–67% B in 2 minutes followed by 0.1 minutes of column flushing and 0.4 minutes of equilibration. The flow rate applied was 2.2 ml/min, the column oven temperature was 60 °C, fluorescence detection was performed, and 5 μl of sample were injected.

operating at extreme backpressure and column formats, which can withhold pressures of up to several hundred bars. Figure 1.5.5 shows an example for ultrafast N-glycan analysis via HILIC. Released and labeled N-glycans of a monoclonal antibody can be separated within a gradient time of only two minutes.

Another example is the ultrafast separation of monoclonal antibody charge variants, which can be a potent tool for candidate screening in early stages of the biopharmaceutical process and product pipeline. Figure 1.5.6 shows the ultrafast charge variant separation via pH-gradient-based CEX chromatography of the mAb trastuzumab with a gradient time of only three minutes.

In case the system capabilities are not sufficient to deal with the backpressure caused by the run parameters desired, there are still ways to maintain the pressure at a reasonable level. One way is to switch to mobile phases of lower viscosity. This obviously only makes sense if the desired selectivity can be maintained. Alternatively, the mobile phase viscosity can be lowered by increasing the column oven temperature, which in many cases does not only reduce the viscosity of the mobile phase and hence the pressure burden but also leads to earlier elution, allowing for an even shorter run time. In cases where gradient separation yields high chromatographic resolution, the gradient time can simply be shortened by making a steeper gradient. The overall efficiency will decrease, but this can be accepted as long as a sufficient separation can be obtained. Considering these suggestions, methods that are no longer than 2–4 minutes can be developed, which, albeit being short, still offer sufficient separation.

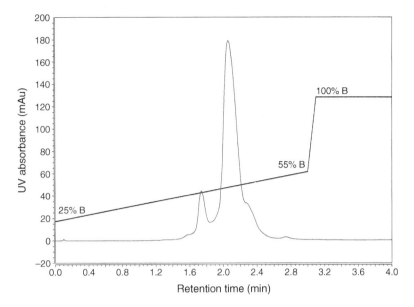

Figure 1.5.6 Separation of trastuzumab charge variants within only three minutes of gradient time. UV trace and separation gradient are indicated. The column was a Thermo Fisher Scientific MAbPac SCX-10 RS with dimensions of 2.1 × 5 mm and a particle size of 5 μm. Separation is based on a pH-gradient performed with Thermo Fisher Scientific CX-1 buffers. The flow rate applied was 1 ml/min, the column oven temperature was 60 °C, and 2.4 μl of drug product were injected on column. Detection was performed at a wavelength of 280 nm.

1.5.3 Optimizing the Sensitivity of an HPLC Method

When optimizing sensitivity, it is important to keep in mind how it relates to other metrics such as speed and selectivity, and also that it potentially comes at the cost of reduced robustness.

Even though not directly related to the applied HPLC method itself, sample preparation is the first step of the analysis workflow ultimately influencing sensitivity of the method. Sample pretreatment such as solid-phase extraction (SPE) or buffer exchange using molecular weight cut-off (MWCO) filters can be used to remove certain components of the sample matrix, which could interfere with the separation or detection [3]. Matrix effects are especially known to have an effect on the sensitivity of LC–MS-based applications [4]. Importantly, sample preparation procedures can also have detrimental effects on sensitivity. An example would be spin filters or tubes irreversibly retaining analytes such as proteins, which can lead to reduced sample concentration and ultimately poor or biased results.

A very effective way to improve the sensitivity of a HPLC method is to change the column dimensions. Reducing the internal diameter of the separation column drastically decreases the peak volume and thereby the dilution of the analyte. This is also beneficial in case of MS detection as a reduced column dimension goes along with lower flow rates, which make it more compatible with electrospray ionization.

Another positive effect of using low internal-diameter columns is the better conduction of temperature. Not having a homogeneous temperature across the separation results in peak broadening and thus reduced sensitivity. Nevertheless, the low flow rate used for small-dimensional columns significantly affects the analysis time and might also have a negative effect on the robustness of the method as it often entails a more complex and fragile instrumental setup. An important consideration, which is also related to the used column (-dimension), is the sample volume injected. Especially for analytes present at low concentration, it is important to inject the maximum amount possible while not exceeding the column-capacity as this would again result in peak broadening [5] (exception: sample focusing in gradient elution).

Additionally, the detector used has a tremendous impact on the sensitivity of the applied HPLC method. In most applications for biomolecules, UV-based detection is used. As defined by the Beer–Lambert law, UV detection is a concentration-dependent detection method, which also relies on the volume and path length of the detector cell. In general, the volume of the detector should never be higher than a 10th of the expected peak volume. Additionally, the detection rate should be high enough to provide at least 25 data points across a peak allowing for accurate quantification.

As a rule of thumb, an improved signal-to-noise ratio and thus lower limits of detection and quantification can be achieved by applying adequate sample preparation as well as by downscaling the separation method. In any case, the user should always be aiming for narrow peaks in combination with a suitable detection strategy to allow for sensitive and accurate HPLC analysis of biomolecules.

1.5.4 Multidimensional Separations (See also Chapter 1.1)

Multidimensional separation modes are commonly used to obtain increased resolving power in HPLC. By combining different separation modes, analytes coeluting in the first dimension can often be sufficiently separated in the following dimension. Thus, the most important considerations are the orthogonality of the methods used as well as their compatibility. This means that different separation mechanisms are combined in a way to result in increased peak capacity, while analytes after the first dimension are ideally within conditions that allow for direct injection into the second dimension separation. Common combinations of separation modes are summarized in Table 1.5.1.

One important decision that needs to be made is whether the two-dimensional (2D) separation is done online or offline. In online 2D-LC, analytes eluting from the first dimension are directly transferred into the second dimension, increasing the overall peak capacity without substantially increasing the analysis time and it can easily be automated. However, for online applications, the analyte diluent after the first dimension has to be compatible with the required starting conditions of the second dimension. Offline 2D-LC does not require direct compatibility and allows for more flexibility. Samples are collected from the first dimension

Table 1.5.1 Commonly combined LC–separation methods and their orthogonality and compatibility.

Mode	RP/RP	IEC/RP	IEC/SEC	SEC/RP	HILIC/RP	IEC/HIC	HIC/RP
Orthogonality	✓✓	✓✓✓	✓✓	✓✓✓	✓	✓✓✓	✓✓
Compatibility	✓✓	✓✓✓	✓✓	✓✓✓	✓✓	✓✓	✓✓

Notes: HIC, hydrophobic interaction chromatography; HILIC, hydrophilic interaction liquid chromatography; IEC, ion exchange chromatography; RP, reversed phase; SEC, size exclusion chromatography.

before subjecting them to the next separation mode. Therefore, even though more tedious and therefore error-prone, samples can be for instance buffer-exchanged or concentrated before continuing the analysis. Additionally, methods can be individually optimized to gain maximum peak capacity.

Fractions of samples derived from the first-dimensional separation can be collected in different ways. In heart-cutting 2D-LC peaks of interest are collected in a targeted manner, while in comprehensive 2D-LC the first-dimensional run gets fully sampled to be further analyzed [6]. In any way, multidimensional separation methods have been proven particularly useful when analyzing highly complex samples such as cell lysates or if the analytes of interest are present at considerably different levels of abundance. An interesting example is the application of high pH–low pH 2D-LC–MS^E for the analysis of host-cell protein (HCP) contaminants in purified monoclonal antibody samples [7]. High pH reversed-phase (RP) fractionation prior to low pH RP-LC coupled to mass spectrometry has been shown to significantly increase the output when analyzing highly complex cell lysate samples [8, 9]. In case of HCP analysis, however, complexity of the sample is not necessarily the problem. In fact, the purified antibodies are present at much greater abundance than the contaminants, which has a negative impact on HCP detection when using mass spectrometry.

Taken together, when selecting compatible and orthogonal multidimensional LC methods, increased peak capacity and greater selectivity can be achieved allowing for the analysis of a multitude of complex biological samples.

1.5.5 Considerations for MS Detection (See also Chapter 1.3)

Some modes are intrinsically well suited for mass spectrometric detection, while others are not. RP or HILIC chromatography, for example, operate without the need for nonvolatile ions in the mobile phases, and usually employed organic solvents and additives such as formic acid greatly enhance analyte desolvation and ionization in the LC–MS interface. Other chromatographic modes like SEC, IEC, or HIC chromatography require mobile phases of certain ionic strength and thus can pose extreme challenges to MS detection. The only way to directly couple such separation modes is the exclusive use of volatile salts like ammonium

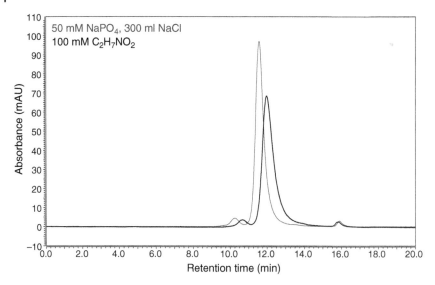

Figure 1.5.7 SEC separation of a monoclonal antibody-based protein with two different mobile phases. Separation column was a Thermo Fisher Scientific MAbPac SEC-1 column with dimensions of 4 × 300 mm and a particle size of 5 µm. The flow rate was 0.2 ml/min, temperature was 25°C, separation was performed under isocratic conditions, and 6 µl of sample were injected.

acetate, ammonium formate, and ammonium bicarbonate. However, volatile salts might not always result in equal separation performance of biomolecules, which is why online MS detection can require trade-offs in regards to separation or peak shape. Figure 1.5.7 shows the separation of monomers and aggregates of a monoclonal antibody-based biopharmaceutical with conventional SEC buffers and with MS-compatible ammonium acetate. In both cases, monomers (RT 11.5–12 minutes) and aggregates (RT ~10 minutes) are separated, but the efficiency and peak shape is clearly superior when using $NaPO_4$ with NaCl.

Even though volatile salts are in general MS friendly, high ionic strength still compromises MS detection, and therefore the salt concentration should be kept as low as experimentally possible at all times. Such interferences are especially problematic when modes like salt-gradient-based IEC or HIC are interfaced, which operate directly via a salt gradient. Size exclusion chromatography as well as pH-gradient based IEC can in contrast operate at ionic strengths of <100 mM, allowing for the acquisition of high-quality MS data. Another consideration is that the flow rate applied has to be kept low to allow for efficient desolvation and ionization. This is especially crucial for separation modes operating with aqueous mobile phases since water, when compared to e.g. organic solvents, has a relatively low volatility. A general rule of thumb is that the flow rate should not exceed 0.5 ml/min to guarantee full desolvation. However, in cases where higher flow rates are required, a reduction is possible using a post-column flow splitter before MS analysis. Something to consider with regard to reproducibility is that the concentration or pH of the mobile phases may change over time when volatile ingredients are used. If, for example, a separation cannot be reproduced after having used the same mobile phases for several days, it is advisable to prepare fresh ones. One phenomenon, which can also be problematic in case of HPLC–MS employing salty mobile phases, is the occurrence of salt adducts. This can, on the one hand, complicate MS data analysis and also decrease sensitivity but it can usually be avoided by the utilization of adequate source and ion-transfer parameters offered by today's state-of-the art mass spectrometers.

1.5.6 Conclusions and Future Prospects

As summarized in Figure 1.5.8, gathering information about the analyte(s) as well as having a well-defined experimental aim allows for tailored optimization of the

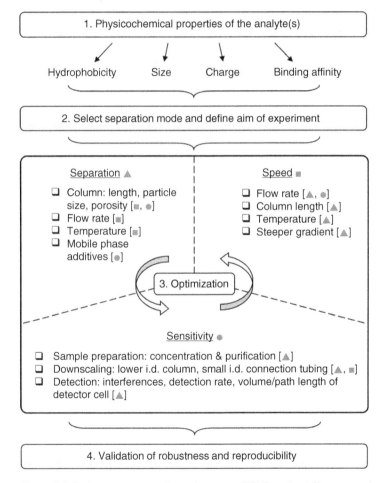

Figure 1.5.8 Important steps in setting up an HPLC method. Shown are the three main sections that allow for optimization: separation, speed, and sensitivity of the method. Symbols (sensitivity, ●x025CF; speed, ■x025A0; separation, ▲x025B2) show how those metrics are affecting each other (e.g. changing the flow rate to improve the separation affects the speed (■x025A0) of the method).

applied HPLC method. However, since separation/selectivity, speed, and sensitivity are interconnected, the analyst has to be aware of how the optimization of a single parameter will likely affect all others.

The analysis of biomolecules constitutes a large field ranging from small molecules, such as metabolites, to larger structures like nucleic acids or proteins, protein complexes, or even intact viruses, each representing unique analytical challenges. Next to academia pushing basic research, large industries like the pharmaceutical sector are the main driving force for continuous developments of more robust, sensitive, fast, selective, and more automated HPLC methods for biomolecules to monitor processes and to ensure product quality. In this regard, automation is gaining increasing importance. Using process analytical technology (PAT) frameworks will ultimately allow for real-time monitoring and analysis of critical quality and performance attributes. Employing faster methods will allow for high-throughput screening during process development.

References

1 Brusotti, G. et al. (2018). Advances on size exclusion chromatography and applications on the analysis of protein biopharmaceuticals and protein aggregates: a mini review. *Chromatographia* 81 (1): 3–23.
2 Fekete, S. et al. (2016). Hydrophobic interaction chromatography for the characterization of monoclonal antibodies and related products. *Journal of Pharmaceutical and Biomedical Analysis* 130: 3–18.
3 Buszewski, B. and Szultka, M. (2012). Past, present, and future of solid phase extraction: a review. *Critical Reviews in Analytical Chemistry* 42 (3): 198–213.
4 Trufelli, H. et al. (2011). An overview of matrix effects in liquid chromatography–mass spectrometry. *Mass Spectrometry Reviews* 30 (3): 491–509.
5 Zhang, J. et al. (2012). Optimization for speed and sensitivity in capillary high performance liquid chromatography. The importance of column diameter in online monitoring of serotonin by microdialysis. *Journal of Chromatography. A* 1251: 54–62.
6 Brandão, P.F., Duarte, A.C., and Duarte, R.M.B.O. (2019). Comprehensive multidimensional liquid chromatography for advancing environmental and natural products research. *TrAC Trends in Analytical Chemistry* 116: 186–197.
7 Farrell, A. et al. (2015). Quantitative host cell protein analysis using two dimensional data independent LC–MS(E). *Analytical Chemistry* 87 (18): 9186–9193.
8 Delmotte, N. et al. (2007). Two-dimensional reversed-phase × ion-pair reversed-phase HPLC: an alternative approach to high-resolution peptide

separation for shotgun proteome analysis. *Journal of Proteome Research* 6 (11): 4363–4373.

9 Yang, F. et al. (2012). High-pH reversed-phase chromatography with fraction concatenation for 2D proteomic analysis. *Expert Review of Proteomics* 9 (2): 129–134.

1.6

Optimization Strategies in Packed-Column Supercritical Fluid Chromatography (SFC)

Caroline West

University of Orleans, ICOA, CNRS UMR 7311, Orleans, France

Supercritical fluid chromatography (SFC) is now mostly conducted with mobile phases comprising pressurized carbon dioxide and a cosolvent, whatever the exact state of the fluid. As packed-column SFC is largely dominating the scene over capillary SFC, the present chapter will only be related to the former. Modern packed-column SFC is a very powerful toolbox comprising many different optimization parameters with different action levels: some of them, like stationary-phase and mobile-phase composition, have strong influence on retention, selectivity, and peak shapes, thereby strongly affecting resolution, while others (column temperature, internal pressure, and mobile-phase flow rate) are fine-tuning parameters. The way of developing and optimizing a method will then usually follow this order: first, selection of a chromatographic system comprising stationary-phase and mobile-phase nature (cosolvent and possible additives) with scouting experiments; second, optimization of mobile-phase composition (isocratic or gradient), temperature, pressure, and flow rate, possibly with design of experiments. Effective optimization requires a good understanding of all effects related to these parameters.

Looking at typical SFC instruments (Figure 1.6.1), the different parts of the instrument are mostly identical to modern high performance liquid chromatography (HPLC) instruments: pumping system, sample injector, column oven to fit several columns and control operating temperature, UV or diode-array detection, then mass spectrometric detection. The pumping system must be adapted to be able to pump the gaseous carbon dioxide as a cold liquid and then mix it to the liquid cosolvent. A make-up fluid is sometimes necessary prior to introduction of the fluid into the mass spectrometer. Finally, a backpressure regulator is used to maintain the desired operating pressure. It may be placed, as in Figure 1.6.1, after a splitter diverts the flow to the mass spectrometer, or inline between the UV and mass spectrometry detectors.

In Figure 1.6.1, the system parts that will need to be optimized are indicated in color, and in suggested order of optimization: firstly, the choice of stationary phases (SP); secondly, the choice of mobile-phase composition (MP); and thirdly,

Optimization in HPLC: Concepts and Strategies, First Edition. Edited by Stavros Kromidas.
© 2021 WILEY-VCH GmbH. Published 2021 by WILEY-VCH GmbH.

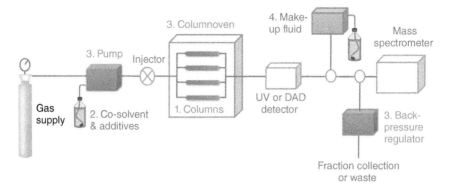

Figure 1.6.1 Packed-column SFC instrument with the different features to optimize in the course of a method development.

the temperature, pressure, and flow rate. They will be described in this order in the following.

Note that make-up fluid and mass spectrometric detection may also require some optimization and will be only briefly evoked in Section 1.6.4 as this chapter is mostly concerned with chromatographic optimization.

1.6.1 Selecting a Stationary Phase Allowing for Adequate Retention and Desired Selectivity

1.6.1.1 Selecting a Stationary Phase for Chiral Separations

Chiral SFC has been a long-time favorite, particularly for pharmaceutical applications. The packed columns available for chiral SFC are mostly the same as those employed in HPLC, with a few exceptions of HPLC columns that are normally operated in aqueous media and should preferably not be operated in water-free mobile phase (like protein-based chiral stationary phases (CSP)).

Polysaccharide-based stationary phases are by far the most commonly used in SFC, as they are highly versatile, providing enantioresolution for a large range of analytes, whether they are neutral, acidic, or basic. In addition, there are now a large number of commercially available polysaccharide CSP, with different chiral selectors, mostly based on amylose or cellulose modified with aromatic groups, and providing interesting complementarity. Some of the most recent versions of these polysaccharide stationary phases are no longer coated on silica support, which made them somewhat susceptible to damage with strong solvents, but immobilized on the silica support, allowing for greater possibilities in mobile-phase optimization. Recent developments have also brought small-particle-size-stationary phases, now based on sub-2-µm fully porous silica particles, offering improved column efficiency. Aside from their versatility, another reason for the long-term preference for polysaccharide CSP is that chiral SFC analyses are often developed for the purpose of transferring the method to preparative scale. When purification of stereoisomers is desired, the high loadability of polysaccharide stationary phases allows for high productivity (mass of product purified according to time and/or mass of stationary phase) [1]. Immobilized polysaccharides are also interesting in this respect because strong solvents may be used in the mobile phase to further improve the mass of product injected in each run [2].

Secondly, brush-type stationary phases (like the Pirkle-type phases and *Cinchona* alkaloid-type phases) usually offer complementary selectivity to polysaccharide CSP, and they may comprise different possibilities for interacting with the analytes, like ionic interactions. They also have the added interest of enantiomeric or diastereomeric stationary phases being available. This is rather useful when a particular elution order of the analyte enantiomers is desired (minor peak eluting first for enantiomeric ratio measurement, or preferred enantiomer eluting first for purification purposes). Thus, replacing one enantiomeric stationary phase by the other (i.e. (R, R) replacing (S, S)) will simply reverse the elution order of the enantiomeric analytes (i.e. $R > S$ or $S < R$). An excellent example of such use can be found in [3]. Most recent versions of some of these CSP are also now available with fully porous sub-2-μm particles, again favoring higher column efficiency.

Synthetic polymer CSP were often cited in the past but are now rarely cited. Macrocyclic glycopeptide CSP, on the other hand, are increasingly used. They come with a variety of antibiotic chiral selectors anchored to silica support, most of them bearing ionizable functions that may vary in ionization state depending on mobile-phase apparent pH [4]. Changing mobile-phase additives (acids or bases) is thus a strong optimization parameter on such CSP. Recently, commercialized phases are now available on sub-3-μm superficially porous silica particles to improve column efficiency and favor fast analyses. Indeed, superficially porous particles generate less pressure drop along the column than fully porous ones, allowing for faster flow rates.

A summary of the remarkable features of frequently used CSP is presented in Figure 1.6.2.

Even considering all these elements, chromatographers are well aware that column selection for chiral separations is mostly a trial-and-error process, with little contribution of the chromatographer's experience being involved, because chiral separations are mostly unpredictable. The usual procedure is then to have a set of columns selected for their complementarity of separation behaviors, and to do a screening step of all these columns to identify a starting point for further method optimization. Typically, a first screening set of 4–8 columns would be used, which may be extended with a second set of columns, should the first set remain unsuccessful in providing a debut resolution. The first screening step also often involves mobile-phase variation, as will be further discussed below.

When screening for chiral separations, it is now customary to employ short columns (typically 50-mm column length). A debut resolution (like $R_s = 0.6$) could then be transformed into a higher resolution with an identical stationary

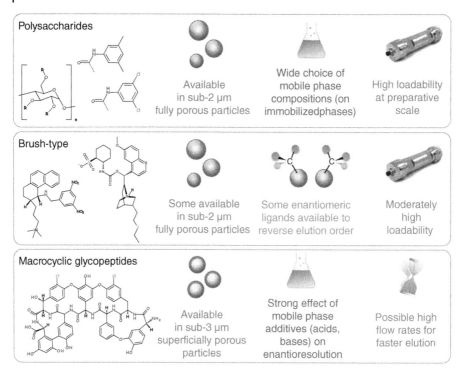

Figure 1.6.2 Enantioselective stationary phases principally used in chiral SFC and their significant features.

phase in a longer column, thanks to improved efficiency, or after mobile-phase optimization.

1.6.1.2 Selecting a Stationary Phase for Achiral Separations

The process of selecting a column for achiral separations is often similar to the one described above for chiral separations, although it would not always need to be. Some basic understanding of the interactions that a stationary phase can establish with the analytes is necessary for a clever, informed choice of stationary phases to test.

The stationary phases available for achiral SFC are actually very numerous because all the columns available for achiral HPLC, whether intended to be used in reversed-phase (RP), normal-phase (NP), or hydrophilic interaction (HILIC) modes, can be operated in SFC conditions. In addition, some manufacturers have designed stationary phases specifically for SFC use. In this case, the ligands are sometimes original, different from anything available in HsPLC. A typical example is that of the famous 2-ethylpyridine-bonded silica stationary phase that is now offered in a variety of similar ligands (4-ethylpyridine, pyridyl-amide, or picolylamine, to name a few). Other stationary phases are based on ligands that are mostly identical to those of HPLC phases but may include some modifications to offer a better

1.6.1 Selecting a Stationary Phase Allowing for Adequate Retention and Desired Selectivity

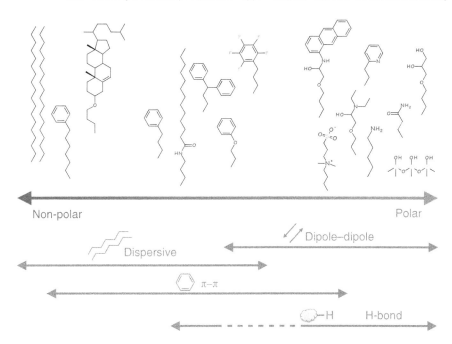

Figure 1.6.3 Typical stationary phases employed in achiral SFC ranked according to their polarity and the possible interactions they can establish with the analytes.

robustness when employed in carbon dioxide–solvent mobile phases. Indeed, it was shown that the stationary phase may progressively lose retentive capability [5], possibly due to the reaction between silanol groups and the alcohol molecules of the mobile-phase cosolvent [6]. This reaction may be prevented by (i) different ligands (with longer, more polar spacer arms) as was proposed by some manufacturers [6] or (ii) use of a small proportion of water in the mobile phase (typically 2%) to reverse the reaction of methyl-silyl formation.

As appears in Figure 1.6.3, the available stationary phases can be simply ranked according to their polarity, going from the least polar alkyl-bonded stationary phases normally used in RP-HPLC, to the most polar bare silica gel employed in NP-HPLC and HILIC.

Figure 1.6.3 also highlights the type of interactions that may be expected from each stationary phase [7, 8]. From these, the typical analytes that should be well retained and separated on them can be deduced. Hydrophobic compounds (i.e. nonpolar lipids, polynuclear aromatic hydrocarbons) will be best retained on nonpolar phases and separated according to the length of alkyl chain, number of double bonds, and/or number or aromatic rings. Analytes possessing polar functions will be retained on polar phases and separated according to the number of hydrogen-bonding groups. Some analytes may be equally well retained on polar and nonpolar stationary phases, but the separation will differ. For instance, lipids can usually be retained on both types of stationary phases, but a polar stationary phase will provide a class separation (based on polar head), while a nonpolar stationary phase will provide intraclass separation (based on hydrophobic tail).

In cases when you have no idea what would be the best choice for your sample, either because it is mostly unknown, or because of its high complexity, it is advisable to have a set of complementary stationary phases to use for initial screening experiments. Four to six adequately selected columns are usually sufficient to solve most problems, where "adequately selected" means that they should have different polarities and provide a different blend of interactions to retain the analytes. Note that stationary phases that are situated in the middle of the polarity axis are more versatile and more likely to solve a variety of problems. A typical example is that of polar aromatic ligands or that of non-endcapped C18 stationary phases, which have been found useful for a large variety of samples [9].

In the end, when no totally satisfactory solution is found, there is always the possibility to couple different columns in series to take advantage of complementary selectivities [10, 11]. This is facilitated by the low fluid viscosity generating lower pressure drops than in HPLC. This process has been fully theorized by Lynen and coworkers [12] with the stationary phase-optimized-selectivity (SOS) concept.

Finally, it is also worth mentioning that, although they are usually more expensive than achiral stationary phases, chiral stationary phases may also be useful for achiral separations, particularly when isomers must be resolved.

Regarding particle size and types, SFC has followed the same trend as HPLC in progressively reducing particle sizes down to sub-2-μm particle diameter. As mentioned above, sub-3-μm superficially porous particles, which also provide improved column efficiency, have the additional benefit of reduced pressure drops. This can be put to advantage with (i) increased flow rates to fasten the analyses, (ii) increased column lengths to further increase column efficiency or use different columns for complementary selectivities, or (iii) use of operating conditions yielding higher internal pressure (high backpressure, low oven temperature, or high proportion of cosolvent).

The combined use of these high-efficiency stationary phases and technologically advanced instruments (notably with reduced dead volumes) has opened the way for ultra-high-performance supercritical fluid chromatography (UHPSFC), similar to the evolution of HPLC into UHPLC.

There is no general consensus on column dimensions. Sub-2-μm particle stationary phases are generally not available in long column lengths or wide internal diameters, due to packing issues. Unlike UHPLC practice, it is however not recommended to reduce the total column length below 100 mm and the internal diameter below 3.0 mm because extra-column volumes, which are still higher in UHPSFC instruments than in UHPLC instruments, would be responsible for reduced efficiency. When larger particle sizes are employed, it is still customary to employ longer and larger column lengths and internal diameters (typically 150–250 mm column length and 4.6-mm internal diameter). While column efficiency would generally be lower than the ones measured on small dimensions–small particle columns, there is still the advantage of less constraints on the pressure drop, allowing for more flexibility in the optimization of all operating parameters impacting the internal pressure (vide infra).

1.6.2 Optimizing Mobile Phase to Elute all Analytes

1.6.2.1 Nature of the Cosolvent

Carbon dioxide is now the most employed fluid in SFC mobile phases. As CO_2 has a low polarity, there are two ways to tune its elution strength: in early days of SFC, it was mostly done with changes in internal pressure (as will be further explained in Section 1.6.3), while nowadays it is mostly done through the introduction of a cosolvent (also called a modifier). Although nearly all solvents are miscible to CO_2, because CO_2 is rather nonpolar, the cosolvent should preferably be polar. Therefore, the most employed cosolvents are short-chain alcohols like methanol, ethanol, or isopropanol. Acetonitrile is less useful than in HPLC, because it is not a good solvent for protic analytes, which is not an issue in HPLC where water is fulfilling that function. However, acetonitrile is still useful in SFC for nonprotic analytes and in ternary mixtures (CO_2–alcohol–acetonitrile) to fine-tune selectivity [13]. Among the alcohols, methanol is by far the most employed [9] in achiral separations. Isopropanol is equally employed for chiral separations because the change in mobile-phase composition often causes significant changes in enantioselectivity and methanol and isopropanol are most complementary [14]. Ethanol is most similar to methanol and should be favored when a greener separation method is desired as its toxicity is much less than that of methanol [15].

Water and CO_2 are not miscible in all proportions, but water may be used in ternary compositions (CO_2-alcohol or acetonitrile-water). In small proportions (typically less than 10% in the cosolvent), it works as an additive (see Section 1.6.2.3). When high proportions of water are desired, due to the limited miscibility of CO_2 and water, the conditions of pressure and temperature will also need to be adjusted to avoid phase separation between a gaseous CO_2 and liquid water-solvent phase.

For achiral separations, the screening of stationary phases can often be done with a single mobile-phase composition, while the nature of mobile phase would be explored in a second stage. Indeed, the nature of the solvent will provide selectivity changes but is often of secondary importance to the stationary phase. On the contrary, when developing a chiral separation method, both stationary phases and solvent should be examined in the first screening steps. It was often observed that no enantioresolution would occur on one CSP with one cosolvent, while baseline resolution could be obtained on the same CSP with a different cosolvent.

The choice of cosolvent has a different impact when the method must be transposed to preparative scale. Indeed, analyte solubility must be sufficient to ensure that a large quantity of it will be injected in each run to favor a high productivity. In that case, a small proportion of a strong solvent (like 10% dichloromethane in methanol) is sometimes sufficient to ensure improved solubility with moderate impact on retention and enantioselectivity [2]. Besides, as the solvent will be evaporated from the purified fractions retrieved, the most volatile solvents should be favored to limit evaporation time and energy consumption.

1.6.2.2 Proportion of Cosolvent

The proportion of cosolvent usually varies between 2% and 50%. Current trends are moving toward higher values of this proportion in gradient operation, even as high as 100% at the end of the gradient. Clearly, the fluid is not in a supercritical state when the proportion of cosolvent increases (the exact proportion at the phase "transition" will depend on temperature and pressure settings). The fluid is then sometimes called "subcritical" or, when the proportion of cosolvent is larger than the proportion of CO_2, a "liquid with enhanced fluidity." Because the properties of the fluid are all continuous and because the way to operate the technique is not any different whatever be the exact state of the fluid, no distinction is made in the following and the term "SFC" is used throughout.

Apart from some rare exceptions, when increasing the proportion of cosolvent, the retention time usually decreases (Figure 1.6.4, arrow (a)).

There are several reasons for that. Firstly, increasing the solvent percentage causes increased viscosity of the mobile phase, which, in turn, increases the internal pressure and thereby increases the elution strength. Secondly, because the cosolvent has different interaction capabilities from carbon dioxide, the interactions with the analytes are generally favored, again increasing elution strength. Finally, the mobile-phase components adsorb on the stationary-phase surface, reducing strong analyte–SP interactions. Plotting the logarithm of retention factors according to the logarithm of solvent percentage in the usual operating range yields near-linear curves (Figure 1.6.4), which makes it possible to estimate the changes based on preliminary experiments. It also means that a plot of retention vs. cosolvent percentage basically decreases in a logarithmic fashion. A consequence is that the first percentages of cosolvent introduced in CO_2 have a tremendous effect on reducing retention, while further increases will have less and less effect.

The interactions between the analytes and mobile-phase molecules will naturally vary between the analytes. For instance, analyte molecules with hydrogen-bonding capability will better interact with methanol. Because not all analytes are affected to the same extent by the proportion of cosolvent, the slopes of the retention vs. solvent percentage curves will be different. As a result, selectivity may also vary when varying the percentage of cosolvent, and may even result in reversals of elution order (Figure 1.6.4, arrow (b)).

Because mobile-phase viscosity increases upon the introduction of cosolvent, it was long believed that chromatographic efficiency would be impaired. On the contrary, because analyte solubility in the mobile phase is improved, column efficiency is mostly improved by the introduction of the cosolvent (Figure 1.6.4, arrow (c)). Also, the improved solubility and reduced analyte–SP interactions generally improve peak symmetry.

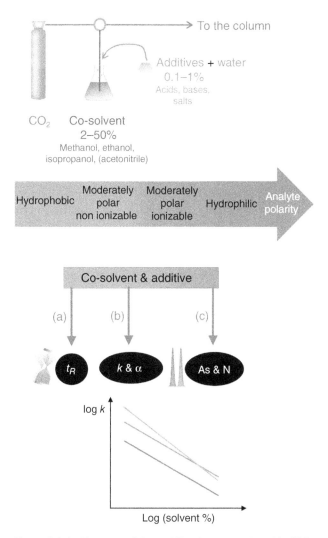

Figure 1.6.4 Features of the mobile-phases employed in SFC: composition in accordance with analyte polarity, multiple effects induced and relation between the percentage of cosolvent and logarithm of retention factor.

Isocratic and gradient methods are equally used depending on the sample complexity. When the sample includes a wide range of analyte polarities, a gradient method is advisable. For instance, on a polar stationary phase, a gradient allows to achieve sufficient retention of the least polar analytes (low percentage of cosolvent) and elution of the most polar analytes (large percentage of cosolvent) in a reasonable analysis time.

When first approaching a new sample, it is then a good idea to start screening for columns with a gradient elution (for instance, 2–50% cosolvent in 10 minutes) to ensure that most analytes will elute from the column. When the column providing adequate selectivity is selected, the mobile-phase composition will then be adjusted to elute the analytes in the desired time frame, and to obtain the desired resolution. Gradient methods are best optimized with a design of experiments, also allowing to include other parameters in the optimization process.

The final aim of the method developed is significant to the choice of isocratic or gradient method and the optimal composition. When the method should be transferred to quality control laboratories, an isocratic or simple (one-step, linear) gradient should be favored. Note that transferring a method to a different laboratory with the same instrument is possible and provides equally good results to those usually observed in HPLC [16]. Interlaboratory validation with different instruments is under way but has not yet been proven.

When the analytical method is developed in the aim to transfer it at the preparative scale, an isocratic method should be preferred because it allows for stacked injections (starting the next injection before the end of the previous one) that permit improved productivity and reduce solvent consumption. The optimal proportion of cosolvent would be in the 15–30% range, where the solubility of the analytes would permit an adequate quantity to be loaded in each injection while the volume of solvent to evaporate in the purified fraction would remain manageable.

1.6.2.3 Use of Additives

Aside from CO_2 and the cosolvent, the mobile phase also often comprises a third component introduced in small quantities and named "additive." The additive is often an acid (e.g. formic acid), a base (e.g. diethylamine or ammonia), or a salt (e.g. ammonium formate or ammonium acetate) [9].

As a rule of thumb, acidic additives are preferred for acidic analytes, while basic additives are used for basic analytes as they would work as competitors for adsorption sites of the stationary phase. However, strong acids are also useful to elute basic analytes in forming ion pairs. Ammonium formate and acetate are found very versatile, providing good results for all classes of analytes. They may however be undesirable in wide gradients with UV detection as baseline drift may be deleterious to sensitivity. Water is also increasingly used as an additive to favor the solubility of polar analytes, often employed in conjunction with a salt, where it would also limit precipitation issues.

The primary function of the additive is to favor the elution of polar neutral, acidic, or basic analytes with reduced retention times and improved peak shapes. The effect results principally from ion suppression, improved solvation of the analyte, and adsorption on the stationary phase [17, 18]. Additives usually have a moderate effect on achiral selectivity but principally affect chromatographic quality. However, in chiral separations, they usually have a strong effect on enantioselectivity. As a result, the initial screening of chiral stationary phases and mobile-phase compositions in chiral method development should preferably include a range of additives (typically, one acid, one base, one composition without additive).

Typical concentrations observed range in the 0.1–1% acids or bases in the cosolvent (for liquid additives) or 10–20 mM (for salts and powders). Water is rather introduced at 2–5% in the cosolvent.

In Figure 1.6.4, the choice of mobile-phase composition is indicated in accordance with analyte polarity. Most hydrophobic analytes like hydrocarbons and long-chain esters can usually be eluted with CO_2 only, or a very small concentration of cosolvent. For polar, nonionizable species (alcohols, amides, etc.), a larger proportion of cosolvent is necessary, but additives are not wanted. Ionizable analytes (acids and mostly bases) require the introduction of additives. Finally, most hydrophilic analytes (like nucleosides, sugars) would be best eluted with some water in the cosolvent.

1.6.2.4 Sample Diluent

While the dilution solvent must dissolve the sample fully, it should also be compatible with the mobile-phase composition. It was observed that the more polar and protic solvents like methanol negatively affect the peak shapes for the least retained analytes [19]. Nonprotic solvents like acetonitrile, methyl-*tert*-butyl-ether, or dichloromethane were found to provide improved peak shapes, especially when the injected volume is large. A portion of water in the sample diluent (like 50%) is also possible for the most polar analytes. In that case, it is preferable to include some portion of water in the mobile-phase composition as well; otherwise, the water molecules from the sample diluent will progressively adsorb on the stationary phase and impair the method reproducibility.

1.6.3 Optimizing Temperature, Pressure, and Flow Rate

The last three chromatographic optimization parameters are best considered in a joint fashion because some of their effects are related, namely the effect they have on internal pressure.

1.6.3.1 Understanding the Effects of Temperature, Pressure, and Flow Rate on your Chromatograms

The effect of pressure is quite straightforward: as a general rule, increasing backpressure in the system will always cause increased internal pressure, which will cause increased viscosity of the fluid, which should always cause increased elution strength of the mobile phase, thereby reducing the overall retention. However, because the change of elution strength will again not affect each analyte in the same manner, separation factors can be affected by the change in pressure. However, the effect of pressure is only significant when the fluid is sufficiently compressible, which happens when the proportion of cosolvent in the mobile phase is low (typically below 20%). In low-compressibility fluids (typically above 20% cosolvent), approaching the liquid state, varying the pressure has less effects on the chromatograms.

The effect of temperature is much more complex. When the fluid is close to a liquid, that is to say when the proportion of cosolvent is high (typically above 20%), the effect of temperature will very much resemble the effects observed in reversed-phase liquid chromatography. Namely, increasing temperature will cause decreased retention and Van't Hoff plots will approach linear tendencies. On the opposite, when the fluid is far from the liquid state, with low proportions of cosolvent and highly compressible conditions, increasing the temperature will cause decreased densities, thereby decreasing elution strength. Thus, increasing the temperature in that case will cause generally increased retention, but as not all analytes will be affected to the same extent, separation factors may also vary. Depending on the range of mobile-phase composition and temperatures explored, U-shaped trends of retention vs. temperature may also be observed. In a mobile-phase gradient elution, early-eluting analytes (eluted in high-compressibility fluid) may experience different effects from late-eluting analytes (eluted in low-compressibility fluid).

Meanwhile, diffusion coefficients will also vary when pressure and temperature vary, which will result in different chromatographic efficiency that cause different resolution. However, the knowledge on these changes is still rather scarce, as most past studies on diffusion coefficients were done in pure CO_2, that is to say in operating conditions that are not relevant of the current practice of SFC.

SFC users often neglect the interest of flow rate as an optimization parameter.

(i) Most chromatographers would vary flow rate in the interest of analysis time, thus mostly trying to increase flow rate as much as possible, within the limits of pressure allowed by the instrument (Figure 1.6.5, arrow (a)).

(ii) Most chromatographers are aware that changing the flow rate will also have an effect on efficiency, as indicated by Van Deemter curves (Figure 1.6.5, arrow (c)). Compared to HPLC Van Deemter curves, keeping the column dimensions and particle size of stationary phases identical, three major observations can be done. Firstly, the optimum point is nearly identical in terms of reduced plate height, meaning that identical particle size will yield comparable chromatographic efficiency in HPLC and SFC when each one is operated at their optimum flow rate. Secondly, this optimum flow rate is obtained at larger values in SFC compared to HPLC, meaning that equally good efficiency is achieved with usually faster analyses in SFC (typically 2–3 times faster). And thirdly, the C-term of Van Deemter equation is lower in SFC conditions compared to HPLC conditions. As a consequence, if the flow rate is increased above the optimum value, the efficiency will be less degraded in SFC than is observed in HPLC. Furthermore, when using sub-2-μm fully porous particles, or when using sub-3-μm superficially porous particles, the curve is nearly flat, making it hard to actually identify the optimum point and providing a wide range of flow rate values for which chromatographic efficiency is nearly unchanged.

(iii) The third effect of flow rate in SFC is the one most neglected because it is unlike HPLC. Increasing the flow rate will increase the internal pressure in the chromatographic system. As described for pressure and temperature effects, increasing pressure for a compressible fluid will cause changes in retention and separation factors, which is never observed in HPLC at usual operating pressures (Figure 1.6.5, arrow (b)).

1.6.3 Optimizing Temperature, Pressure, and Flow Rate | 99

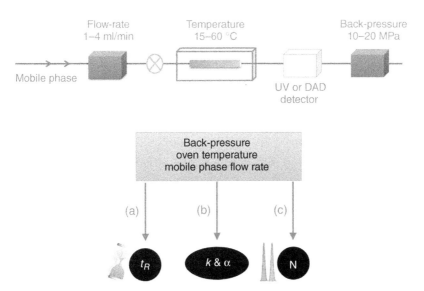

Figure 1.6.5 The effects of back-pressure, oven temperature and mobile-phase flow rate, and the typical range in which they may be varied.

1.6.3.2 Optimizing Temperature, Pressure, and Flow Rate Concomitantly

As they are rather fine-tuning parameters (much less influent than stationary and mobile phases), the best way to optimize these three is certainly to use a three-parameter design of experiments (DoE). Optimization of one factor at a time is naturally also possible but is unlikely to provide the desired optimum point. Because both column equilibration and analyses are fast in SFC, a complete DoE can be achieved in a very short time (typically a few hours) to define the best compromise of all parameters. The range of each parameter that can be explored naturally depends not only on the instrument but also on previous choices made on column dimensions (length, internal diameter, and particle size) and on mobile-phase composition, especially the proportion of cosolvent that is affecting the internal pressure, particularly at the end of a gradient program.

Less compressible fluids, obtained at high back pressure or flow rate and low temperature, will allow for more robust operating conditions (less sensitive to small variations in operating conditions) and improved sensitivity with UV detection. Indeed, UV baseline noise is essentially caused by the variations in refraction index of the mobile phase when the back-pressure control valve is regulating the system pressure through successive opening and closing. For less compressible fluids, the variation of refraction index is smaller; thus, the baseline noise is also reduced.

(i) Backpressure is typically varied between 10 and 20 MPa. Depending on column dimensions and other operating conditions, the pressure drop typically varies between 1 and 25 MPa. Consequently, the inlet pressure would vary from 11 to 45 MPa. All instruments allow for an inlet pressure of 40 MPa, with some of them allowing for higher values of 60 or 66 MPa. Along a mobile-phase gradient, the internal pressure will increase with analysis time; thus, reversed backpressure gradients are also possible to avoid reaching the upper pressure limit of the pumping system [20].

(ii) Depending on the instrument, temperature may vary between 5 and 90 °C. Most analyses are rather conducted in a narrower range of 15–60 °C. Cooling the fluid is often interesting for chiral separations because lower-temperature values make for more rigid stationary phases and analytes, which is often favorable to enantiorecognition. Temperature is not conveniently varied in the course of an analysis as equilibration is longer than backpressure, so it is preferably maintained constant throughout one run.

(iii) Again depending on the instrument, the flow rate may vary from 0.1 to 10 ml/min, but most analyses are conducted with a flow rate ranging from 1 to 4 ml/min. Similar to pressure, when short analysis time is desired, consider using a reverse-flow rate gradient starting at the fastest flow rate possible and ending at low-flow rate to avoid reaching the upper pressure limit at the end of the mobile-phase gradient [21].

1.6.4 Considerations on SFC–MS Coupling

Nowadays, SFC hyphenation to mass spectrometry (MS) has been made very easy by instrument manufacturers. This reflects in the increasing proportion of published papers where MS is used, while UV detection was the favorite choice in the past [9]. There are basically two principal ways of hyphenating these devices [22]. The first one is to simply maintain the whole mobile-phase flow eluting from the column and going through the UV detector and backpressure regulator entering the MS entirely (without any splitting). Naturally, this causes some limitations on the flow rate, as the MS ionization sources would not be capable to handle the large gas expansion caused by CO_2 depressurization when the flow rate is high. So this configuration is preferably employed with moderate flow rates (typically no higher than 2 ml/min). The second option is to split the flow, which is usually done after the column and UV detector but before entering the backpressure regulator (as in Figure 1.6.1). Then one portion of the flow is directed to the BPR, while the other one is directed to the MS. In doing so, the exact proportion of the fluid entering the MS is unknown, and will vary depending on operating conditions and depending on the pressure set by the MS interface.

In both cases, it is customary to introduce a make-up fluid prior to entering the MS (and preferably before the BPR). This can serve two purposes: one is to avoid analyte precipitation caused by the cold CO_2 depressurization. The other one is to introduce ionization enhancers, like acidic additives, that may serve as proton donors for electrospray ionization. The make-up fluid composition and flow rate should preferably be optimized after chromatographic optimization to maximize sensitivity.

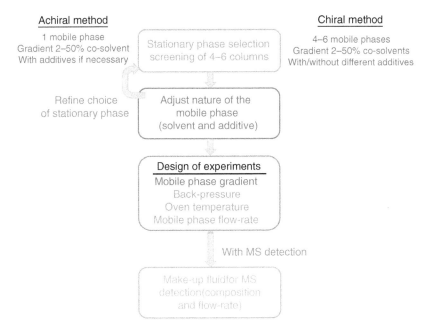

Figure 1.6.6 Proposed method development process in achiral and chiral SFC.

1.6.5 Summary of Method Optimization

Figure 1.6.6 summarizes the proposed optimization process.

When a stationary phase can be selected based on some knowledge of the sample, the first step of screening for columns can be avoided. When a screening step is necessary, an adequate selection of stationary phases to offer complementary selectivities is essential. Gradient elution is recommended at the initial stages of method development. A single mobile-phase composition to allow for the elution of a broad range of analytes (typically methanol comprising ammonium formate or acetate) should ensure that the information retrieved from the screening step is not biased by inadequate elution strength. For chiral separations, the screening step should include mobile-phase variation with and without additives. When the stationary phases selected fail to produce adequate retention and a debut separation, then an orthogonal stationary phase must be looked for. When one promising stationary phase is found (adequate retention range, some selectivity), it is also useful at this stage to investigate other stationary phases of close polarity that may offer somewhat different selectivity. When the analytes are not eluted from any of the columns tested, consider increasing the gradient up to 100% cosolvent. If the analytes are still not eluted in such conditions, then there is a solubility issue and SFC is inadequate; change of technique is required.

After stationary-phase selection, the nature of mobile phase can be optimized (changes in cosolvent and additive) to obtain satisfying chromatographic quality. The isocratic proportion of cosolvent or gradient range is preferably optimized in the third step, with a design of experiments possibly, including secondary optimization parameters (backpressure, oven temperature, and flow rate).

In the end, when the separation is still not perfect, looking for another stationary phase with close selectivity can produce a separation with close attributes but slightly different resolution.

Unlike reversed-phase HPLC, there is currently no software available to predict the outcome of an SFC separation based on a few scouting experiments. The fact that most effects described above are nonlinear and that the parameters are strongly interacting is an additional difficulty. We can however hope for such solutions in near future. Also, retention prediction that would help identify unknown analytes is yet imperfect, but some work is under way to achieve this goal in a near future.

1.6.6 SFC as a Second Dimension in Two-Dimensional Chromatography

There are currently no commercial systems, including SFC as a first or second dimension in a two-dimensional separation, only custom-made systems. There are however numerous experiments in the literature to prove the interest of such an approach [23–26]. SFC is most orthogonal to HPLC; thus, the interest of associating them to increase peak capacity is certain. As SFC can operate in different modes depending on the choice of stationary phase (reversed-phase-like with a nonpolar stationary phase, normal phase with a polar stationary phase), it may be orthogonal to reversed-phase, normal-phase, or hydrophilic interaction liquid chromatographic modes. Interfacing the two is the major issue as the solvent plug from the first dimension should preferably not contain too much water, due to miscibility issues of CO_2 and water, and because adsorption of water on the stationary phase in the SFC dimension will progressively modify the chromatography until a stable state is reached. This problem can be avoided if the SFC mobile phase already contains a small proportion of water, or with the use of a trapping column in the interface to eliminate the solvent from the first dimension prior to injecting the fractions in the second dimension.

1.6.7 Further Reading

Several recent papers have addressed the subtleties of SFC method development [27–29]. Interested readers will find there are more elements to understand the fundamental effects observed in carbon dioxide–based mobile phases. A primer is also available from Agilent, written by Terry Berger, one of the major contributors to the developments of packed-column SFC. All SFC manufacturers make application notes available from their websites, which can also be useful to give the users a head-start with new samples.

References

1 Francotte, E.R. (2001). Enantioselective chromatography as a powerful alternative for the preparation of drug enantiomers. *Journal of Chromatography. A* 906 (1-2): 379–397.
2 Miller, L. (2012). Evaluation of non-traditional modifiers for analytical and preparative enantioseparations using supercritical fluid chromatography. *Journal of Chromatography. A* 1256: 261–266.
3 Mazzoccanti, G., Ismail, O.H., D'Acquarica, I. et al. (2017). Cannabis through the looking glass: chemo- and enantio-selective separation of phytocannabinoids by enantioselective ultra high performance supercritical fluid chromatography. *Chemical Communications* 53 (91): 12262–12265.
4 Khater, S. and West, C. (2019). Characterization of three macrocyclic glycopeptide stationary phases in supercritical fluid chromatography. *Journal of Chromatography. A* 1604: 460485.
5 Ebinger, K. and Weller, H.N. (2014). Comparative assessment of achiral stationary phases for high throughput analysis in supercritical fluid chromatography. *Journal of Chromatography. A* 1332: 73–81.
6 Fairchild, J.N., Brousmiche, D.W., Hill, J.F. et al. (2015). Chromatographic evidence of silyl ether formation (SEF) in supercritical fluid chromatography. *Analytical Chemistry* 87 (3): 1735–1742.
7 West, C. and Lesellier, E. (2008). A unified classification of stationary phases for packed column supercritical fluid chromatography. *Journal of Chromatography. A* 1191 (1–2): 21–39.
8 West, C., Lemasson, E., Bertin, S. et al. (2016). An improved classification of stationary phases for ultra-high performance supercritical fluid chromatography. *Journal of Chromatography. A* 1440: 212–228.
9 West, C. (2018). Current trends in supercritical fluid chromatography. *Analytical and Bioanalytical Chemistry* 410 (25): 6441–6457.
10 West, C., Lemasson, E., Bertin, S. et al. (2018). Interest of achiral-achiral tandem columns for impurity profiling of synthetic drugs with supercritical fluid chromatography. *Journal of Chromatography. A* 1534: 161–169.
11 Welch, C.J., Biba, M., Gouker, J.R. et al. (2007). Solving multicomponent chiral separation challenges using a new SFC tandem column screening tool. *Chirality* 19 (3): 184–189.
12 Delahaye, S. and Lynen, F. (2014). Implementing stationary-phase optimized selectivity in supercritical fluid chromatography. *Analytical Chemistry* 86 (24): 12220–12228.
13 Brunelli, C., Zhao, Y., Brown, M.-H., and Sandra, P. (2008). Pharmaceutical analysis by supercritical fluid chromatography: Optimization of the mobile phase composition on a 2-ethylpyridine column. *Journal of Separation Science* 31 (8): 1299–1306.
14 De Klerck, K., Vander Heyden, Y., and Mangelings, D. (2014). Generic chiral method development in supercritical fluid chromatography and

ultra-performance supercritical fluid chromatography. *Journal of Chromatography. A* 1363: 311–322.

15 Alder, C.M., Hayler, J.D., Henderson, R.K. et al. (2016). Updating and further expanding GSK's solvent sustainability guide. *Green Chemistry* 18 (13): 3879–3890.

16 Dispas, A., Marini, R., Desfontaine, V. et al. (2018). First inter-laboratory study of a supercritical fluid chromatography method for the determination of pharmaceutical impurities. *Journal of Pharmaceutical and Biomedical Analysis* 161: 414–424.

17 West, C., Melin, J., Ansouri, H., and Mengue Metogo, M. (2017). Unravelling the effects of mobile phase additives in supercritical fluid chromatography. Part I: Polarity and acidity of the mobile phase. *Journal of Chromatography. A* 1492: 136–143.

18 West, C. and Lemasson, E. (2019). Unravelling the effects of mobile phase additives in supercritical fluid chromatography. Part II: Adsorption on the stationary phase. *Journal of Chromatography. A* https://linkinghub.elsevier.com/retrieve/pii/S0021967319301335.

19 Desfontaine, V., Tarafder, A., Hill, J. et al. (2017). A systematic investigation of sample diluents in modern supercritical fluid chromatography. *Journal of Chromatography. A* 1511: 122–131.

20 Taguchi, K., Fukusaki, E., and Bamba, T. (2014). Simultaneous analysis for water- and fat-soluble vitamins by a novel single chromatography technique unifying supercritical fluid chromatography and liquid chromatography. *Journal of Chromatography. A* 1362: 270–277.

21 Raimbault, A., Noireau, A., and West, C. (2019). Analysis of free amino acids with unified chromatography-mass spectrometry—application to food supplements. *Journal of Chromatography. A* 460772.

22 Tarafder, A. (2018). Designs and methods for interfacing SFC with MS. *Journal of Chromatography B* 1091: 1–13.

23 Sarrut, M., Corgier, A., Crétier, G. et al. (2015). Potential and limitations of on-line comprehensive reversed phase liquid chromatography × supercritical fluid chromatography for the separation of neutral compounds: an approach to separate an aqueous extract of bio-oil. *Journal of Chromatography. A* 1402: 124–133.

24 Iguiniz, M., Corbel, E., Roques, N., and Heinisch, S. (2018). On-line coupling of achiral reversed phase liquid chromatography and chiral supercritical fluid chromatography for the analysis of pharmaceutical compounds. *Journal of Pharmaceutical and Biomedical Analysis* 159: 237–244.

25 Sun, M., Sandahl, M., and Turner, C. (2018). Comprehensive on-line two-dimensional liquid chromatography × supercritical fluid chromatography with trapping column-assisted modulation for depolymerised lignin analysis. *Journal of Chromatography. A* 1541: 21–30.

26 François, I., Pereira A dos, S., and Sandra, P. (2010). Considerations on comprehensive and off-line supercritical fluid chromatography × reversed-phase

liquid chromatography for the analysis of triacylglycerols in fish oil. *Journal of Separation Science* 33 (10): 1504–1512.

27 Nováková, L., Grand-Guillaume Perrenoud, A., Francois, I. et al. (2014). Modern analytical supercritical fluid chromatography using columns packed with sub-2 μm particles. *Analytica Chimica Acta* 824: 18–35.

28 Lesellier, E. and West, C. (2015). The many faces of packed column supercritical fluid chromatography – a critical review. *Journal of Chromatography. A* 1382: 2–46.

29 Tarafder, A. (2016). Metamorphosis of supercritical fluid chromatography to SFC: an overview. *TrAC Trends in Analytical Chemistry* 81: 3–10.

1.7

Strategies for Enantioselective (Chiral) Separations
Markus Juza

Purification Technology Expert, Corden Pharma Switzerland LLC, Eichenweg 1, CH-4410, Liestal, Switzerland

Chiral[1] chromatographic methods like high-pressure liquid chromatography (HPLC) and supercritical fluid chromatography (SFC) have become an important tool in the determination of enantiomeric compositions in different areas of contemporary research and industrial applications (see Figure 1.7.1).

Chirality is not restricted to carbon atoms with four different substituents, but can also be found at sulfur, phosphorus, and many more atoms in the periodic table. Beside these chiral centers, structures like spiro compounds, allenes, or atropisomers exist as enantiomers that can be analyzed via chiral HPLC.

An example with no chiral atom is Norketotifen, a compound currently under development for treatment of allergen-induced allergic rhinitis[2]. The enantiomers of Norketotifen do not possess asymmetric carbon atoms as is typical of optical isomers. Rather, the enantiomers result from molecular asymmetry due to hindered interconversion of the seven-membered ring and could be separated on a cyclodextrine[3] chiral stationary phase (CSP) (Figure 1.7.2).

Separation of enantiomers via chromatographic methods has a history of more than 60 years in academia and industry. Today, it can be considered as a mature and robust technology. Many of the early issues (memory effects, stability issues, irreproducibility) have been overcome or understood during the past 20 years. Nevertheless, users still have the impression that enantioselective methods like HPLC or SFC require specialist knowledge to run them in a routine environment.

The following paragraphs give a guideline on how to start and develop chiral separation methods and will try to answer questions that will arise during setup and

---

1 Throughout this chapter the term "chiral" HPLC or "chiral" SFC will be used as a synonym for enantioselective chromatographic methods, even though the term is scientifically not correct. "Chiral" implies that there exist two different chemical entities that are not superimposable. Separations of enantiomers are based on chiral selectors, but chiral columns do not exist.
2 http://emergotherapeutics.com/about/.
3 EP 1 218 007 B1; conditions described for enantiomer separation of rac-Norketotifen are: Merck Chiradex 5 µm, mobile phase 95:5; pH 4.0 sodium phosphate/acetonitrile, flow rate 1 ml/min, chart speed 0.3 cm/min, 254 nm, Rt_1 7.5 minutes ((S)-Norketotifen fumarate), Rt_2 13.0 minutes ((R)-Norketotifen fumarate).

Optimization in HPLC: Concepts and Strategies, First Edition. Edited by Stavros Kromidas.
© 2021 WILEY-VCH GmbH. Published 2021 by WILEY-VCH GmbH.

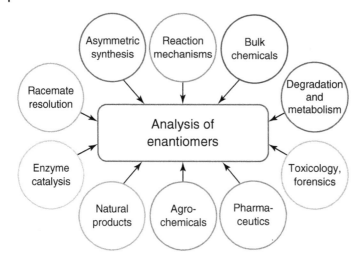

Figure 1.7.1 Application areas of enantioselective chromatography.

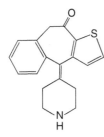

Figure 1.7.2 Structure of rac-Norketotifen.

running such methods. Like any other chromatographic method, chiral separations need, in general, optimization or fine-tuning of parameters to achieve optimum results in regard to resolution and selectivity. Such optimizations follow rules that are comparable to those known from reversed- and normal-phase chromatography and require no specific expertise.

1.7.1 How to Start?

The first step for any development or optimization of a chiral separation should be the definition of the separation problem. If just two enantiomers of a high-purity racemate have to be separated, the task will be relatively simple: any combination of CSP and mobile phase that results in a baseline separation will fulfill the requirements, regardless of retention times, peak shapes, and solubility. Many chiral methods are used to measure the enantiomeric excess (ee) of an intermediate or active pharmaceutical ingredient (API) – a simple two-component mixture problem; very obviously, there is no need for a method with 45-minute runtime.

In a scenario where several stereogenic centers are present in a molecule and resolution of all stereoisomers and impurities at low limit of quantification (LOQ) are required, the screening and development call for high efficiency (i.e. small particle sizes) and sufficient selectivity and enantioselectivity to resolve all peaks of interest. The same will be true when a number of racemates and achiral components can be present in the samples. Here, method development will not only focus on enantioselectivity but also maximize the peak capacity to accommodate all peaks in the chromatogram.

In case the matrix for the racemate(s) is on aqueous base, e.g. in case of pharmacokinetic studies, metabolites, or degradants in soil, a reversed-phase chiral method will be preferred over a normal-phase separation (and likewise if a hyphenation to an MS to increase the LOD is required). A similar situation may arise from the fact that not all analytical laboratories can and like to use normal-phase chromatography. However, compounds like acid chlorides or anhydrides can very quick react in an aqueous mobile phase and are in general not suited for reversed-phase chromatography.

In the end, the method must be "fit for purpose."

The definition "fit for purpose" might differ significantly if the method is used only in one laboratory for a few times or if it is intended for method qualification, validation, or transfer to other laboratories. In the end, the user (or customer) has to define the parameters, like robustness, linearity, accuracy, and LOQ just as for any other analytical method, make a benefit–cost analysis, and decide how often the individual plans to run the separation and if an existing method can fulfill these criteria or if a totally new development is justified.

1.7.2 Particle Size

Next to the wide chemical variety of the CSPs, the particle size technology used routinely has moved from 10 or 5 to 3 µm, which has been introduced to the market by several producers during the last decade allowing for fast screening and development of ultra-high-performance liquid chromatographic separations and greatly facilitating high-throughput screening. In spite of a certain difficulty in producing small-particle polysaccharide phases, columns packed with 3.0-µm particles (and smaller) are now available. These advances in column technology aim to reduce mass transfer resistance and improve efficiency, paving the way to shorter columns and shortening analysis times. This development is of major importance as it is known that the maximum column efficiency is inversely proportional to the square of the particle diameter.

Therefore, new developments of chiral separations should be performed on 3-µm-particle size CSPs on short columns rather than the traditional 250 mm × 4.6-mm ID column dimensions. Also, column packed with 10-µm particle chiral phases should no longer be used for the same reasons 10-µm achiral phases are no longer are used – they belong to a generation of stationary phases that have reached the end of their lifecycle and might only be useful for some semipreparative HPLC experiments.

1.7.3 Chiral Polysaccharide Stationary Phases as First Choice

Nowadays, polysaccharide CSPs are the predominant type of CSPs employed in chiral HPLC and SFC due to a number of reasons:

- Their wide applications window
- Screening conditions can be standardized and require no detailed knowledge of interaction mechanisms and conditions that must be avoided
- Availability of numerous different chiral selectors either in coated or in immobilized versions, and
- Large loading capability useful for preparative scale.

It has been claimed that today more than 99% or all chiral separations are performed on polysaccharide CSPs [1]. They have taken over the role of many other CSPs developed during the 1980 and 1990s, which also explains why today rarely new chiral phases are commercialized that do not rely on polysaccharides[4].

Polysaccharide CSPs can be derived from cellulose or amylose. Almost all of those CSPs are synthesized by reaction of a substituted aromatic isocyanate with the cellulose/amylose chains. The degree of substitution is always below the theoretical maximum of 3. The obtained polymers are coated on spherical silica particles ranging from 1.7 to 20 μm with c. 15–20 wt%. Unfortunately, these polymers used to have a major disadvantage. They could be used under very nonpolar conditions (eluents with large amounts alkanes, such as hexane or heptane) or under very polar conditions as encountered in reversed-phase chromatography or the so-called polar organic mode employing alcohols (like methanol or ethanol) or acetonitrile. The medium-polarity solvents, like ethers, ester, and many more, destroy nonimmobilized CSPs by dissolving the chiral polymer.

To increase chemical stability of the polymers, various procedures for their immobilization have been described. This immobilization procedure changes not only the chemical properties of the CSP, but can also influence significantly the enantioselectivity of the phase. The properties of immobilized CSPs will be discussed in one of the following paragraphs.

To find an enantiomer separation, a screening of various CSPs with different chiral selectors and eluent systems is necessary. Such a screening should at least include five CSPs with different chiral selectors.

Derivatized polysaccharide stationary phases exhibit both polar and hydrophobic properties, and combine weak hydrophobic interactions with polar interactions. This enables successful HPLC separations of complex samples on these CSPs. They can be used in normal-phase, polar organic, or in reversed-phase applications, and are therefore in some regards similar to diol, cyano, or amide stationary phases known from achiral HPLC (see Box 1.7.1).

---

4 This statement does not imply that polysaccharide CSPs are the only useful type of CSP and that there are no alternatives.

1.7.3 Chiral Polysaccharide Stationary Phases as First Choice | 111

A larger part of the "mystery of chiral HPLC" originated from the arcane art of finding an eluent system that led to retention and enantioselectivity, and did not destroy the column after one or two runs.

Box 1.7.1 Box: Understanding Polarity of Achiral Stationary Phases, Polysaccharide CSPs, and Eluent Systems

Eluents and Eluent systems for successful enantiomer separations using polysaccharide-derived stationary phases – a key to the "mystery of chiral HPLC"

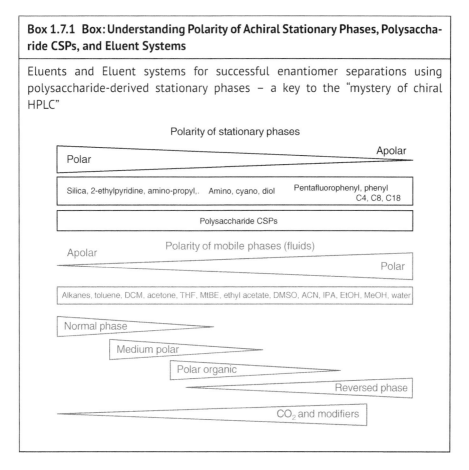

For a successful working with polysaccharide-CSPs, it is of key importance to understand that enantioselectivity originates from an interaction of chiral selector, analyte, and the selected solvent system; all three must fit together to achieve robust and reproducible enantiomer separations. Regrettably, classical RP eluents are in many cases not the best-suited mobile phases for good chiral separations.

The systematic screening of polysaccharide stationary phases to find a separation can be best understood using a "Lego®-approach." The available columns can be tested in various modules that differ in polarity. In principle, all polysaccharide CSPs can be tested in the normal-phase module, polar organic module, reversed-phase module, and the polar organic SFC module (see Box 1.7.2). All of them "click" together, even though some suppliers have special editions of the phases for reversed phase and SFC conditions, especially for coated selectors.

Box 1.7.2 Box: Polarity Modules for Screening NP Module, PO Module, MP Module, RP Module, SFC PO Module, and SFC MP Module

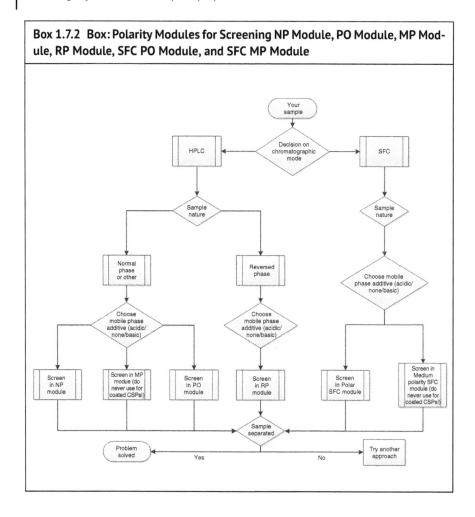

Immobilized polysaccharide CSPs offer a special feature: Those can be tested using a medium-polarity module that is not compatible with coated polysaccharide CSPs – they are dissolved by the medium-polarity solvents. They also can be tested in a medium-polarity super critical fluid module in which the nonpolar solvent n-heptane is replaced by the nonpolar fluid carbon dioxide. This medium-polarity module has also been named "convergent chromatography range" [2] as it bridges the normal-phase range and the reversed-phase range of solvents. Supercritical CO_2 is miscible with the entire eluotropic series of solvents, opening up a wide range of solvent selectivities to develop a separation.

Besides compatibility with all organic solvents, immobilized polysaccharide CSPs have a number of attractive properties compared to the coated polysaccharide CSPs. The immobilization of the phase makes them immune to "memory effects" originating from prolonged use of acidic or basic modifiers, allows regeneration of these CSPs, and makes it possible to use sample solvents like DCM or DMSO in which some compounds can show extreme solubilities, see Table 1.7.1.

Table 1.7.1 Comparison of coated and immobilized polysaccharide CSPs

	Coated CSPs	Immobilized CSPs	
Robustness	Restricted solvent selection possible	High solvent flexibility	@
Reproducibility	Memory effects possible	Easy column regeneration	@
Loading capacity	High loading capacity	Loading capacity improved due to analyte solubility	@
Enantioselectivity	Broad range of chiral separations	Different range of chiral separations	@

Immobilization of the polysaccharide phases is achieved via radical chemistry or UV treatment, which, besides changing the physical properties of the treated phase, also changes the enantioselectivity because the three-dimensional structure of the polysaccharide helix is altered irreversibly. Immobilized CSPs therefore have in some applications significantly different selectivity compared to the original non-immobilized polysaccharide phase.

1.7.4 Screening Coated and Immobilized Polysaccharide CSPs in Normal-Phase and Polar Organic Mode

The primary screening for coated and immobilized polysaccharide CSPs should start with four different eluents, two of them typical normal-phase chromatographic conditions, i.e. alkane with an alcohol modifier, two of them for the polar organic mode. The polar organic mode is often overlooked as alternative to normal-phase conditions, because such polar eluents are not very promising candidates for separations on silica or aluminum oxide and are not MPLC or TLC solvents one would select in a first screening. It should be noted that the coating of the silica particles by the polymer drastically changes the surface properties of the silica core (see Box 1.7.1).

Eluent mixtures known from normal-phase separations, like heptane/ethyl acetate or dichloromethane/methanol, will dissolve the coated chiral selectors very quickly and will destroy the column **irretrievably**. The same is true for solvents of medium polarity such as tetrahydrofuran, MTBE, acetone, and many more.

Therefore, the screening should start with the following conditions:

(a) Alkane/2-propanol	80 : 20
(b) Alkane/ethanol	80 : 20
(c) MeOH	100
(d) ACN	100

The alkane eluent part can be *n*-hexane, *n*-heptane, or even *iso*-hexane; *n*-heptane should be preferred over *n*-hexane as it is not metabolized to the toxic 2,5-hexane dione [3].

Acidic additives (e.g. TFA) or basic additives (e.g. DEA) are necessary if acidic or basic compounds are to be separated. If the peak shapes with TFA (acidic) or DEA (basic) are not acceptable, further acidic or basic additives can be tested.

The equilibration volume between the four solvent combinations should be at least five column volumes, since the solvents modify the supramolecular structure of the polysaccharides, which takes a certain amount of time. Only then can reproducible results be obtained.

Figure 1.7.3 shows an example for an enantiomer separation in the normal-phase mode; the racemate is separated in less than five minutes under optimized isocratic conditions. The compound AHC 2082782 is used for treatment of endoparasites in warm-blooded animals [4].

Figure 1.7.4 gives a systematic scheme for finding and optimizing separations in normal-phase mode and polar organic mode for coated and immobilized polysaccharide CSPs.

If the retention times in normal phase mode are too short, more alkane should be added. However, solvent systems with very low amounts of alcohol (e.g. less than 2%) should be avoided. Such mixtures are sensitive to errors in preparation and in summertime evaporation of the eluent can lead to a changing composition of the eluent, resulting in varying retention times and irreproducible results. It has also been reported that this type of eluent can draw moisture from air humidity, which can have significant impact on the reproducibility of enantiomer separations [5].

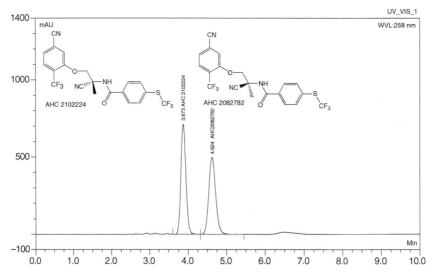

Figure 1.7.3 Enantiomer separation of AHC 2102224 and AHC 2082728 on Chiralpak AD-H. Eluent: *n*-hexane/ethanol (50 : 50); column: 250 mm × 4.6 mm ID, flow: 1.00 ml/min, detection: 258 nm, $T = 30\,°C$.

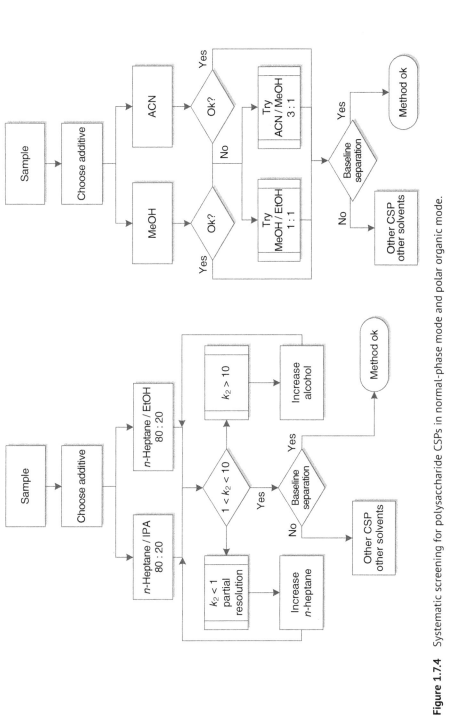

Figure 1.7.4 Systematic screening for polysaccharide CSPs in normal-phase mode and polar organic mode.

If the retention times are too long or if the analyte is not eluted, the retention time can be shortened as follows:

(a) Alkane/2-propanol	80 : 20	⇒more 2-propanol
(b) Alkane/ethanol	80 : 20	⇒more ethanol
(c) MeOH	100	⇒different solvent
(d) ACN	100	⇒different solvent

If there is no retention using polar solvents (MeOH or ACN), experiments on the tested column should be aborted. If there is retention, a secondary screening can be successful:

(a′) EtOH/MeOH	50 : 50
(b′) ACN/MeOH	75 : 25

These combinations of polar organic solvents in the secondary screening should not be underestimated; they can lead to unforeseen high enantioselectivities and enantiomer separations that cannot be explained by rational approaches, but depend on the minute changes in the three-dimensional structure of the polysaccharidic backbones that lead to enantiomer separation, but cannot be predicted.

It must be stressed that in cases where no peak is observed, the compound is with high probability still on the column and must be removed by flushing the column with a suitable solvent. If this simple precaution is not followed, the compound (or enantiomer) can elute in one of the next chromatograms, resulting in three peaks or two peaks of dissimilar area for a racemate.

A typical example for the optimization of an enantiomer separation in the normal phase mode is given in Figure 1.7.5. The amount of polar alcohol influences strongly the retention time of the nicotinoly-derivative [6] enantiomers, peak resolution, and enantioselectivity. By systematic variation of the eluent composition, a good compromise between overall retention time and baseline separation can be achieved. Best conditions are alkane/EtOH 70 : 30% under which the enantiomer pair is well separated under 20 minutes of run time.

1.7.5 Screening Coated and Immobilized Polysaccharide CSPs in Reversed-Phase Mode

There are a number of possible advantages using reversed-phase conditions, such as good solubility of polar compounds, simplified sample preparation for serum, plasma or soil samples, and the use of cheaper solvents.

Acetonitrile or methanol is often used as an organic modifier under RP conditions. The retention times of an analyte *can* increase with increasing proportion of modifiers. Often, the same proportion of acetonitrile gives shorter retention time than the same amount of methanol.

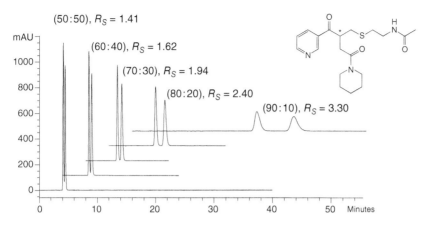

Figure 1.7.5 Enantiomer separation of a nicotinoyl derivative on Chiralcel OD-H. Eluent: hexane/ethanol, various mixing ratios; column: 250 mm × 4.6 mm ID, flow: 1.00 ml/min, detection: 225 nm, $T = 25\,°C$.

When using polysaccharide CSPs, it is important to have the analyte in neutral form to maximize the interaction between CSP and the analyte. For a neutral analyte, acidic and basic additives have only a minimal impact. An acidic mobile phase must be used for an acidic analyte. For a basic analyte, two parameters must be considered: the choice of the buffer system and a suitable pH. A borate buffer with ~pH 9 is a good starting point.

Therefore, the screening should start with the following conditions:

ACN/buffer[a]	40 : 60
MeOH/buffer[a]	60 : 40

a) No buffer needed for neutral water soluble analytes.

Acidic additives (e.g. HCOOH) or basic additives (e.g. NH_4HCO_3, DEA) are necessary in the dissolved sample and the eluent when acidic or basic compounds are to be separated. If the peak shapes with formic acid (acidic) or ammonium hydrogen carbonate (basic) are not acceptable, further acidic or basic additives can be tested.

Like in the normal-phase module or the polar organic-phase module, the equilibration volume between the solvent combinations should be at least five column volumes.

It should be noted that coated polysaccharide CSPs, when used under reversed-phase conditions, should not revert to other modes. In many cases, the three-dimensional structure of the phase is changed on contact with water and reconditioning may succeed – or not. For this reason, columns dedicated to aqueous eluents, like Chiralpak AD-RH, must only be used in RP conditions successfully.

If the retention times are too short, then less organic modifier is added.

If the retention times are too long or if the analyte is not eluted, the retention time can be shortened as follows:

ACN/buffer 40 : 60	⇒ more acetonitrile
MeOH/buffer 60 : 40	⇒ more methanol

A flow chart for a systematic screening under RP conditions is given in Figure 1.7.6. In case a partial resolution is obtained, other typical reversed-phase

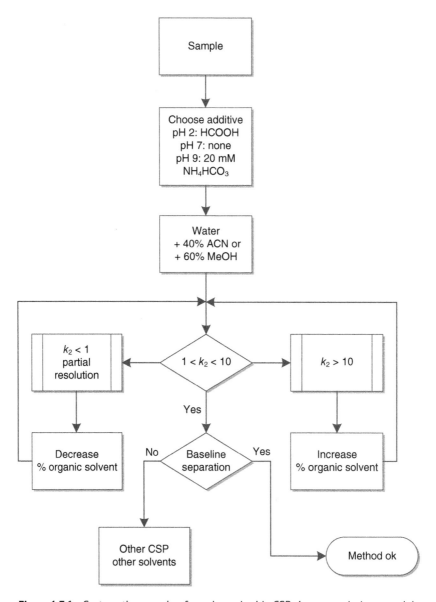

Figure 1.7.6 Systematic screening for polysaccharide CSPs in reversed-phase module.

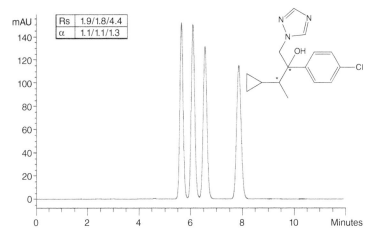

Figure 1.7.7 Enantiomer separation of cyproconazole on CHIRAL ART Cellulose-SC. Eluent: water/ACN (48 : 52); column: 250 mm × 4.6 mm ID, flow: 1.20 ml/min, detection: 220 nm, $T = 25\,°C$ (by courtesy of YMC).

additives can be tested. Acidic additives can be TFA, formic acid, or phosphoric acid; basic additives can include DEA, TEA, and ammonia.

An example of a typical reversed-phase separation is shown in Figure 1.7.7. The mixture of four stereoisomers is separated in less than 10 minutes under optimized isocratic reversed-phase conditions. Cyproconazole is used against powdery mildew, rust on cereals and apple scab, and applied by air or on the ground to cereal crops, coffee, sugar beet, fruit trees, and grapes.

1.7.6 Screening Immobilized Polysaccharide CSPs in Medium-Polarity Mode

The screening of immobilized polysaccharide CSPs in the medium-polarity module should start with three different eluents. These medium-polarity eluent mixtures **must exclusively be used on immobilized** polysaccharide CSPs – they are not compatible with coated polysaccharide CSPs!

(a) Alkane/CH_2Cl_2/EtOH	50 : 50 : 1	
(b) MtBE/EtOH	98 : 2	
(c) Alkane/Ethyl acetate	50 : 50	

Acidic additives (e.g. TFA) or basic additives (e.g. DEA) are necessary, as with the coated phases, if acidic or basic compounds are to be separated. If the peak shapes with TFA (acidic) or DEA (basic) are not acceptable, further acidic or basic additives can be tested.

The equilibration between the three solvent combinations should be longer than that for the coated phases, since the solvents modify the supramolecular structure of the polysaccharides, which takes more time, especially for immobilized polysaccharide phases. Only then can reproducible results be obtained.

Figure 1.7.8 gives a systematic scheme for finding and optimizing separations using immobilized polysaccharide CSPs in the medium-polarity module.

If the retention times are too short, more alkane should be added (also with MtBE).

If the retention times are too long or if the analyte is not eluted, the retention time can be shortened as follows:

(a) Alkane/CH_2Cl_2/EtOH	50 : 50 : 1	⇒	more DCM or EtOH, decrease alkane
(b) MtBE/EtOH	98 : 2	⇒	more ethanol or another modifier, e.g. THF
(c) Alkane/ethyl acetate	50 : 50	⇒	decrease alkane and/or add alcohol

An example for the systematic screening of immobilized phases in normal-phase (NP) module, medium-polarity (MP) module, and polar organic (PO) module for the racemate hydroxychloroquine (HCQ) is shown in Figure 1.7.9. The racemate can be separated in less than two minutes under optimized isocratic medium-polarity conditions. HCQ is a medication used to prevent and treat malaria; other uses include treatment of rheumatoid arthritis, lupus, etc.

1.7.7 Screening Coated and Immobilized Polysaccharide CSPs under Polar Organic Supercritical Fluid Chromatography Conditions

The principles of SFC are similar to those of liquid chromatography (see also Chapter 1.6); however, SFC typically uses carbon dioxide as the main mobile phase. SFC is essentially a normal-phase chromatographic technique with inherent high speed and efficiency due to its mobile phase. As a high-pressure liquid, or supercritical fluid, CO_2 is an excellent solvent due to its low viscosity and high diffusivity. SFC excels at separating and purifying chiral compounds because it's faster, uses much less solvent, and overall it is a less expensive and "greener" method than HPLC for chiral separations.

As far as enantioselectivity is concerned, like in HPLC, it is impossible to predict which solvent will provide the most favorable separation conditions for a given racemate.

Alcohols (methanol, ethanol, and 2-propanol) or acetonitrile are typically used as first-choice solvents in combination with CO_2. All polysaccharide CSPs, whether coated or immobilized, are suitable. It should be noted that at higher percentages of polar organic modifiers, the supercritical fluid eluent may become a subcritical fluid. However, this change of physical state rarely leads to chromatographic problems, but can cause a change in retention behavior.

Samples dissolvable in alcohol are most applicable for SFC, but also ACN can be used as sample solvent.

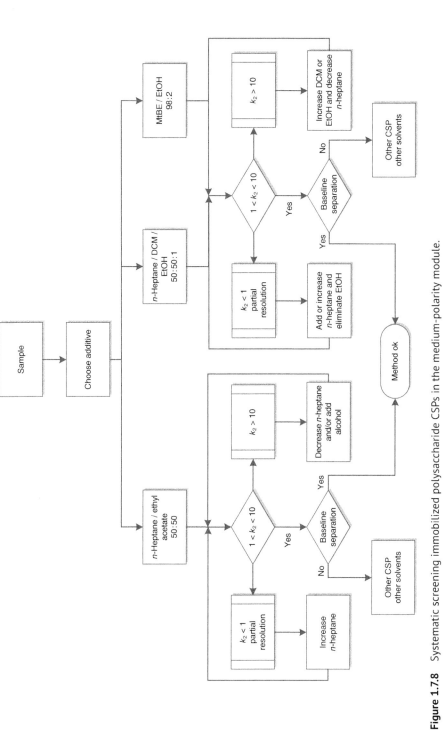

Figure 1.7.8 Systematic screening immobilized polysaccharide CSPs in the medium-polarity module.

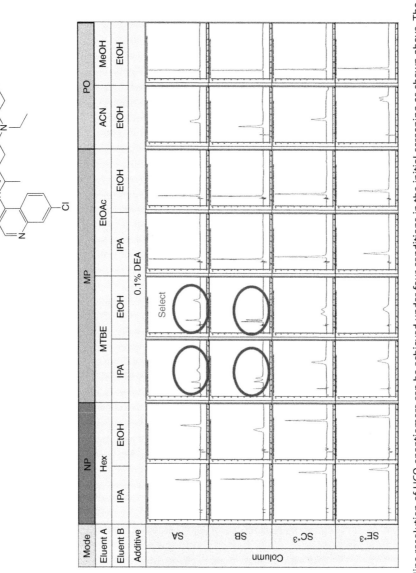

Figure 1.7.9 Baseline resolution of HCQ enantiomers can be achieved under four conditions in the initial screening as shown above. The combination of MTBE/EtOH (90 : 10) containing 0.1% DEA on CHIRAL ART Amylose-SA (3 µm, 50 × 3.0 mm ID, flow rate: 0.85 ml/min, detection at 344 nm, $T = 25\,°C$) was selected as the most favorable condition based on retention and resolution – best results shown in circle $R_s > 1.5$ (by courtesy of YMC) [7].

The screening should start with the following conditions:

(a)	CO_2/MeOH	80 : 20
(b)	CO_2/ethanol	80 : 20
(c)	CO_2/2-propanol	80 : 20
(d)	CO_2/ACN	70 : 30

Acidic additives (e.g. TFA) or basic additives (e.g. DEA) are necessary if acidic or basic compounds are to be separated. If the peak shapes with TFA (acidic) or DEA (basic) are not acceptable, further acidic or basic additives can be tested. It should be noted that CO_2 is slightly acidic and that not in all cases an acidic modifier will be required for acidic analytes.

The equilibration volume between the four solvent combinations should be at least five column volumes, since the solvents modify the supramolecular structure of the polysaccharides, which takes a certain amount of time. Only then can reproducible results be obtained.

Figure 1.7.10 shows the enantiomer separation of p-amino-benzoyl-(L)-(−)-glutamic acid, called "L-ABGA," in polar organic SFC under subcritical conditions on a phase with particle size of 3 µm. The sample (enriched in the desired S enantiomer (ratio D:L 37 : 63) is separated in less than 0.5 minutes under optimized isocratic conditions. L-ABGA is further reacted with a pteridine moiety leading to the nature identical and enantiomerically pure L-folic acid.

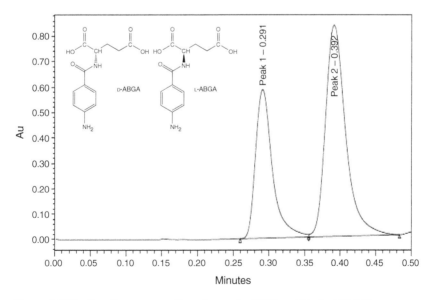

Figure 1.7.10 Enantiomer separation of D- and L-ABGA on Chiralcel OJ-3 $Rt_{D\text{-}ABGA}$ 0.29 minutes, $Rt_{L\text{-}ABGA}$ 0.39 minutes. Eluent: CO_2/ethanol (60 : 40); column: 50 mm × 4.6 mm ID, flow: 2.00 ml/min, $p = 150$ bar, detection: 280 nm, $T = 35\,°C$.

The retention times of the enantiomers are reduced by c. 50% compared to the optimized isocratic normal-phase conditions using the same stationary phase (Chiralcel OJ-3, 50 × 4.6 mm, mobile phase: *n*-heptane/ethanol/TFA (70/30/0.1, v/v/v), flow: 1.00 ml/min) [8].

Figure 1.7.11 gives a systematic scheme for finding and optimizing separations in polar organic mode for coated and immobilized polysaccharide CSPs in SFC.

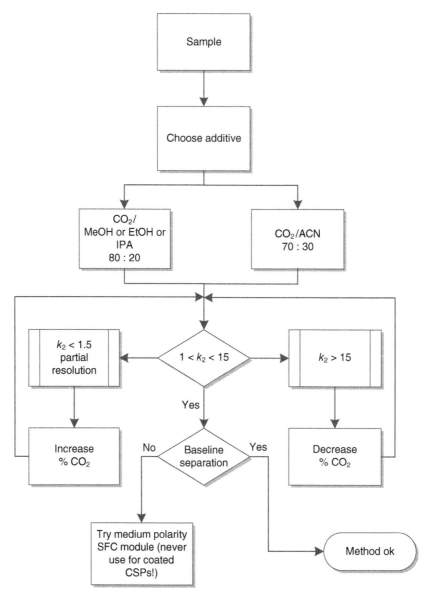

Figure 1.7.11 Systematic screening polysaccharide CSPs in the polar organic SFC module.

If the retention times in the polar organic SFC mode are too short, more CO_2 should be added.

If the retention times are too long or if the analyte is not eluted, the retention time can be shortened as follows:

(a), (b), (c) CO_2/alcohol	80 : 20	⇒more alcohol, i.e. decrease CO_2
(d) CO_2/ACN	70 : 30	⇒more acetonitrile, i.e. decrease CO_2

If there is no retention using polar organic solvents experiments on the tested column should be aborted, a screening on immobilized polysaccharide CSPs in the medium-polarity module should be initiated.

1.7.8 Screening Immobilized Polysaccharide CSPs in Medium-Polarity Supercritical Fluid Chromatography Conditions

The screening of immobilized polysaccharide CSPs in the medium-polarity module should start with four different eluents. These medium-polarity eluent mixtures must exclusively be used on immobilized polysaccharide CSPs – **they are not compatible with coated polysaccharide CSPs**!

Three of these eluent systems are mixtures of organic solvents with methanol (dichloromethane/methanol 90 : 10, ethyl acetate/methanol 90 : 10, and MtBE/methanol 80 : 20) that are added to the supercritical carbon dioxide. The fourth medium-polarity eluent that can be added to the mobile phase is THF. The later eluent system has properties similar to those used for curing polymers used for seals on HPLC and SFC equipment, and can cause leakages of some seals. Therefore, these conditions should be tested prior to using them in a screening to ensure proper operation of the system with CO_2/THF as eluent.

The screening should start with the following conditions:

(a) CO_2/THF	75 : 25
(b) CO_2/DCM/MeOH	80 : 18 : 2
(c) CO_2/ethyl acetate/MeOH	80 : 18 : 2
(d) CO_2/MtBE/MeOH	75 : 20 : 5

Acidic additives (e.g. TFA) or basic additives (e.g. DEA) are necessary, as with the coated phases, if acidic or basic compounds are to be separated. If the peak shapes with TFA (acidic) or DEA (basic) are not acceptable, further acidic or basic additives can be tested.

Figure 1.7.12 gives a systematic scheme for finding and optimizing separations in medium polar mode for immobilized polysaccharide CSPs in SFC.

If the retention times in the medium polar SFC mode are too short, more CO_2 should be added, or the amount of methanol should be decreased.

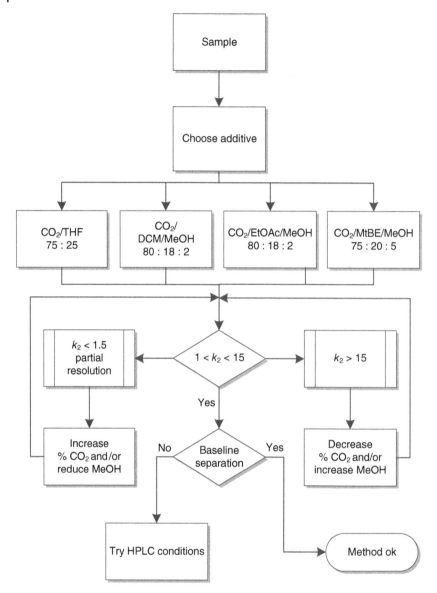

Figure 1.7.12 Systematic screening of immobilized polysaccharide CSPs in the medium polar SFC module.

If the retention times are too long or if the analyte is not eluted, the retention time can be shortened as follows:

(a) CO_2/THF	70 : 30	⇒ more THF, i.e. decrease CO_2
(b) CO_2/MP-solvent/MeOH	80 : X : Y	⇒ more MeOH, i.e. decrease CO_2

1.7.9 SFC First?

This question arises, of course, only if both instrument types are available in the laboratory. Screening under SFC conditions will be faster as the choices of mobile-phase modifiers are more restricted and due to the faster diffusion in SFC the chromatogram runtime is much shorter. An extensive screening of five to six different CSPs can therefore be realized overnight or even faster. However, there will remain a number of racemates unresolved after an SFC screening that must be tested via chiral HPLC.

Also in cases where the sample nature requires reversed-phase conditions, SFC will not be very attractive based on the simple fact that no reversed-phase module is available in SFC, even though water can be used as modifier together with, e.g. alcohols in low percentage in SFC.

1.7.10 Are There Rules for Predicting Which CSP Is Suited for My Separation Problem?

Unfortunately, the answer is very short: no. There are as of today no guidelines or rules of thumb which polysaccharide stationary phase will resolve the enantiomers of any given racemate. This remains one of the significant differences to reversed-phase, normal-phase, or HILIC separations of nonchiral compounds.

The only reliable approach is screening different CSPs and mobile-phase combinations until a suitable combination is found for a new chiral structure.

1.7.11 Which Are the Most Promising Polysaccharide CSPs?

The answer to this key question will differ, depending on whom you will ask. Every supplier of course must stress that "their" CSPs will be the best choice. Literature surveys on published enantiomer separations most likely will cover mostly academic examples published in the past in one or two journals. And, of course, every year several new stationary phases are commercialized that will shift the focus of tested CSPs to those newcomers to prove their superiority. Therefore, it will only be possible to give an answer to this question on a temporary base – very similar like answering the question "which is the best car?"

Based on recommendations of three major producers of CSPs, the answer for the year 2021 seems to be the following (see Table 1.7.2).

Table 1.7.2 Recommendation for CSPs to be included in a screening.

Polysaccharide	Bonding	Substitution	Supplier/producer		
			YMC	Phenomenex	Daicel
A	i	Tris(3,5-dimethyl-phenylcarbamate)	CHIRAL ART Amylose-SA	Lux i-Amylose-1	Chiralpak IA
A	n		CHIRAL ART Amylose-C (Neo)	Lux Amylose-1	Chiralpak AD
A	i	Tris(3-chlor-5-methylphenyl-carbamate)		Lux i-Amylose-3	Chiralpak IG
C	i	Tris(3,5-dimethylphenyl-carbamate)	CHIRAL ART Cellulose-SB		Chiralpak IB (-N)
C	i	Tris(3,5-dichlorphenyl-carbamate)	CHIRAL ART Cellulose-SC	Lux i-Cellulose-5	Chiralpak IC
C	n	Tris(4-methylbenzoate)	CHIRAL ART Cellulose-SJ	Lux Cellulose-3	Chiralcel OJ

A, amylose; C, cellulose; i, immobilized; n, not immobilized.

Both amylose and cellulose derivatives play a role; however, there is a trend to use preferably immobilized stationary phases in a first screening. Besides symmetrically substituted derivatives (with a 3,5-dimethyl phenyl carbamate structure), also phases with substituents containing chlorine (3-chlor-5-methylphenyl-carbamate and 3,5-dichlorphenyl-carbamate) or a noncarbamate derivatization chemistry (4-methylbenzoate) seem to give orthogonal enantioselectivity that may not be found elsewhere. A similar observation was made for SFC by C. West [9] who proposed a retention map to compare CSPs using multivariate data analyses. Also, for SFC, the most orthogonal selectivities can be found for ester derivatives, nonhalogenated and halogenated carbamate derivatives.

1.7.12 Are some CSPs Comparable?

From Table 1.7.2, also another problem (or mystery) of chiral separations is obvious. Every provider of CSPs uses their own naming or nomenclature. The "jungle" of trade names makes it almost impossible to discern identical and different chiral selectors. Box 1.7.3 provides an overview of the different polysaccharide substituents, trade names, and suppliers.

CSPs, based on the identical selector, should provide in general a similar or comparable enantioselectivity. In other words, there will be little benefit to include two or three columns with the same selector in a screening; results will be very comparable, even though peak shapes, resolution, and retention times can differ, depending to the origin of the polysaccharide-starting material, the derivatization chemistry, coating procedure, carbon load, and immobilization procedure.

Box 1.7.3 gives a systematic overview of coated and immobilized CSPs, using the polysaccharide backbone as common feature to grouping the CSPs.

Box 1.7.3 Box: Survey of Polysaccharide CSPs Based on Backbone and Derivatization		
Finding equivalent selectors – cellulose derivatives		
Selectors based on cellulose	**Coated**	**Immobilized**
Tris(4-methylbenzoate)	[USP L80/L107],Chiralcel OJ, Eurocel 03, CHIRIS-IOJ, Lux Cellulose-3, Reprosil Chiral-JM, Enantiocel-C3, Chiral Cellulose T-MB, Chromega Chiral CCJ	Chiralpak IJ, Refelect I-Cellulose J, CHIRAL ART-SJ
Tris(cinnamate)	Chiralcel OK	

(Continued)

Box 1.7.3 (Continued)

Selectors based on cellulose	Coated	Immobilized
Tris(3,5-dimethyl-phenylcarbamate)	[USP L40/L93], Chiralcel OD, Nucleocel Delta, Eurocel 01, CelluCoat, Lux Cellulose-1, RegisCell, CHIRIS-IOD, Epitomize CSP-1C, Reflection CL, Chiral-OM, Cellulose DMP, Reprosil Chiral-OM, CHIRAL ART Cellulose-C, Enantiocel-C1, ACQUITY Trefoil CEL1, Chiral Cellulose T-DPC, Chromega Chiral CCO, Sepapak-1	Chiralpak IB, Reflect I-Cellulose B, CHIRAL ART Cellulose-SB
Tris(phenylcarbamate)	[USP L70], Eurocel 04, CHIRIS-IOB, Chiralcel OC, Reprosil Chiral-CM	
Tris(4-methylphenyl-carbamate)	Chiralcel OG	
Tris(4-chlorphenyl-carbamate)	Chiralcel OF, Reprosil Chiral-FM	
Tris(3-chlor-4-methyl-phenylcarbamate)	Chiralcel OZ, Lux Cellulose-2, Epitomize CSP-1Z, Chromega Chiral CC2, Reprosil Chiral ZM, Enantiocel-C2, ACQUITY Trefoil CEL2, Sepapak-2	CHIRAL ART Cellulose-SZ
Tris(4-chlor-3-methyl-phenylcarbamate)	Chiralcel OX, Lux Cellulose-4, Chromega Chiral CC4, Sepapak-4	
Tris(4-fluoro-3-methyl-phenylcarbamate)	Chromega Chiral CCO F4	
Tris(4-fluoro-3-trifluormethyl-phenylcarbamate)	Chromega Chiral CCO F4T3	
Tris(3,5-dichlor-phenylcarbamate)	Sepapak-5	[USP L119], Chiralpak IC, CHIRAL ART Cellulose-SC, Lux i-Cellulose-5, Reflect I-Cellulose C
Tris(2-fluor-5-methyl-phenylcarbamate)	Chromega Chiral CCO F2	
Tris(3-chlor-4-methylphenyl-carbamate) and tris (3,5-dichlorphenyl-carbamate)	Chromega Chiral CCC	

Finding equivalent selectors – amylose derivatives		
Selectors based on amylose	**Coated**	**Immobilized**
Tris(3,5-dimethyl-phenylcarbamate)	[USP L51], Chiralpak AD, AmyCoate, Nucleocel ALPHA, Europak 01, RegisPack, CHIRIS-IAD, Epitomize CSP-1A, MultoHigh Chiral-AM, Reprosil Chiral-AM, CHIRAL, ART Amylose-C, Enantiopak-A1, ACQUITY Trefoil AMY1, Lux Amylose-1, Chiral Amylose T-DPC, Chromega Chiral CCA	[USP L99], Chiralpak IA, Lux i-Amylose-1, CHIRAL ART Amylose-SA, Reflect I-Amylose A
Tris(3,5-dimethylphenyl-carbamate) and tris(4-methyl-phenylcarbamate)	Chromega Chiral CCX	
Tris[(S)-methylbenzyl-carbamate]	[USP L90], Chiralpak AS-V, Reprosil Chiral-AS, Chromega Chiral CCS	Chiralpak IH
Tris(3-chlor-4-methyl-phenylcarbamate)	Chiralpak AZ, Epitomize CSP-1K, Reprosil Chiral-ZA	Chiralpak IF
Tris(3-chlor-4-methyl-phenylcarbamate) and tris(4-methylphenyl-carbamate)	Chromega Chiral CCU	
Tris(3-chlor-5-methyl-phenylcarbamate)		Chiralpak IG, Lux 1-Amylose-3
Tris(4-fluor-3-methyl-phenylcarbamate)	Chromega Chiral CCA F4	
Tris(5-chlor-2-methyl-phenylcarbamate)	Chiralpak AY, Chromega Chiral CC3, RegisPack CLA-1, Reprosil Chiral-YM, Lux Amylose-2, Sepapak-3	
Tris(3-chlorphenyl-carbamate)		Chiralpak ID
Tris(3,5-dichlorophenyl-carbamate)		Chiralpak IE, CHIRAL ART Amylose-SE

USP refers to the United States Pharmacopeia; Chiralcel® and Chiralpak® are products of Daicel Chemical Industries/Chiral Technologies; Eurocel of Knauer; Lux® of Phenomenex; CelloCoat and AmyCoate of Kromasil®; RegisCell®, RegisPack®, and Reflect™ of Regis Technologies Inc.; CHIRAL ART of YMC Europe GMBH; Sepapak of Sepaserve; Astec® Cellulose DMP of Merck Millipore; Reprosil Chiral of Dr. Maisch GmbH; Chromega of ES Industries; Epitomize of Orochem Technologies; Nucleocel of Macherey & Nagel; Chiral Cellulose and Amylose of Silicycle; ACQUITY® Trefoil of Waters; Enantiocel and Enantiopak of Column Tek.

1.7.13 "No-Go's," Pitfalls, and Peculiarities in Chiral HPLC and SFC

Like in achiral HPLC or SFC, the history of a column used should be traceable. A column contaminated by dirty samples will not be able to yield reproducible results. Also, columns stored for a long time should be equilibrated and tested before they are used for any screening or development work. If the column does not pass the test, it should be discarded. HPLC columns are consumables and should be treated as such.

Chiral columns should always be used in the conditions they were designed for, e.g. columns conditioned by the supplier for reversed-phase applications (e.g. "Chiralpak-RH" columns) should be employed with aqueous eluents and not others.

It should be noted that there are cases in which a partial separation cannot be optimized in a way that a full baseline separation is achieved. Rather than trying to optimize a bad separation for weeks on an old column, a new column with the same chiral selector should be tested.

As outlined in the paragraphs before, a thorough screening must be based on a number of different stationary phases. There will be no significant benefit to include two or even more chemically identical selectors in a screening – all will yield similar or identical results.

Chiral chromatographic separations are controlled by thermodynamic equilibria between enantiomers adsorbed on/in the CSP and dissolved in the mobile phase. In almost all cases, the higher the temperature of the separation is, the lower the enantioselectivity becomes. So lowering the temperature will lead to an improved enantiomer separation, while increasing the temperature will decrease enantioselectivity [10].

The elution order of enantiomers is always a result of the interaction of CSP and mobile phase. Changing the mobile phase or the stationary phase can easily invert the elution order [11]. Elution orders cannot be predicted on polysaccharide CSPs, the less stable intermediate diastereomeric complex between stationary phase and the analyte will be the one for the first eluted enantiomer, while the more stable complex will lead to the elution of the second eluted enantiomer.

The nature of the sample should also be transparent and traceable. Unfortunately it occurs that an enantiomer separation is tried on a sample that is already enantiomercially pure – even the best screening cannot find an enantiomer separation for one single enantiomer. Special care should be given to crystalline compounds – not all enantiomers crystallize as racemates – also conglomerates of pure enantiomers crystallizing from racemic solutions have been reported that can lead to significant confusion when analyzed [12].

Samples with several components can also be a source for errors. A real enantiomer separation of a racemate will *always* result in a peak area ratio of 50 : 50 (in UV detection). Any deviation from the expected value indicates the coelution of an impurity beneath one of the peaks. This may even be true for reference samples from worldwide known suppliers. Racemic trans-stilbene-oxide samples often contain up to 4% of the achiral meso-form (i.e. cis stilbene oxide) that can coelute with one of the enantiomers, leading to area ratios of 52% : 48%.

It also should be noted that not all samples are stable regarding their stereochemical orientation. Some compounds may racemize (or enantiomerize) during chiral chromatography leading to a plateau between the two peaks of the enantiomers [13]. These plateaus are temperature dependent, but also in some cases the right choice of mobile phase may suppress the online racemization.

1.7.14 Gradients in Chiral Chromatography

CSPs, in which the chiral ligand is covalently bound, are fully compatible with gradient elution. But when would one use gradient elution in chiral HPLC? First, gradients can be used to speed up the screening process with a generic method that covers a wider range of sample types. Selectivity observed in gradient runs can later be optimized for isocratic operation. The second reason for using gradients is to ensure that contaminants are not trapped on the column when impure samples are injected. In these cases, gradients can be run to remove strongly bound matrix components from the column.

Typical RP eluents, which include combinations of acetonitrile, methanol, 2-propanol, or THF with an aqueous solvent, can be used with a gradient elution of 5–80% organic component [14]. Gradients can also be employed in SFC increasing the elution strength of the mobile phase by increasing the amount of polar coeluent.

1.7.15 Alternative Strategies to Chiral HPLC and SFC on Polysaccharide CSPs

So far only selectors based on polysaccharide CSPs have been described, for the main reason that these are in general the most promising candidates for finding an enantiomer separation. Numerous other CSPs are on the market that can be an alternative to polysaccharides. An overview of those and guidelines for screening and optimization have been given in one of the earlier editions of this book [15] and can be found in the literature for protein-based CSPs [16], cyclodextrins [17], Pirkle-type CSPs [18], crown-ether-based CSPs [19], and CSPs based on macrocyclic antibiotics [20, 21]. A recent example for an enantiomer separation on a nonpolysaccharide CSP is given in Figure 1.7.2.

The most remarkable development in recent years are ion-exchange CSPs [22], which can be anionic, cationic, or zwitterionic. While anion exchangers are appropriate for chiral separation of acidic compounds, their chiral selectors are based on cinchona alkaloids derived from quinine and quinidine. Cation-exchanger selectors are useful for basic analytes, which are structurally based on chiral sulfonic or carboxylic acid compounds as selectors. Zwitterionic selectors were introduced by Lindner and coworkers by combining cation- and anion-exchange moieties in one

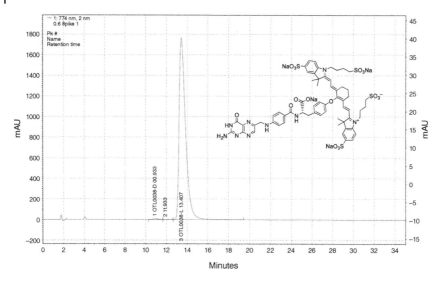

Figure 1.7.13 Enantiomer separation of OTL38 on Chiralpak ZWIX(+): 150 × 3.0 mm, 3 μm, flow: 1 ml/min, 100 mM formic acid/50 mM DEA in 49/49/2 ACN/MeOH/H$_2$O, 35 °C, 774 nm.

single chiral selector and have been commercialized by Daicel under the trade name ZWIX. Those CSPs can be applied for the enantiomer separation of acid, basic, and amphoteric compounds [23]. The chiral mechanism of recognition is mainly based on ionic interactions between the charged analytes and the opposite charged groups of the CSPs. Hydrogen bonds as well as π–π interactions can also influence complex formation. The ion pairing of solvent controls the adsorption and retention of the analytes [24]. Polar-organic and reversed-phase elution modes are the preferential elution modes for this type of CSPs. The retention and enantioselectivity are affected by the pH and the nature and concentration of acid or base added to the mobile phase [25].

An example is given in Figure 1.7.13 for the fluorescent marker OTL38, which comprises a near-infrared dye and a ligand that binds to the receptors overexpressed on cancer cells. These markers illuminate the cancerous lesions, literally lighting the way for the resection of malignant tissue. OTL38 targets folate receptors commonly found on many cancers, including ovarian and lung. The enantiomers are separated on a ZWIX(+) column in polar ionic mode [26].

There are also numerous reports of derivatizing enantiomers with enantiomercially pure reagents leading to diastereomers that have different achiral properties, i.e. compounds that can be separated on achiral phases [27]. These approaches have mostly been abandoned as the time needed to find a complete derivatisation, optimizing the conditions without racemization of the sample, and finding a nonchiral separation are outweighed by the short time needed to find an enantiomer separation on a CSP.

However, there remain a number of separation problems that cannot be solved via chiral HPLC or SFC. Mostly these compounds are either very volatile, or very polar and ionic. Volatile chiral compounds can be separated via enantioselective

GC employing predominantly chiral selectors derived from cyclodextrins or amino acids [28], or via addition of chiral additives in capillary electrochromatography that complex the enantiomers that are then separated due to their differences in their respective electrophoretic mobility [29].

1.7.16 How Can I Solve Enantiomer Separation Problems Without Going to the Laboratory?

Most webpages of column producers contain "application databases," which are an excellent source for information on enantiomer separation of a racemate with a previously described structure or trade name. Several thousands of such applications can be found. Numerous enantiomer separations have been reported over the last 40 years in the scientific literature, in bachelor, master, and PhD theses, in patents, or in pharmacopendial methods. Even without access to sci-finder or similar platforms, such information can be retrieved within a short time from the world wide web (www).

An example is the chiral herbicide fenoxaprop-p-ethyl (see Figure 1.7.14), which is used as a selective, systemic herbicide against annual weed-leaved weeds (grasses), especially in cereal cultivation. How to find a combination of stationary and mobile phase for the separation of the racemate in soil samples?

Using a search engine with the phrase "chiral herbicide fenoxaprop-ethyl," a number of hits appear that are of interest:

1) One of them has the title "Environmental Fate of Chiral Herbicide Fenoxaprop-ethyl in Water-Sediment" [30]. In the section Material and Methods, one finds a subsection entitled "Enantioselective and quantitative HPLC-MS/MS analysis" that leads to supplemental data in which one can find the experimental conditions used by the authors: Column: Chiralpak IC (250 × 4.6 mm), eluent: methanol/water/formic acid, 85 : 15 : 0.1 (v/v/v), flow rate 0.5 ml/min, Rt_1 24.1 minutes ((R)-enantiomer), Rt_2 25.7 minutes ((S)-enantiomer).
2) The second promising entry bears the title "Enantioselective Environmental Behavior of the Chiral Herbicide Fenoxaprop-ethyl and Its Chiral Metabolite Fenoxaprop in Soil." The conditions for the enantiomer separation given are "an amylose tri-(3,5-dimethylphenylcarbamate)" [31] – even without knowledge of the experimental conditions, it is a good starting point to know a second possible CSP for this reversed-phase application. For polysaccharide CSPs with this structure, see Box 1.7.3.
3) The third hit leads to an application note from Regis. The stationary phase is a (S,S)-DACH-DNB, mobile phase is hexane/IPA, 95 : 5 (v/v), retention time is

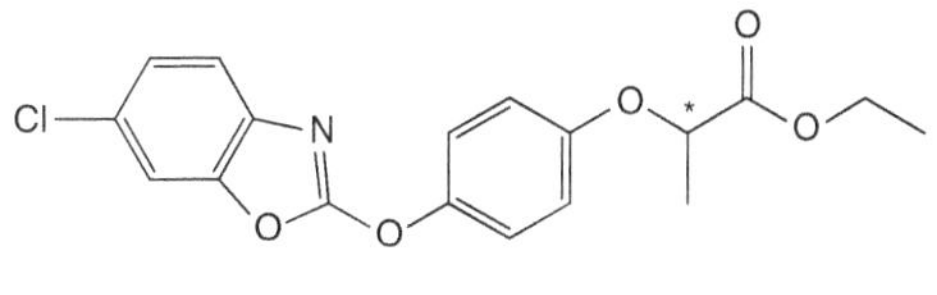

Figure 1.7.14 Finding separations for rac fenoxaprop-p-ethyl.

given as less than 25 minutes with complete baseline separation [32]. This method may be suitable for the racemate, but will be difficult to apply to soil samples without prior sample work-up and extraction into an organic solvent.

These three randomly picked examples are only an illustration of what can be found on the www by using search engines and suitable key phrases.

Many problems regarding enantiomer separations have already been solved – one just has to find the conditions under which a racemate was analyzed. For this purpose, the database ChirBase (http://chirbase.u-3mrs.fr/) was created. Today it contains a collection of 220 000 chiral HPLC/SFC separations extracted from literature and patents. Chemical structure search (Exact, Substructure, Similarity) can be performed as well as simple or complex searches on other fields such as chemical structure, IUPAC name, trade name, specific optical rotation, and absolute configuration. Besides literature references, it contains chromatographic conditions: CSP, mobile phase, temperature, flow rate, detection, elution order, retention times, enantioselectivity, resolution.

1.7.17 The Future of Chiral Separations – Fast Chiral Separations (cUHPLC and cSFC)?

What can we expect in the next 10 years for chiral separations? As in any other chromatographic technique, there will be two main driving factors:

- New CSPs will come ready for the market and many of them will have small particle sizes, i.e. <2 µm.
- The analytical instrument market will focus on speed and reproducibility opening avenues to fast and ultrafast (i.e. less than a minute) separations for enantiomers.

Both developments will result in very fast chiral UHPLC (cUHPLC) and chiral SFC (cSFC) separations. Some developments for new stationary phases in academia have recently been reviewed [33]. The touchstone for these developments will be the acceptance rate in laboratories.

Despite the many advantages short analysis times will bring, the methods will have to be qualified and validated and will require investments into new equipment, education, and require acceptance on management level. There is also the real possibility that inertia and habits will slow down the step to the next evolution level in chiral separations.

On the other side, the economic benefits of quick access to information on enantiomeric composition in routine analytics in industry (raw materials, intermediates, API, DP), as well as in catalyst screening, the potential of real-time monitoring of product formation in asymmetric synthesis, and requirements arising from high-throughput experimentation approaches in R&D environments will be a strong driving force for implementing high-speed chiral separations.

Another need for ultrafast enantiomer analytics will come from two-dimensional LC methods (achiral–chiral separations) for complex matrices, ranging from pharmacokinetic and pharmacodynamic profiles of the individual enantiomers and metabolites in clinical development, drug monitoring, and toxicology, to the extraction of plant and soil samples in agrochemical degradation and metabolism studies, just to name a few. For 2D separations, a segment of the first dimension (e.g. part of an achiral HPLC run) is transferred to a second (chiral) column and analyzed in this second dimension for enantiomer separation. The transfer from one column to the next is often referred to as "heart-cutting."

An example of a two-dimensional liquid chromatography–mass spectrometry achiral–chiral separation has been reported by Daali et al. [34]. The authors developed a bioanalytical method for the simultaneous determination of paracetamol and enantiomers of ketorolac in human plasma. Separation was first achieved in a reversed-phase C18 column under gradient RP conditions. The run between 8.9 and 9.9 minutes, corresponding to phenacetin (internal standard) and racemic Ketorolac, was transferred via a six-port switching valve to an amylose tris-(dimethylphenylcarbamate) derived CSP on which the Ketorolac enantiomers were subsequently separated using an isocratic mobile phase composed of ACN/0.1% formic acid 50 : 50 (v/v). The total run time was less than 18 minutes. By this innovative strategy, the lifetime of chiral columns could be prolonged avoiding damages due to the sample matrix (Figure 1.7.15).

The recent progress in the field of ultrafast enantioselective chromatography [35] provides significantly faster separations with excellent selectivity, peak shape, and repeatability by combining achiral and chiral narrow-bore columns in the first dimension with columns packed with highly efficient chiral selectors (sub-2-µm fully porous or 2.7-µm fused-core particles) in the second dimension [36]. The second dimension can also be run on a chiral column under SFC conditions.

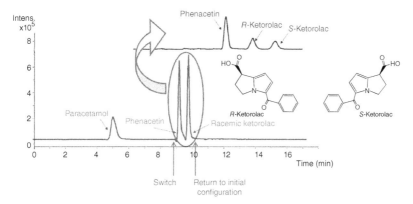

Figure 1.7.15 Separation of spiked plasma (0.5 mg/ml of racemic ketorolac and 2.5 mg/ml paracetamol). Achiral (first dimension) column: Discovery C18 (150 × 2.1 mm ID, d_p = 5 mm) with guard column, eluent: gradient water/0.1% formic acid/ACN, flow: 0.2 ml/min; chiral (second dimension) column Chiralpak AD-RH (150 × 2.1 mm ID, d_p = 5 mm), eluent: water/0.1% formic acid/ACN (50 : 50), flow 0.15 ml/min, detection: MS/ESI. Source: Kuntheavy et al. [34]. © 2009 Elsevier.

Automated evaluation of chiral screenings will become one of the next challenges, especially the pharmaceutical industry has to meet. In environments generating literally thousands of new compounds per year, a larger percentage of those will require enantiomer separation on analytical and semipreparative level to speed up the drug development cycle.

Data storage and data retrieval will require customized software solutions connecting structural data to experimental results. It will be interesting to see if artificial intelligence will help the users to find currently unknown rules for predicting success rates of enantiomer separations on a given set of stationary phases. Once these rules have been established, screening protocols will become much smarter and faster. Until then, we have to rely on the systematic trial-and-error approach outlined in the paragraphs above.

References

1 Ali, I., Saleem, K., Hussain, I. et al. (2009). Polysaccharides chiral stationary phases in liquid chromatography. *Separation and Purification Reviews* 38: 97–147.
2 Tarafder, A., Hill, J.F., and Baynham, M. (2014). Convergence chromatography, versus SFC - What's in a name? *Chromatography*: 34–36.
3 Mckee, R.H., Adenuga, M.D., and Carrillo, J.-C. (2015). Characterization of the toxicological hazards of hydrocarbon solvents. *Critical Reviews in Toxicology* 45 (4): 273–365. https://doi.org/10.3109/10408444.2015.1016216.
4 P. Durray, N. Gauvry, T. Göbel, F. Pautart, (2005). Process for the preparation of enantiomers of amidoacetonitrile compounds from their racemates. EP 04026510, filed 07 November 2005 and issued 18 May 2006.
5 Hintermann, L. (2007). Peak separation by adventitious or added water in normal-phase chiral HPLC. *The Journal of Organic Chemistry* 72: 9790–9793.
6 S. Saito, M. Nagai, T. Morino, T. Tomiyoshi, T. Nishikiori, A. Kuwahara, T. Sato, T. Harada (1998), Novel compound having effect of promoting neuron differentiation. WO patent 1999005091A1, filed 24 November 1998 and issued 04 February 1999.
7 Adapted from D. Eßer, N. Shoji, T. Sato, C. Yokoyama, T. Takai, 2016 Poster Chirality - Fast Method Screening for Separation of Enantiomers in HPLC and SFC Utilising Novel Polysaccharides Type Chiral Stationary Phases Based on Small Particles with permission from YMC.
8 Frühauf, D. and Juza, M. (2012). Development, optimisation and validation of a sub-minute analytical enantioselective high pressure liquid chromatographic separation for a folic acid precursor. *Journal of Chromatography. A* 1269: 242.
9 West, C. (2013). Chirality Seminar, Basel.
10 Matarashvili, I., Kobidze, G., Chelidze, A. et al. (2019). The effect of temperature on the separation of enantiomers with coated and covalently immobilized polysaccharide-based chiral stationary phases. *Journal of Chromatography A* 1599: 172–179.

11 Vanthuyne, N. and Roussel, C. (2013). Chiroptical detectors for the study of unusual phenomena in chiral chromatography. *Topics in Current Chemistry* 340: 107–105.

12 Roussel, C., Vanthuyne, N., Jobert, J.-L. et al. (2007). HPLC on chiral support with polarimetric detection: application to conglomerate discovery. *Chirality* 19 (6): 497–502.

13 Trapp, O. (2013). Interconversion of stereochemically labile enantiomers (enantiomerization). In: *Differentiation of Enantiomers II. Topics in Current Chemistry*, vol. 341 (ed. V. Schurig). Cham: Springer.

14 U. Schlund, D. Eßer, A. Bräutigam, *Whitepaper Chiral LC & SFC Method Development, YMC*, 2019.

15 Juza, M. (2006). Optimization of chiral separations in HPLC. In: *HPLC Made to Measure - A Practical Handbook for Optimization*, vol. 427 (ed. S. Kromidas). Weinheim: Wiley-VCH.

16 Moldoveanu, S.C. and David, V. (2017). Stationary phases and columns for immunoaffinity type separations. In: *Selection of the HPLC Method in Chemical Analysis*, 387–392. Amsterdam, The Netherlands: Elsevier.

17 Xiao, Y., Ng, S.C., Tan, T.T.Y., and Wang, Y. (2012). Recent development of cyclodextrin chiral stationary phases and their applications in chromatography. *Journal of Chromatography. A* 1269: 52–68.

18 Fernandes, C., Tiritan, M.E., and Pinto, M. (2013). Small molecules as chromatographic tools for HPLC enantiomeric resolution: Pirkle-type chiral stationary phases evolution. *Chromatographia* 76: 871–897.

19 Hyun, M.H. (2016). Liquid chromatographic enantioseparations on crown ether-based chiral stationary phases. *Journal of Chromatography. A* 1467: 19–32.

20 Armstrong, D.W., Tang, Y., Chen, S. et al. (1994). Macrocyclic antibiotics as a new class of chiral selectors for liquid chromatography. *Analytical Chemistry* 66: 1473.

21 Ward, T.J. and Farris, A.B. (2001). Chiral separations using the macrocyclic antibiotics: a review. *Journal of Chromatography A* 906: 73–89.

22 Mahalingam, S.M., Kularatne, S.A., Myers, C.H. et al. (2018). Evaluation of novel tumor-targeted near-infrared probe for fluorescence-guided surgery of cancer. *Journal of Medicinal Chemistry* 61: 9637–9646.

23 Hoffmann, C.V., Pell, R., Lämmerhofer, M., and Lindner, W. (2008). Synergistic effects on enantioselectivity of zwitterionic chiral stationary phases for separations of chiral acids, bases and amino acids by HPLC. *Analytical Chemistry* 80: 8780–8789. https://doi.org/10.1021/ac801384f.

24 Lämmerhofer, M. (2010). Chiral recognition by enantioselective liquid chromatography: mechanisms and modern chiral stationary phases. *Journal of Chromatography. A* 1217: 814–856. https://doi.org/10.1016/j.chroma.2009.10.022.

25 Lämmerhofer, M. (2014). Liquid chromatographic enantiomer separation with special focus on zwitterionic chiral ion-exchangers. *Analytical and Bioanalytical Chemistry* 406: 6095–6103.

26 Mahalingam, S.M., Kularatne, S.A., Myers, C.H. et al. (2018). Evaluation of novel tumor-targeted near-infrared probe for fluorescence-guided surgery of cancer. *Medicinal Chemistry* 61: 9637–9646.

27 Sun, X.X., Sun, L.Z., and Aboul-Enein, H.Y. (2001). Chiral derivatization reagents for drug enantioseparation by high-performance liquid chromatography based upon pre-column derivatization and formation of diastereomers: enantioselectivity and related structure. *Biomedical Chromatography* 15: 116–132. https://doi.org/10.1002/bmc.41.

28 Schurig, V. and Juza, M. (2014). Analytical separation of enantiomers by gas chromatography on chiral stationary phases. *Advances in Chromatography* 52: 117–168.

29 Gübitz, G. and Schmid, M.G. (2001). Chiral separation by chromatographic and electromigration techniques. A review. *Biopharmaceutics & Drug Disposition* 22: 291–336. https://doi.org/10.1002/bdd.279.

30 Jing, X., Yao, G., Liu, D. et al. (2016). Environmental fate of chiral herbicide fenoxaprop-ethyl in water-sediment microcosms. *Scientific Reports* 6: 26797. https://doi.org/10.1038/srep26797.

31 Zhang, Y., Liu, D., Diao, J. et al. (2010). *Journal of Agricultural and Food Chemistry* 58 (24): 12878–12884, doi: https://doi.org/10.1021/jf103537a.

32 Gasparrini, F., Misiti, D., Villani, C. et al. (2002). Efficient HPLC separation of Herbicide enantiomers on the new DAC-DNB chiral stationary phase. In: *The application notebook*, 29.

33 Teixeira, J., Tiritan, M.E., Pinto, M.M.M., and Fernandes, C. (2019). Chiral stationary phases for liquid chromatography: recent developments. *Molecules* 24: 865–903.

34 Ing-Lorenzini, K.R., Desmeules, J.A., Besson, M. et al. (2009). Two-dimensional liquid chromatography–ion trap mass spectrometry for the simultaneous determination of ketorolac enantiomers and paracetamol in human plasma: application to a pharmacokinetic study. *Journal of Chromatography A* 1216 (18) 3851-3856. doi: https://doi.org/10.1016/j.chroma.2009.02.071.

35 Stoll, D.R. and Maloney, T.D. (2017). Recent advances in two-dimensional liquid chromatography for pharmaceutical and biopharmaceutical analysis. *LCGC North America* 35: 680–687.

36 Barhate, C.L., Regalado, E.L., Contrella, N.D. et al. (2017). Ultrafast chiral chromatography as the second dimension in two-dimensional liquid chromatography experiments. *Analytical Chemistry* 89 (6): 3545–3553. https://doi.org/10.1021/acs.analchem.6b04834.

1.8

Optimization Strategies Based on the Structure of the Analytes

Christoph A. Fleckenstein

BASF SE, Industrial Petrochemicals Europe, Carl-Bosch-Strasse 38, 67056, Ludwigshafen am Rhein, Germany

1.8.1 Introduction

Prediction of the polarity or solution behavior of organic compounds is a general problem when developing high performance liquid chromatography (HPLC) optimization strategies. Of course, it is general expert knowledge that to achieve good separation, analytes require good interaction with both the stationary phase and the mobile phase (here: the solvent) – following the basic principle of "like dissolves like."

In this context, it needs to be mentioned that there are multiple so-called specific and nonspecific interactions between molecules that partly act simultaneously. The essential types of interactive forces between molecules are compiled in Table 1.8.1.

Within the context of liquid chromatography, one attempts to concentrate on the general term "polarity" when discussing the topic of interactions. At a first glance, this term appears to be handy, because there is a plethora of "polarity tables" for existing solvents [1, 2].

The assessment of the polarity of analyte molecules is far more difficult. Normally, there are no tables to be found. Based on the number and complexity of the analytes, this is simply impossible. Starting to dig deeper into this topic, one soon gets caught in a jungle of theoretical considerations that are – discretely looked at – absolutely valid. They contribute to a better understanding of the nature of polarity and interactions of solvents and molecules (analytes).

Among others, this is associated with the fact that the term "polarity" is not defined in a standardized way. The term "polarity" rather includes all interactions between molecules (usually solvent and solvate molecule). There is no convenient dimension or scale so far that manages to combine all interactions in a meaningful and pragmatic way at the same time. Hence, a plurality of scales that describe polarity has been developed, which are applied in the different areas of natural sciences – meanwhile there are more than 184 (!) different scales for polarity [3].

The multiplicity of scales for polarity of solvents is a clear sign, too, that scientists have elaborated manifold to analyze solvents, to calibrate them in different ways,

Optimization in HPLC: Concepts and Strategies, First Edition. Edited by Stavros Kromidas.
© 2021 WILEY-VCH GmbH. Published 2021 by WILEY-VCH GmbH.

Table 1.8.1 Interactive forces between molecules.

Nonspecific interactions	Specific interactions

• Van der Waals forces
  o London forces (interaction between two polarizable molecules)
  o Debye forces (interaction between a dipole and a polarizable molecule)
  o Keesom forces (interaction between two dipoles)
	• H-Donor/acceptor interactions (hydrogen bonds)

• Ion-dipole forces (Coulomb)	• e^--Pair donor/acceptor interactions

or to calculate some of their properties via theoretical considerations. Although lots of these data and findings for solvents can be found in scientific literature, this is normally not the case for analyte molecules, particularly not in the case of new and so far unknown substances.

So the question is: if we struggle to get access to existing data and info for an analyte molecule, how can one still manage pragmatically to get a "feeling" for polarity, possible interactions with the chemical environment or chemical behavior?

It was shown that already very basic considerations based on the fundamentals of organic chemistry may lead to an extremely good assessment of the properties of a molecule.

Some may hastily throw the towel referring to the point that organic chemistry is known to be very complex and coming from the analytical community one may usually not possess an extremely deep expertise in organic chemistry. That is by far wrong!

The following chapter demonstrates that for our purpose, thus a pragmatic assessment of the character of a molecule, it is already sufficient to consider few, basic features to get a purposeful picture of a molecule and predict differences between two molecules.

In this chapter, we will first deal with selected crucial functional groups of organic molecules and their effects on polarity and water solubility. Furthermore, we will learn that hydrogen bonds and the possibility to predict formation of the same are essential to anticipate solubilities. The influence of pH value on particular functional moieties and on the solubility of a molecule will be briefly discussed. We will then answer the question how to approach the assessment of large, complex multifunctional molecules in an unafraid way. Finally, we will learn how one can assess solubility of a molecule beyond looking at functional groups, but based on the octanol–water-coefficient ($\log P$) or the Hansen parameters and how we can get access to the respective data.

1.8.2 The Impact of Functional Moieties

Functional groups decisively influence the physicochemical properties of molecules. This is illustrated quite impactfully in Table 1.8.2.

Table 1.8.2 Impact of functional moieties.

	Butane	Butyric acid	Butyl amine	γ-Amino-butyric acid
Aggregate state	Gas	Liquid	Liquid	Solid
Water solubility [20 °C]	Poor (61 mg/l)	Good (completely)	Good (completely)	Excellent (>1000 g/l) (formation of zwitter ions in H_2O)

Introduction of simple functional moieties heavily influences the initially gaseous and very hydrophobic molecule butane in respect of water solubility and the intermolecular cohesion.

Organic chemistry knows a plethora of functional groups. Rapidly, one may lose the overview and even well-versed organic chemists generally do not recognize every existing exotic functional moiety sufficiently. However, this is even not necessary for the assessments in our case. As a general rule, it is enough to know a very limited selection of functional groups and to be able to estimate their influence on the properties of a molecule. This selection covers functional moieties that are to be found in almost all organic molecules.

Compounds without a discrete functional group (such as alkanes, alkenes, or alkines as well as alicyclic aromatic hydrocarbons, hence compounds that only consist of carbon and hydrogen) hardly exhibit any polarity. They appear hydrophobic and exhibit decreased solubility in water. This assessment is rather pragmatic and generally held, because, in a more detailed view, one would realize that C—C multiple bonds indeed influence the polarity of a molecule. However, the influence of C—C multiple bonds is negligible in relation to the strength of other functional groups.

By far, most functional groups more or less increase the polarity of molecules. The most important functional moieties and their properties are depicted in Table 1.8.3.

Functional groups, which have the ability to form hydrogen bonds, are of extraordinary importance regarding the influence in the solubility in water and lower alcohols. Hydrogen bonds (H-bonds) are of an electrostatic nature and stronger than usual dipole–dipole interactions. Typically, H-bonds exhibit a binding strength of 17–63 kJ/mol [4]. Hydrogen bonds rank among "specific interactions." They can only occur in presence of certain functional groups within the two molecules or parts, which are interlinked via H-bonds.

1.8.3 Hydrogen Bonds

The term "hydrogen bonds" refers to an interaction force between a hydrogen atom, which is covalently bound to a heteroatom (H-donor), and a free electron pair of

Table 1.8.3 Overview over selected functional moieties and their properties.

Functional group		Specific polarity properties	pH-specific behavior	Oxidation behavior
Hydroxy – Primary (min. R = H, R′=H) – Secondary (R = H, R′ = C, R″=C) – Tertiary (R + R′ + R″ = C)	R′—C(R)(R″)—OH	– H-bonding possible – Increased water solubility – High polarity		– Oxidizable toward aldehyde/carboxylic acid – Oxidizable toward ketone – Not oxidizable
Carbonyl – Aldehyde (R = H, R′ = C) – Ketone (R^1 = C, R^2 = C)	R′—C(=O)—R	Formation of hydrogen bonding dependent on concrete structure		– Oxidizable toward carboxylic acid – Not oxidizable
Carboxyl	R—C(=O)—OH	– H-bonding possible – Increased water solubility – High polarity	Acidic	– Not oxidizable
Amino – Primary (R = C, R′ = R″ = H) – Secondary (R = R′ = C, R″ = H) – Tertiary (R = R′ = R″ = C)	R—N(R′)—R″	– H-bonding possible – Increased water solubility – High polarity – Only H-bonding acceptor – Slightly increased water solubility	Basic	Oxidizable

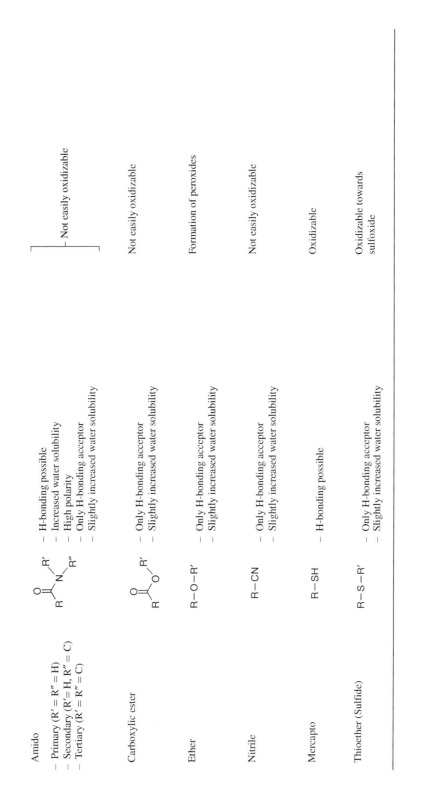

another heteroatom (H-acceptor). Heteroatoms in organic chemistry are mostly the elements O and N. These are significantly more electronegative than hydrogen, whereby the hydrogen bond is clearly polarized.

The donor–acceptor interactions of hydrogen bonds are depicted in Scheme 1.8.1.

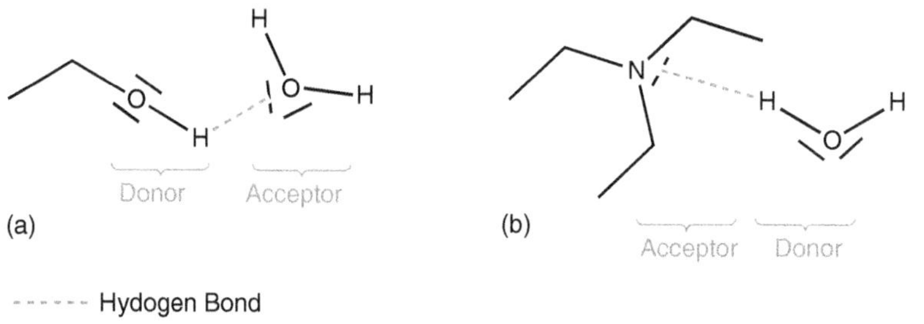

(a) Donor Acceptor   (b) Acceptor Donor

------ Hydogen Bond

Scheme 1.8.1 Examples for hydrogen bonds. (a) Between water and ethanol. (b) Between water and triethyl amine.

It should be noted that certain functional groups can simultaneously act as donors or acceptors. Triethyl amine (Scheme 1.8.1b), e.g. can serve as an acceptor only, whereas the hydroxy group in the ethanol molecule (Scheme 1.8.1a) can act as both a donor and an acceptor due to the free electron pairs of the hydroxy group.

Functional moieties that can serve as both donor and acceptor contribute significantly higher to an increase of water solubility than those that can only act as an acceptor. As an example, let us compare the two amines, tributyl amine and dihexyl amine. Both components exhibit the same total formula. Dihexyl amine is a secondary amine; its amino moiety can act as hydrogen donor and acceptor. The water solubility of dihexyl amine at ambient temperature is approx. 0.3 g/l. Tributyl amine on the other hand exhibits a ter. amino moiety, which can only serve as hydrogen acceptor. Its water solubility at ambient temperature is approx. 0.05 g/l, which means it is nearly one order of magnitude less water soluble than the secondary amine.

It is worth mentioning in the frame of application-oriented analytical assessments that hydrogen bonds do not only play a role within organic compounds or solvents.

Also, metal surfaces, in particular steel, are capable of forming hydrogen bonds to the matrix due to their superficial hydroxy groups.

Metals, especially steels, are frequently used in chromatography as capillaries, but also for other devices. Hydrogen bonds between metal surfaces and analytes can cause undesirable effects like physisorption and, as a consequence, "ghost peaks," etc. Materials like PEEK or even TEFLON are much less prone to forming hydrogen bonds to the matrix, hence being often advantageous.

1.8.4 Influence of Water Solubility by Hydrate Formation of Aldehydes and Ketones

In our previous considerations regarding solubility of molecules in water or polar protic solvents, we assumed that the influencing functional groups to be static,

1.8.4 Influence of Water Solubility by Hydrate Formation of Aldehydes and Ketones

thus invariable. In most cases, this assumption is completely sufficient. However, it is known that – under certain conditions – aldehydes and ketones tend to form hydrates in an aqueous environment. This means that a water molecule adds to the carbonyl group formally generating a geminal diol, i.e. a carbon atom bearing two hydroxy groups (Scheme 1.8.2).

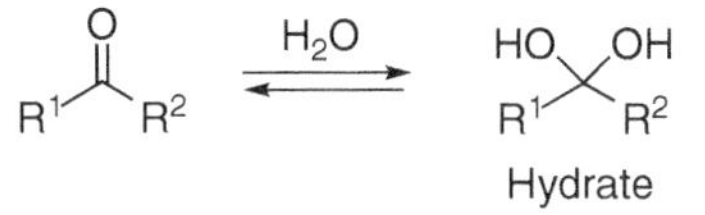

Scheme 1.8.2 Hydrate formation of carbonyl compounds.

This, of course, leads to a significant change of solubility in water and polar solvents, because in place of a carbonyl moiety, which can solely serve as a hydrogen acceptor, there are now two hydroxy groups, which act as H-donors and acceptors. Hydrate formation is usually prohibited according to the Erlenmeyer rule. However, it happens even preferred, the more electron deficient the two rests (R^1, R^2) flanking the carbonyl group are.

Acetone as an example – here are two electron-density pushing methyl rests flanking the carbonyl group – exists in water with <0,1% as a hydrate. Formaldehyde, which does not possess electron-density pushing flanking substituents, can be found >99% as a hydrate in water.

The relevance of hydrate formation can be illustrated by means of the following real practical example: Within the synthesis of an active ingredient, acetyl pyridine (**A**) is radically monochlorinated (Scheme 1.8.3). Besides the monochlorinated product of value, the double-chlorinated side component **C** accrues inevitably. After subsequent workup, purity control is performed. The method of choice, RP-HPLC, provides useful results. However, the observant chromatographer will realize some inexplicable phenomena.

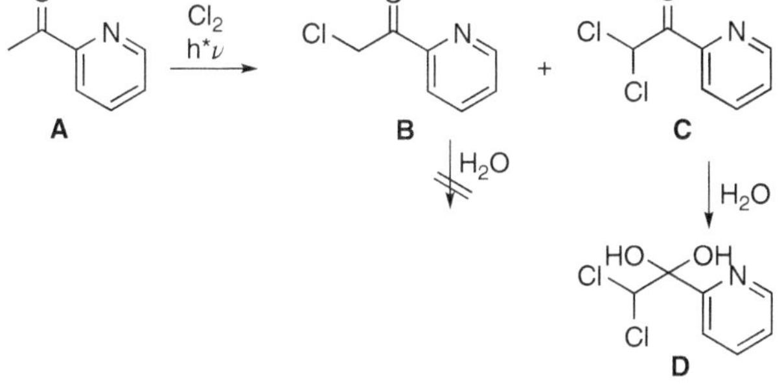

Scheme 1.8.3 Radical chlorination of acetyl pyridine.

Compound **C** elutes in an unusual manner faster than the other derivatives **A** and **B**, which seem chemically very similar. Shape and symmetry of the peak caused by **C** are broader and less ideal than the later eluting peaks of the derivatives **A** and **B**. All these findings cannot be explained from a purely analytical point of view. However, if one looks at the acetyl pyridine derivatives from an organic chemical point of view, one can notice that, due to the higher degree of chlorination at compound **C**, the electron-density pull at the carbonyl group is increased exactly to an extent so that hydrate formation is preferred under aqueous conditions.

In contrast to **A** and **B**, compound **C** exists not in its carbonyl but in its hydrate structure **D** during the process of chromatography at an RP column.

1.8.5 Does "Polar" Equal "Hydrophilic"?

The terms "polar" and "hydrophilic" are all too often misleadingly used as synonyms. It is useful for our discussions in the area of chromatography to distinguish both terms from each other. Hydrophilic means to have a high affinity for water. Usually via formation of hydrogen bonds, hydrophilic molecules can interact extremely well with water or lower alcohols such as methanol or ethanol and dissolve in the same. Polar molecules however are simply molecules characterized by an electrical dipole moment. Water however exhibits both features: It is polar and based on its extremely high ability to form hydrogen bonds it is hydrophilic.

In the first instance, the terms "polar" and "hydrophilic" do not provide sufficient evidence to draw conclusions about the quality of solubility.

In any case, it is advisable not to choose the strongest or most polar solvent for polar substances, but the most suitable one. The solubility is all the better, the more alike the interaction forces are between the solvent molecules and the dissolved compound. This is illustrated briefly as follows.

The two solvents, water and dimethyl formamide (DMF), are known to be very strong and polar solvents. Just looking at the solubility of sucrose, water turns out to be the better solvent. Sucrose exhibits eight hydroxy groups, which can form hydrogen bonds with both of the solvents. Water is the more polar molecule. It possesses hydrogen donor as well as – acceptor properties. The solubility of sucrose in water is reported with 68.7 g/100 ml solution at 30 °C [5].

DMF is an alternative polar solvent; however, it only possesses hydrogen-acceptor properties. The solubility of sucrose in DMF is lower than that in water. It is reported with just 14.1 g/100 ml solution at 30 °C [6].

But why is DMF (Scheme 1.8.4) at all hands known as an outstanding solvent for organic molecules?

In elutropic series DMF ranks directly behind water, but in front of methanol in terms of polarity. This is all the more surprising because as a polar aprotic solvent, DMF does not possess the capability to act as a hydrogen donor.

DMF exhibits an extremely high dipole moment in the tightest space. Moreover, this comparatively small molecule acts as e^--pair donor as well as acceptor. Hence, it can interact extremely well with other electron pair–donating molecules such as amines, alcohols, ethers, but also with electron pair acceptors such as carbonyls and nitro- or sulfonic compounds.

1.8.5 Does "Polar" Equal "Hydrophilic"?

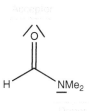

Scheme 1.8.4 Dimethyl formamide.

Table 1.8.4 Comparison of ethyl acetate and 1,4-dioxane.

		Dipole moment (D)	Water solubility (20 °C)	Log P
Ethyl acetate		1.78	8.3 g/100 ml	0.71
1,4-Dioxane		0.45	Infinitely miscible	−0.27

This combination in the tightest space makes DMF a unique solvent. That also explains why DMF can dissolve numerous organic impurities and elute them, for example, from a chromatography column, whereas this cannot be achieved with the vast majority of other solvents.

The dipole moment is a measure for the strength of a dipole and therefore the polarity of a molecule. The dipole moment can be determined via the electricity constant of a molecule. The unit that is typically used for the dipole moment is D (Debeye, named after the Dutch physicist Peter Debeye). The dipole moment of the above-mentioned solvent DMF is 3.82 D, the dipole moment of water is "just" 1.84 D. With these numerical values, one can easily quantify the above comparisons regarding DMF and water.

The comparison between the solvents ethyl acetate and 1,4-dioxane shall serve as a further example that the terms "polarity" and "hydrophilicity" should not be used synonymously.

As depicted in Table 1.8.4, ethyl acetate is the more polar molecule compared to 1,4-dioxane. Its dipole moment is 1.78 D, the dipole moment of 1,4-dioxane merely is 0.45 D.

One reason for the comparatively low dipole moment of dioxane is due to the fact that dioxane exists in various conformers (e.g. chair and boat formation). Not all of these conformers contribute toward an increased dipole moment. However, water solubility of both solvents behaves independently from their polarity: Whereas 1,4 dioxane is infinitely water miscible (soluble), ethyl acetate exhibits only a comparatively moderate solubility in water with 8.3 g/100 ml.

Another more academic and less relevant to practice but still impressive example to compare the terms "dipole moment" and "water solubility" is the molecule 1,2,3,4,5,6-hexafluoro cyclohexane [7]. This molecule, which has been synthesized for the first time only very recently, exhibits a dipole moment of 6.2 D. For an organic and nonionic compound, this is an extremely high dipole moment; it ranks among the highest dipole moments that are known for organic nonionic molecules. It may be surprising, but it is hardly water soluble. The molecule possesses no positions to form any hydrogen bonds, which would be a good base for water solubility.

This example shows that structural elements, being able to form hydrogen bonds, are significantly more important for hydrophilicity (or water solubility) than a high polarity of the molecule.

1.8.6 Peroxide Formation of Ethers

The ether group deserves special attention among the functional organic groups. Ether moieties can be found in numerous organic active ingredient molecules, be it acyclic or cyclic compounds. Ether functionalities are also to be found in various solvents such as THF or dioxane, but in alkoxylate-based surfactants or dispersing agents, e.g. polyethylene glycols as well.

In general, the ether group is known to show high stability. This statement is correct regarding nucleophilic or electrophilic attack, e.g. hydrolysis or similar. However, ether groups show certain reactivity for radical attacks and tend to form so-called hydroperoxides in presence of oxygen.

The reaction mechanism for formation of a hydroperoxide is illustrated in Scheme 1.8.5, using THF as an example for an ether. The first step is, usually induced by UV light, cleavage of an α-H-atom in the neighborhood of the ether moiety. The resulting C-radical recombines with oxygen, which is a diradical, forming a THF-peroxide radical. This THF-peroxide radical now reacts with another THF molecule to form THF-hydroperoxide and a new THF radical. Each reaction cycle forms a new THF radical. Hence, the reaction is a classical chain reaction that will move on reacting until all oxygen is depleted. In general, cyclic ethers are more prone to formation of peroxide than acyclic ethers.

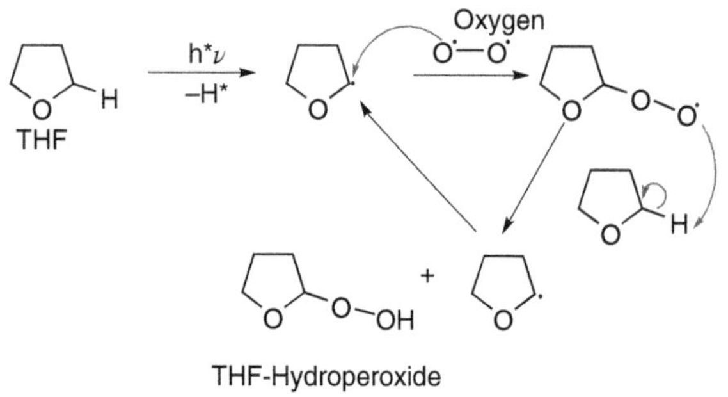

Scheme 1.8.5 Formation of ether hydroperoxides.

Ether-hydroperoxides are very instable compounds that usually cannot be isolated without decomposition. Decomposition, however, that may happen explosively, is only of subsequent interest in the field of analytics. More important for analytical chemistry is the fact that ether compounds may degrade via hydroperoxide formation. On the other hand, the hydroperoxides formed can act as excellent oxidizing agents for oxidizable functional groups such as primary or secondary hydroxy groups or aldehydes. Thereby, in the analytical environment, new peaks may appear whose reference substances have not been present in the original sample. One may keep in mind, that in the case of ethers used as solvents, the same are present in multiple excess compared to the analyte. This significantly enhances the chance for undesirable side reactions. In the same way, ether hydroperoxides may modify stationary phases in chromatography via oxidation [8].

Formation of peroxides occurs in presence of UV light and oxygen. The latter is hardly avoidable looking at the entire analytical process chain, i.e. sampling, sample preparation, chromatographic separation. However, an effective protection against UV light is a good tool against formation of peroxides. Hence, use of brown glass vials and bottles for ether-containing solvents and analytes is highly recommended. Brown glass devices are almost impermeable to UV radiation. Prior to use of ethereal solvents such as THF or dioxane, it is recommended to perform a rapid test for peroxides. As a rapid test, one can use commercially available potassium-iodide–starch paper (it turns deeply blue-violet color in presence of oxidizing agents) or specific test sticks.

To protect ethereal solvents from unwanted formation and accumulation of hydroperoxides, they are commonly marketed as stabilized. This means that the manufacturers add some ppm of a radical scavenging stabilizer. THF, for example, gets often stabilized with ~300 ppm BHT (Butylhydroxytoluene; 2,6-di-*tert*-butyl-*p*-cresol). This reduces the formation of peroxides effectively. The use of stabilizing additives however often entails some disadvantages in chromatography.

Some ethers are less prone to formation of peroxides, among them *tert*-butyl-methyl-ether (MTBE) or cyclopentyl-methyl-ether (CPME) [9].

1.8.7 The pH Value in HPLC

It is often of interest to perform chromatographic separations at exactly defined pH values. These can be adjusted precisely and constantly by addition of buffers to the eluent. One of the advantages is that the user can influence - depending on the structure of the analyte – whether the analyte is present in its ionic or nonionic form. Typically, hydrophobic interactions between analyte and stationary phase are stronger if the analyte is present in its nonionic form. In general, this results – at least in combination with the commonly used silica-based modified phases – in narrow and symmetric peaks. In addition, the selective change of an analyte from its ionic toward its nonionic form (or vice versa) helps to move certain analyte components into a different elution window. The pH dependency of RP retention for acids and basis is depicted in Figure 1.8.1.

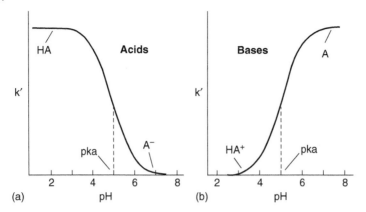

Figure 1.8.1 pH dependency of RP retention for acids and bases. Source: Snyder et al. [10].

Of course, organic chemistry knows numerous functional groups can be ionized or deionized depending on the pH value. However, only for few groups, protonation can be influenced within a pH range that is still acceptable for chromatography (~pH 2–12). Looking at the acidic groups in organic chemistry, there is, for example, the group of sulfonic acids. These, however, belong to the strong acids (their pK_a usually ranks at <−1 (!)). This means that a high degree of dissociation of the sulfonic acid group is unavoidable under typical chromatographic conditions in an aqueous environment. The molecule exists in its anionic structure. Hence, sulfonic acid moieties in analytes provide an extreme increase of molecular polarity and improvement of water solubility.

In the following, the most important acidic and basic groups for HPLC will be discussed from an organic chemical point of view.

1.8.7.1 Acidic Functional Groups

Carboxylic acids rank among the weak acids. Depending on the substituents of the acyl rest, pK_a values of carboxylic acids typically rank between 2 and 5 (see Table 1.8.5). It is known that substituents exhibiting a negative inductive effect, e.g. halogens, facilitate tendency toward dissociation of carboxylic acids, thus decreasing the pK_a value. On the contrary, the pK_a value is increased in the presence of substituents with a positive inductive effect, e.g. alkyl rests. Mesomeric effects act similarly: the negative mesomeric effect of a phenyl ring has a decreasing effect on the pK_a value of a carboxylic acid. It should also be noted that inductive effects are in particular so-called distance effects. This means that they can show their full effect if they are closely located to the functional group considered, here the carboxylic group. The larger the distance, the weaker is the inductive effect. It can be assumed that inductive effects cannot show an effect above more than three neighboring bonds. The "distance effect" of substituents with an inductive effect can be studied when comparing α-chloro-propionic acid (pK_a 2.8; Cl-substituent is closely neighbored) and β-chloropropionic acid (pK_a 4.1; Cl-substituent is further

Table 1.8.5 pK_a-Values of selected carboxylic acids.

Carboxylic acid	Formula	pK_a	Carboxylic acid	Formula	pK_a
Formic acid	HCOOH	3.77	α-Chloropropionic acid	$CH_3CHClCOOH$	2.8
Acetic acid	CH_3COOH	4.76	β-Chloropropionic acid	$ClCH_2CH_2COOH$	4.1
Trimethylacetic acid	$(CH_3)_3CCOOH$	5.05	Benzoic acid	H_5C_6COOH	4.22
Monochloroacetic acid	$ClCH_2COOH$	2.81	Trifluoroacetic acid	F_3CCOOH	0.23

away). Consequently, the α-isomer is more than 10 times more acidic compared to the β-isomer.

Of course, organic chemistry also knows carboxylic acids that belong to the group of strong acids. The (de-)protonation of the same cannot be influenced under pleasant conditions because of their low pK_a value. An example for a strong organic acid is trifluoroacetic acid. The three fluoro substituents that are very closely positioned to the carboxylic moiety make trifluoroacetic acid with a pK_a value of 0.23 (Table 1.8.5) more than 10 000 times more acidic than acetic acid.

These extreme substitution patterns, however, fortunately seldom appear in most of the analyte molecules.

Besides carboxylic acids, in our considerations, one should not forget phenolic hydroxy groups as acidic functional groups. Aliphatic alcohols usually exhibit comparatively high pK_a values, e.g. methanol with a pK_a value of 15.5. Consequently, a meaningful influence of the dissociation is not feasible for aliphatic alcohols under HPLC conditions. This is different for aromatic hydroxy compounds, e.g. phenols, because the negative mesomeric effect of the aromatic ring influences the acidity of the hydroxy group significantly. Phenol, with a pK_a value of 9.99, is about 100 000 times more acidic than methanol. Depending on the substitution pattern of the aromatic ring, aromatic hydroxy groups may in extreme cases even rank in the range of strong acids, e.g. picric acid (2,4,6-trinitro phenol) with a pK_a of 0.29.

1.8.7.2 Basic Functional Groups

Among the basic functional groups, in organic chemistry, amino groups play an important role for our discussions in HPLC analytics. Within amines, one differentiates between primary, secondary, and tertiary amines (Table 1.8.3). The basicity is significantly influenced by the electron density of the free electron pair of the nitrogen atom, because this is able to bind protons as a base. The more electron-rich the nitrogen atom and consequently the free electron pair is, the higher is the basicity of the amine. The main considerations can be made in analogy to carboxylic acids, just with a reversed effect. Hence, substituents with a positive inductive effect lead

Table 1.8.6 pK_a-Values of selected amines (pK_a-values of the corresponding acids).

Amine	Formula	pK_a
Ethylamine	$(CH_3CH_2)NH_2$	10.75
Diethylamine	$(CH_3CH_2)_2NH$	10.94
Triethylamine	$(CH_3CH_2)_3N$	10.79

to an increase of the electron density at the nitrogen atom, thus increasing basicity. The effect of functional groups with a negative inductive effect is vice versa.

Because of this, primary amines (which exhibit one alkyl rest, thus one rest with a positive inductive effect) are less basic compared to secondary amines (which exhibit two alkyl rests, thus two rests with a positive inductive effect). Formally, one would expect that tertiary amines should consequently exhibit the comparatively highest basicity. Nonetheless, this is mostly not the case – at least not under aqueous conditions.

Table 1.8.6 shows that basicity (which can be concluded by the pK_a values of the corresponding acids) is very close together; the basicity of triethylamine even slightly decreases again (which can be concluded from the minimal increasing pK_a value). The background is that two factors play a role, which influence basicity: besides the influence by the substituents of the nitrogen atom, there is also the influence by solvation (so influences by solvents). Thus, primary and secondary amines can be better solvated than triethylamine because their N–H moieties enable easier formation of hydrogen bonds. Because pK_a and pK_b values are usually measured in water (they are only defined in aqueous media), solubility influences basicity of amines, too.

The organic compound lincomycin (Scheme 1.8.6) shall serve as a practical example. When assessing the functional groups of the lincomycin molecule, one realizes that the molecule indeed exhibits quite a number of functional groups (amino, amido, hydroxy, thioether). However, only the amino group (tertiary amine in the heterocyclic ring on the left side) can be ionized in an HPLC-relevant pH range, e.g. via buffering.

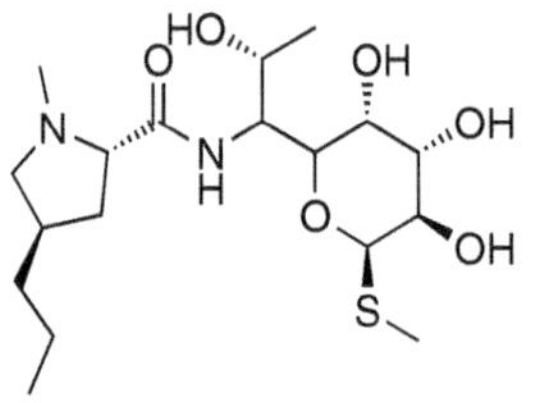

Scheme 1.8.6 Lincomycin.

Depending on the requirements of the analytics, different RP-HPLC chromatographic methods have been developed for lincomycin at various pH ranges. As an example, one can use a mobile phase modified with a phosphate buffer (pH 6). In this case, the amino group is unprotonated [11]. Respective methods elute the analyte lincomycin at longer retention times. Other methods use mobile phases exhibiting a significantly lower pH value. In case of a pH decrease to about 3, e.g. via addition of phosphoric acid [12], lincomycin tends to exist in its protonated salt form. This leads to a shift of the elution window toward shorter retention times. It should be noted that when choosing the optimum pH range, not only the optimization, but also sensitivity of chromatography (elution time, shape of peaks, separation, etc.) should stand in the foreground. It is important as well to consider that the stability of the analyte may vary at different pH ranges. For example, lincomycin easily tends toward hydrolytic degradation already at slightly alkaline conditions; under acidic conditions, it is significantly more stable against hydrolysis [13]. So it is not surprising that most chromatographic methods for lincomycin reported in the literature are run at a pH range between 3 and 7.

1.8.8 General Assessment and Estimation of Solubility of Complex Molecules

The view at single (isolated) functional groups or the comparison of simply functionalized molecules in analogy to the knowledge of the most important functional groups is usually logical and comprehensible for the practitioner. Unfortunately, our daily working environment usually does not only offer such simple structures. Typically, active ingredients for crop protection or pharmaceuticals are often complex and highly functionalized. Even intermediates and fine chemicals may bear numerous functional groups. One may have already noticed this aspect looking at lincomycin in the previous paragraph. Using another example, namely streptomycin (Scheme 1.8.7), we want to approach the problem a little more closely.

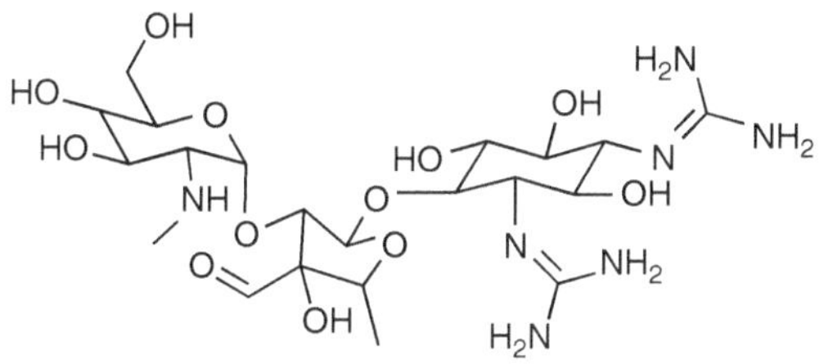

Scheme 1.8.7 Streptomycin.

If one approaches such a complex molecule and tries to predict its chemical behavior or solubility in theory, it is important to overcome the fear one may have of the "organic assessment" of such large molecules. A sentence that is often heard is: "That is too complex for me. My knowledge in organic chemistry is not broad enough."

But instead of burying the head in the sand, we should concentrate on looking holistically at the molecule and recollect what is familiar to us.

Almost none of us know all existing functional groups of organic chemistry by heart. Therefore, a little "Brave the gap" should be acceptable in our context. It helps tremendously to approach the molecule using a kind of tabular structure, to envision what is known, and to try to make a rough assessment. Table 1.8.3 of this chapter may serve as a template. Now it is simply about identifying and counting of the single functional groups.

Looking at streptomycin, the following things (using Table 1.8.3) come to our mind:

- *Hydroxy groups*: 1× primary, 5× secondary, 1× tertiary
    - Hydroxy groups significantly contribute to an increase of water solubility (hydrogen bonds; H-donor and acceptor). No contribution to pH value. Primary and secondary hydroxy groups can be oxidized.
- *Carbonyl groups*: 1× aldehyde
    - H-bond-acceptor. No contribution to pH value. Can be oxidized (formation of carboxylic acid).
- *Ether groups*: 4× ether
    - H-bond-acceptor. No contribution to pH value. Can form peroxides via radical oxidation.
- *Amino groups:* 4× primary, 1× secondary
    - Amino groups significantly contribute to an increase of water solubility (hydrogen bonds; H-donor and acceptor). Contribution to basic pH value.

Using this simple schematic, nearly all functional groups that streptomycin is bearing can be recognized and classified. The attentive viewer has not missed that two of the nitrogen moieties, which are present in the streptomycin molecule, have not been mentioned and assessed. In this case, it is not about amino groups but the so-called imines. This functional group is not mentioned in Table 1.8.3. It shall serve as a good example for the category: "Brave the gap."

The last step is a summarizing conclusion: The molecule streptomycin is

- *Extremely well water soluble*: In total, it exhibits 17 functional groups, which increase water solubility via formation of hydrogen bonds. There is no evidence of a very hydrophobic group.
- *Basic*: In total, it exhibits five functional groups (amino groups), which can act as proton acceptors. The molecule exhibits no acidic moiety. It can be concluded that the secondary amino group provides a higher basicity in direct comparison to the four primary amino groups. In case of an acid–base titration (or even simple salt formation), the secondary amino group can be expected to get protonated first.
- *Better soluble in acidic environment*: In acidic pH range, one or more of the amino groups are expected to get protonated. Consequently, the molecule is ionized. This leads to a further increase of water solubility of streptomycin, which was already assessed to be high.
- *Sensitive to formation of ether hydroperoxides*: In total, the molecule exhibits 4 ether groups with an α-H-atom, which can form ether hydroperoxides. Thus, direct UV radiation should be avoided; brown glass should be chosen as a material for sample vials.

- *Sensitive to oxidation*: Some functional groups (primary + secondary hydroxy groups, aldehyde group) can be oxidized. Thus, care should be taken during sample preparation and chromatography so that they do not provide conditions that favor oxidation. So be careful when not using redox active solvents, e.g. DMSO or conditions, which facilitate formation of oxidizing ether hydroperoxides.

1.8.9 Octanol–Water Coefficient

The octanol/water partition coefficient is established as a simple and pragmatic method to measure and describe the ratio between lipophilicity and hydrophilicity, hence the solubility behavior of a substance in polar and nonpolar solvents [14]. According to the Nernst distribution theorem, the partition coefficient is the ratio of concentrations of a substance in the two immiscible phases of octanol and water in the equilibrium at a specified temperature. The octanol/water partition coefficient is also abbreviated as K_{ow}-value or P-value, respectively.

It was first used in pharmaceutical chemistry and biochemistry to describe the potential of a substance to get dissolved intra- or extracellular (here is the hydrophilic area), but also to pass cell membranes, which are usually significantly lipophilic. For passing the cell membranes, the permeating substance understandably needs lipophilic properties, too.

The value of the octanol/water partition coefficient of a substance is higher than one if the substance dissolves better in lipoid solvents such as n-octanol, it is smaller than one in case it is better soluble in water.

Generally, the octanol/water partition coefficient is reported as the decimal logarithm as $\log P$ (also $\log P_{ow}$):

$$\log P = \log \frac{c_o^{s_i}}{c_w^{s_i}}$$

In this logarithmic equation, the numerator relates to the concentration of a species i in octanol; the denominator shows concentration of a species i in water.

Accordingly, the $\log P$ shows positive values for lipophilic compounds and becomes negative for hydrophilic substances. For example, it is approx. -0.4 (at 25 °C) for formic acid and approx. $+3.4$ (at 25 °C) for 1,4-dichlorobenzene. Hence, $\log P$ offers a fast and good possibility to compare lipophilicity or water solubility between two or more compounds.

The $\log P$ can be of great help especially for analysts in HPLC. The original intention of the $\log P$, namely the description of the behavior of a substance between the polar cytoplasm and the nonpolar cell membrane, which the substance should pass, is found in a very similar way in chromatography: this is the case if we try to estimate the behavior of a substance relative to a solvent (mobile phase) and the column (stationary phase). For most chemical compounds, especially for those that are commercially available, the $\log P$ value can be found in the material safety datasheet (Chapter 1.9 of the MSDS, paragraph: physical and chemical properties). For other compounds, the log *P* is published in chemical journals. Practically measured values of log *P* may not exist for all substances. However, there are various models to predict log *P*, e.g. via quantitative structure activity relationship (QSAR) or via *linear free energy relationships* (LFER). A very good collection of log *P* values, albeit for a fee, is offered by the *Dortmunder Datenbank*.

Fortunately, we can, in everyday life, even circumvent a time-consuming search in databases or journals, at least if we want to get a rough or first estimation of log *P* or the water/fat solubility of a substance.

Most of the popular chemical software drawing programs offer the possibility to have log *P* indicated with no effort. The software programs calculate (in a simplistic way) the log *P* based on the chemical structure using the models mentioned above. As an example, there is a program *Biovia Draw* (formerly known as Accelrys Draw) or *Chemdraw*, which belongs to the Perkin Elmer company. Both programs are even available free of charge in some (simple) versions. The function for prediction of log *P* can be found – dependent on the software version – each under the rubrics "Chemistry → Calculator" or "View → Chemical Properties", respectively. The efficiency of these simulation methods should not be underestimated as we will see below.

In Scheme 1.8.8, there are three relatively simple functionalized compounds depicted. Through our methodically structured approach, namely the prediction of solubility and polarity based on the presence of functional groups, a differentiation of these compounds is not easily possible – even by an experienced specialist. The existing structural features, the possibilities for hydrogen bonding, and donor–acceptor properties are quite similar.

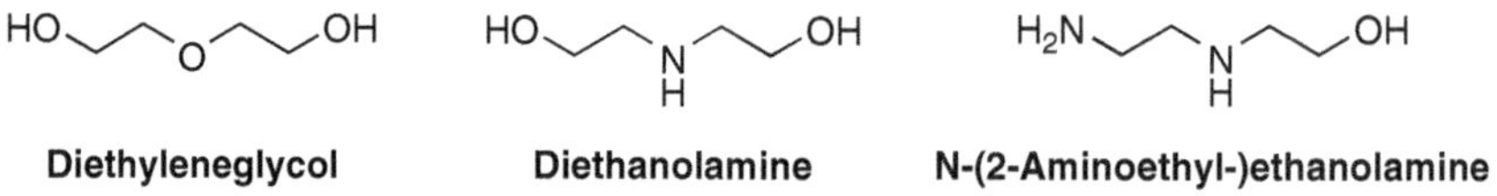

Diethyleneglycol   **Diethanolamine**   **N-(2-Aminoethyl-)ethanolamine**

Scheme 1.8.8 (Amino)alcohols.

Table 1.8.7 shows the log *P* data for the three compounds. In the first column, one can find experimentally based data; the following two columns contain calculated data from *Chem Draw* and *Biovia Draw*.

It stands out that the theoretically simulated values differ only slightly from each other. However, the individual log *P* values clearly show a similar trend. The experimentally gained octanol/water partition coefficient values behave similarly in the order of magnitude, however, show a different sequence of solubility of the compounds among each other. These differences can essentially be explained by the fact that computer simulations strictly apply parameters and increments of the occurring functional groups for their calculations. Effects, which are subsequent to solution of a molecule, e.g. solvation effects, generally cannot be incorporated by these simple programs. In theoretical considerations, the compounds are usually regarded to be pure. Effects caused by impurities for example are not considered in the calculations.

However, it should be noted, that actually experimentally determined log *P* values should be viewed with a sense of proportion, too. For one and the same compound,

Table 1.8.7 Log P-values of selected (amino-)alcohols.

Compound	Log P[a] (experimental)	Log P[b] (in silico)	Log P[c] (in silico)
Diethyleneglycol	−1.52	−0.95	−1.03
Diethanolamine	−2.18	−1.17	−1.25
N-(2-aminoethyl-)ethanolamine	−1.46	−1.55	−1.58

a) Log P experimental (OECD-guideline 117) [15].
b) Log P in silico (Chem Draw).
c) Log P in silico (Biovia Draw).

one can find quite a certain spread of the reported values. A spread of ±0.5 for the same substance is absolutely common.

This is due to the fact that, the purity of the substance and the exact testing conditions are influencing factors, besides the experimental mistake that everyone of us makes when doing analyses.

We can conclude that computer simulation can provide a simple and good addition to the experimental determination of log P. With the help of simulation, especially solubility of very complex molecules can be rapidly assessed. As an example, the theoretically calculated log P value of streptomycin is −6.99 (*Chem Draw*). This means that streptomycin is highly water soluble. It confirms our conclusion that we had previously made for streptomycin based on assessment of functional groups.

At the beginning of the section, we had found that the hydrophilic/hydrophobic interactions that are described with the octanol–water partition coefficient can be found in chromatography, too.

So it is not surprising that meanwhile HPLC is used as an established method for determination of log P. Instead of classical determination of the partition coefficient of a substance in an octanol/water mixture, the retention time of a substance in a validated RP-HPLC analytics allows conclusions to be drawn regarding the log P of the substance [16]. The advantage of this method is the high speed of determination as well as an easy way of automation.

Vice versa, log P can be very helpful for HPLC. Log P is excellent for method optimization due to the fact that separation mechanisms are based to a large extent on the partition between polar and nonpolar phases. The interactions (Table 1.8.1) between stationary and mobile phase as well as the analyte are the same as for the octanol–water partition coefficient. Calculation of intermolecular forces like van der Waals forces, hydrogen bonds, or electrostatic interactions can now be made well for quantitative models and predictions in chromatography.

Even simulative calculation of retention times is possible via log P, particularly good in the field of RP-HPLC [17, 18]. This is known as "in silico chromatography." However, simulation of retention times is not feasible without having knowledge about methods in theoretical computational chemistry. There is no simple tool available for everyone in everyday life of chromatography. In addition, *silico chromatography* usually assumes optimal conditions. Interactions with impurities are neglected as well as solvation effects. Because of that, simulation for gradient elution, where solvation changes in the course of the chromatography, is currently not easily feasible.

1.8.10 Hansen Solubility Parameters

We had learned using the example of the solvent DMF that a good water solubility or high polarity does not say everything about the actual solution power a solvent. Similarly, the octanol–water partition coefficient, which has been discussed previously, ultimately just describes the hydrophilic and lipophilic nature of a substance.

A comparatively simple approach to show the interactions in the case of solubility and to predict solubility is provided by the so-called Hansen solubility parameters [19].

The concept follows the basic principle of "like dissolves like," which means that a molecule is highly soluble if there is good interaction with the solvent, it appeals to the same interactions.

Hansen defines three so-called Hansen parameters for every molecule. These three parameters allow the definition of a dot in a three-dimensional space, the so-called Hansen space. The Hansen parameters are based on the following interaction considerations:

- δ_D, cohesion energy from dispersion forces between molecules
- δ_P, energy from dipolar intermolecular force between molecules
- δ_H, energy from hydrogen bonds between molecules

The closer two substances are in the Hansen space, the higher the probability that they are soluble in one another.

To better understand the Hansen space and to show the relationships of the Hansen parameters more clearly, let us have a look at the solubility of an analyte in various typical solvents. As an example, the well-known painkiller paracetamol (Scheme 1.8.9) is chosen.

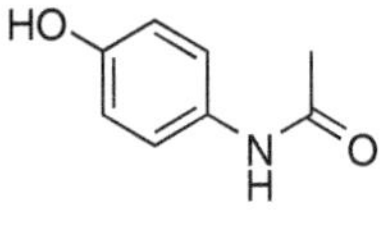

Scheme 1.8.9 Paracetamol.

Looking at paracetamol, one can see that the molecule itself is not too complex. Bearing a hydroxy- and a secondary amido group, paracetamol offers two functional groups with hydrogen donor- as well as acceptor quality. The other residual structural elements, such as the aromatic ring or the methyl group, entail hydrophobic properties to the molecule. Hence, the paracetamol molecule is slightly polar; overall, however, its polarity is fairly balanced. The molecule should impart a certain water solubility.

Table 1.8.8 Hansen parameters of paracetamol and selected solvents as well as the real solubility of paracetamol in these solvents.

Compound	Hansen parameter [19]			Solubility of paracetamol at 30 °C (g/kg solvent) [20]
	δ_D	δ_P	δ_H	
Paracetamol	17.8	10.5	13.9	—
Water	15.5	16.0	42.3	17.4
Ethanol	15.8	8.8	19.4	232.8
1,4-Dioxane	19.0	1.8	7.4	17.1
Chloroform	17.8	3.1	5.7	1.5
Toluene	18.0	1.4	2.0	0.3

These theoretical considerations can be easily verified in reality: The solubility of paracetamol in various solvents is depicted in Table 1.8.8. Paracetamol is only moderately soluble in the very polar solvent water. The active ingredient is more soluble in the less polar ethanol by more than an order of magnitude (232.8 g/kg ethanol). Using an even less polar solvent such as 1,4-dioxane, solubility of paracetamol decreases again. A comparatively poor solubility can be considered in highly nonpolar solvents such as chloroform or toluene.

With the help of the Hansen parameters, this real solubility can be predicted quite well in a qualitative way. As we had already noted above, according to Hansen, two substances can dissolve well into one another if they are nearby in the Hansen space. Figure 1.8.2 depicts the three-dimensional Hansen space with the positioning of paracetamol as well as the five respective solvents. It can be clearly seen here that the solvent ethanol is spatially closest to the paracetamol to be dissolved. Due to its high hydrogen bond energy (δ_H), water is located spatially somewhat more distant from paracetamol in the Hansen space. The residual solvents 1,4-dioxane, chloroform, and toluene are again located spatially somewhat more distant from paracetamol in the Hansen space due to their very low hydrogen bond energy (δ_H).

Meanwhile, Hansen parameters are known for a plethora of solvents, active ingredients, and other compounds or can easily get calculated or determined [21].

Like every model, the Hansen solubility model has limitations, too. As an example, the factors of temperature and molecular size are not considered reasonably. Furthermore, the Hansen concept strictly follows the basic principle of "like dissolves like." However, one can observe that especially for molecules with acidic or basic structural elements, solubility of the molecules can even be improved in an "opposite environment." Carboxylic acids, for example, show often excellent solubility in basic solvents. In this regard, parameter systems for prediction of solubility like the Hansen system are constantly being developed further. As an example, Beerbower et al. have developed a four-parameter system establishing an acidic (δ_a) as well as a basic (δ_b) solubility parameter replacing the Hansen parameter δ_H. This helps

1.8 Optimization Strategies Based on the Structure of the Analytes

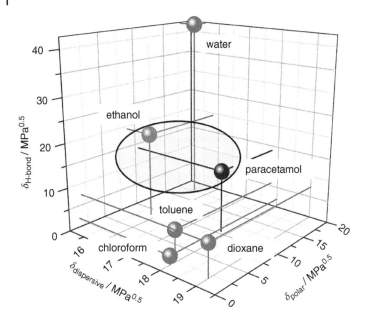

Figure 1.8.2 3-D-illustration of Hansen solubility parameters of paracetamol and selected solvents.

to improve forecast accuracy, electron-donor and -acceptor properties of functional groups are better taken into account at different pH ranges [22].

1.8.11 Conclusion and Outlook

For a long time, scientists and working groups have been dealing with the issue to predict solubility of substances in water or other solvents or solvent mixtures, respectively. Against this background, a variety of polarity- and solubility scales has been developed. With their help, it usually works quite well to classify certain experimentally found results and to explain via post rationalization. However, a precise forecast for more or less unknown systems remains difficult.

Much progress has been made in science in recent years to predict solubility via computerized systems. Most of this progress was made to improve prediction accuracy in aqueous solvent systems [23]. Similarly, computer-based prediction accuracy for the octanol/water partition coefficient is increasing [24].

However, these methods are not yet perfect. In addition to numerous inaccuracies that occur in practical use, it is aggravating that computer-based systems of course make use certain databases and scales, e.g. polarity scales and values [25]. As a rule, however, these scales were developed for pure solvents. The major part of chemistry in nature, laboratory, and thus in our analyses, too, however takes place in solvent mixtures.

One focus of future developments will therefore be in the extension of polarity scales toward solvent mixtures. Another focus will be to better consider the influence of other contaminants.

However, it should not go unmentioned that such a computer-based prediction of solubility properties can be easily obtained. Usually, the experienced use of these methods is only mastered by theoretical chemists. A quick statement and processing by the user, thus in our context the experienced chromatographer, is not yet possible.

For the latter, at least for now, remains the knowledge about organic chemistry and the impacts of the single functional groups of a molecule (as we have discussed in the previous paragraphs) to be an indispensable tool for prediction of substance properties – with a little practice even unbeatably fast!

Acknowledgments

The author would like to thank Prof. Dr. Alfons Drochner, Technical University of Darmstadt, for his active support in creating the figure of the Hansen space.

References

1 Duve, G., Fuchs, O., and Overbeck, H. (1976). *Lösemittel Hoechst*, 6e. Hoechst Aktiengesellschaft.
2 Meyer, V.R. (2009). *Praxis der Hochleistungsflüssigchromatographie*, 77–79. VCH: Weinheim.
3 Katritzky, A.R., Fara, D.C., Yang, H. et al. (2004). *Chemical Reviews* 104: 175–198.
4 Jeffrey, G.A. (1997). *An Introduction to Hydrogen Bonding*. Oxford: Oxford University Press.
5 Browne, C.A. (1912). *Handbook of Sugar Analysis*. New York: John Wiley and Sons.
6 Kononenko, O.K. and Herstein, K.M. (1956). *Industrial and Engineering Chemistry* 1 (1): 87–92.
7 Keddie, N.S., Slawin, A.M.Z., Lebl, T. et al. (2015). *Nature Chemistry* 7 (6): 483–488.
8 Rabel, F.M. (2010). *Normal-Phase Chromatography, Encyclopedia of Chromatography*, 3e, 1601–1603. Taylor & Francis.
9 Watanabe, K., Yamagiwa, N., and Torisawa, Y. (2007). *Organic Process Research and Development* 11 (2): 251–258.
10 Snyder, L.R., Kirkland, J.J., and Glajch, J.L. (1988). *Practical HPLC Method Development*, 2e. New York: John Wiley & Sons.
11 Abualhasan, M.N., Batrawi, N., Sutcliffe, O.B., and Zaid, A.N. (2012). *Scientia Pharmaceutica* 80 (4): 977–986.
12 Dousa, M., Sikac, Z., Halama, M., and Lemr, K. (2006). *Journal of Pharmaceutical and Biomedical Analysis* 40: 981–986.

13 Shantier, S.W., Elimam, M.M., Mohamed, M.A., and Gadkariem, E.A. (2017). *Journal of Innovations in Pharmaceutical and Biological Sciences* 4 (4): 141–144.
14 Sangster, J. (1997). *Octanol-Water Partition Coefficients: Fundamentals and Physical Chemistry*, Solution Chemistry, vol. 2. Chichester: John Wiley & Sons.
15 BASF SE (2020), Safety Datasheet.
16 Valko, K., Du, C.M., Bevan, C. et al. (2001). *Current Medicinal Chemistry* 8 (9): 1137–1146.
17 Hanai, T. (2019). *Current Chromatography* 6 (1): 52–64.
18 Tetko, I.V., Poda, G.I., Ostermann, C., and Mannhold, R. (2009). *QSAR and Combinatorial Science* 28 (8): 845–849.
19 Hansen, C. (2007). *Hansen Solubility Parameters: A User's Handbook*, 2e. Boca Raton: CRC Press.
20 Granberg, R.A. and Rasmuson, A.C. (1999). *Journal of Chemical & Engineering Data* 44 (6): 1391–1395.
21 Scheler, S., Fahr, A., and Liu, X. (2014). *ADMET & DMPK* 2 (4): 199–220.
22 Beerbower, A., Wu, P.L., and Martin, A. (1984). *Journal of Pharmaceutical Sciences* 73: 179–188.
23 Bergström, C.A.S. and Larsson, P. (2018). *International Journal of Pharmaceutics* 540: 185–193.
24 Fan, S., Iorga, B.I., and Beckstein, O. (2020). *Journal of Computer-Aided Molecular Design* 34 (5): 543–560. https://doi.org/10.1007/s10822-019-00267-z
25 Llinas, A. and Avdeef, A. (2019). *Journal of Chemical Information and Modeling* 59 (6): 3036–3040.

1.9
Optimization Opportunities in a Regulated Environment
Stavros Kromidas

Breslauerstr. 3, 66440, Blieskastel, Germany

1.9.1 Introduction

In regulated areas, very little or no change may be made to existing methods. This means that a more or less major challenge is faced if defined specifications are not met. In a routine laboratory, for example quality control or in-process control, the usual requirements for the main component are: A certain value for the coefficient of variation of the peak area (VC, relative standard deviation, RSD) must be achieved. Or a certain value for the resolution between active ingredient (API) and impurity. In the case of a trace analysis method, the limit of quantification (LOQ) or the reporting level should be proofed, the former often defined by the peak-to-noise ratio. Quite seldom, a certain plate number is also required. In the following, I would like to show possibilities how to meet common requirements even if "nothing" may be changed.

1.9.2 Preliminary Remark

Many methods from pharmacopoeias often do not deliver the expected results when applied. It is a well-known fact. I just mention two reasons for this: The robustness of HPLC methods is often not tested extensively enough and/or in a practical way. Furthermore, the chromatographic conditions as described in the original method are not always feasible, just think of different hardware. This is the reason why there is traditionally a lot of leeway in the application of pharmacopoeia methods. So in USP, EP, BP, JP, etc. this leeway is intended to give the routine HPLC laboratory the possibility to react flexibly if necessary. The only one – but eminent! – prerequisite is that the requirements for the method are still fulfilled. This applies, for example, to the criteria of the system suitability test (SST) after a possible change. The freedom given by the pharmacopoeias is not the subject of this book, so I will limit myself here only to a few quotations from USP and EP (I did the highlighting of text passages) with occasional short comments. The point is simply to remind you of the freedom you have.

Optimization in HPLC: Concepts and Strategies, First Edition. Edited by Stavros Kromidas.
© 2021 WILEY-VCH GmbH. Published 2021 by WILEY-VCH GmbH.

- In other circumstances, it may be desirable to use an HPLC column with *different dimensions to those prescribed in the official procedure* (different length, internal diameter, and/or particle size) (Comment: This means that you are practically free to choose all physical parameters of a column)
- No change in the identity of the stationary phase substituent is permitted (for example, no replacement of C18 by C8...) (Comment: I can therefore use a completely different C18 column if necessary, since the requirement "remaining same chemistry" is fulfilled, because "C18" means "same chemistry" for the pharmacopoeias, regardless of the manufacturer. For the selection of a suitable alternative column, see)
- ...chromatographic support, surface modification, and extent of chemical modification must be the same; a change from totally porous particle (TPP) columns to superficially porous particle (SPP) columns is allowed *provided these requirements are met*. (Comment: I can alternatively use a core shell C18 column, which usually allows faster separations and sharp peaks at lower pressure drop)
- If you do make adjustments, which are within a reasonable range, *you do not have to entirely revalidate*, although you must show an improvement in chromatography using reference standards.(Comment: "improvement" is a broad term, so by changing the chromatographic conditions I could create a faster or a more robust method, or by changing, for example, from a 4-mm to a 3-mm column – which is compatible with the allowed difference of 25% in inner diameter – you save about 45% in solvent and waste costs, the peak height also increases noticeably)
- If compliance with the system suitability requirements cannot be achieved, it is often advisable to check the dwell volume or replace the column (Comment: It is advised to take action if necessary)
- ... some modifications of the chromatography conditions may nevertheless be necessary to meet the requirements prescribed for system suitability... (Comment: Note: "required" and not just "allowed")
- ...in such a case, it may be necessary to replace the column by another of the same type (for example, octadecylsilylated silica gel), which shows the desired chromatographic behavior.

Let us now come to reality: For reasons not dealt with here, it is the case in practice that the leeway given by the pharmacopoeias, which is fortunately remarkably large, see above, is rarely used. In case of problems, only the column is replaced or the eluent is replenished. What to do? There is a positive aspect in this context: method details or method parameters that are missing in a method description are logically not given and can therefore be changed. And, fortunately, there are a number of such changes that can lead to an improvement in the result and one can act in accordance with the pharmacopoeia. We will deal with such changes below. Which of these can actually be implemented can only be decided individually. In the following, we will discuss resolution, peak shape, baseline noise, and coefficient of variation. However, before we deal with the most common requirements in a routine laboratory below, the following note: Sometimes it is desirable to modify a "classical" pharmacopoeia method as LC–MS method to specifically characterize an impurity. In this case, the often-used nonvolatile buffer interferes. With the help of 2D-HPLC, the salt can be removed in the second dimension, thus creating an MS-compatible method from a "classical" UV method (see details in Chapters 1.1 and 4.4): The method would remain as prescribed in the first dimension and the requirements would be met.

1.9.3 Resolution

Resolution (R, resolution) is – to put it simply – the distance between two peaks at the peak base. The resolution depends on:

- *The capacity*: strength of the interaction between analyte and stationary phase, measure: retention factor k
- *The selectivity*: the ability of a chromatographic system to distinguish two components, measure: separation factor alpha α
- *The efficiency*: peak shape, measure: plate number N

The retention and separation factor depends on the "chemistry," e.g. pH value, eluent composition, including gradient slope, etc., diluent, stationary phase, and temperature. For gradient methods, often also the dwell or delay volume of the apparatus: this is the volume from the point where the solvents are mixed, i.e. from the mixing valve/mixing chamber to the column head.

If the original method does not contain exact data for a given parameter or the handling practice, the resulting freedom can be used. For example, premixing in case of gradients "yes/no" or just the way how to prepare the mobile phase.

If, with regard to the mobile phase, only: "60/40, v/v, methanol/water" is written, such a note can be interpreted appropriately and the eluent can be prepared as follows: 60 vol% methanol is present and it is made up with water, or both portions are put into the storage vessel at the same time or the pump is mixing the solvents. In the three procedures, different eluent compositions are produced due to volume contraction. The procedure chosen is the one that achieves the required resolution. In case the "chemistry" must not be changed, we now turn to other possibilities to change k, α, and N. We start with changes according to the hardware.

1.9.3.1 Hardware Changes

1.9.3.1.1 Preliminary Remark

It is assumed here that minor hardware changes may be made in the laboratory. If, however, the effort required for a subsequent mandatory performance qualification proves to be disproportionately high, the only remaining possibility would be to change the setting parameters, see below.

First, we should distinguish between isocratic and gradient separations:

Gradient separations

A change in the dwell volume can (it does not have to!) influence, among other things, the resolution, and this may be different in the front and rear sections of the chromatogram. The handles for this: Changing the mixing chamber (geometry, volume), changing the loop volume in the autosampler, changing the diameter and length of the capillaries to the column. I deliberately use the word "change" here and do not give a general recommendation "reduction," because this can go in both directions.

Isocratic separations

Dead volume ("extra column volume," "dispersion volume") is the volume of the apparatus from the autosampler to the detector without column. If this volume is reduced, e.g. by using shorter/thinner capillaries or a smaller cell volume, the peak shape and thus the resolution is improved. This effect becomes more noticeable the smaller the column volume is and the earlier the peaks elute.

1.9.3.1.2 UHPLC Systems

From approximately 800 bar on, the polarizability of (polar) molecules can change, thus also the selectivity and consequently the resolution. This can be achieved, for example, by installing a very thin capillary directly after the column. In the case of gradient separations, hardly any shift of the retention time is observed.

1.9.3.1.3 Column Oven

In HPLC, both diabatic and adiabatic column ovens are used. Furthermore, in the case of an air oven, a distinction should be made between a forced-air and a still-air oven. Depending on the mode of operation, the resolution can change – despite the same temperature indication on the display of different ovens. Additionally, a particular air oven can also be operated in both modes. And analogous to the dwell volume in gradient separations, the resolution can change differently depending on where in the chromatogram the peaks of interest elute, see Figure 1.9.1.

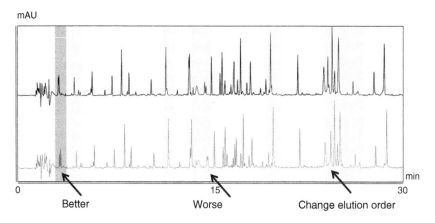

Same HPLC-Separation with a still air (above) and a forced-air oven (below)

Figure 1.9.1 Separation can become worse or better when using different air ovens, elution reversal is also possible; for details, see 1.9.3.1. Source: Michael Heidorn, ThermoScientific © Thermo Fisher Scientific Inc.

1.9.3.2 Improving the Peak Shape

In addition to reducing the dead volume, the following options are available to improve the peak shape:

- If eluent tempering is not explicitly "forbidden," the eluent can be pretempered. This can improve the peak shape, which may lead to the desired resolution.

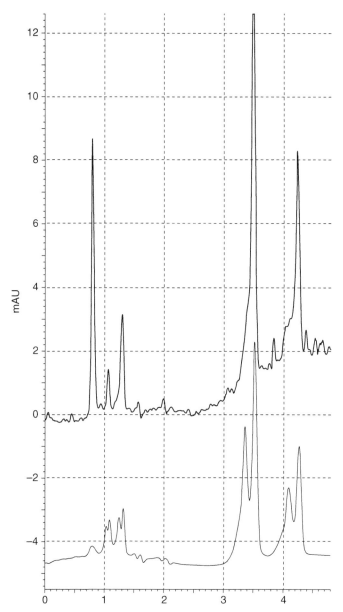

Figure 1.9.2 By turning the column upside down, double peaks are prevented, often the peak shape is improved; see upper chromatogram.

- Inject 10–15 μl air together with the sample – first the sample, then the air from an empty vial. This "air cushion" directly in front of the substance zone prevents the building of a laminar profile of the sample plug on its way from the autosampler to the column. By the way, some UHPLC instruments optionally generate such air segments. As soon as the air bubble reaches the column, it is dissolved in the eluent due to the pressure built up by the column and does not interfere with the sample.
- You can turn the column over, see Figure 1.9.2. This should not be done with UHPLC columns; furthermore, this simple trick only helps with isocratic methods

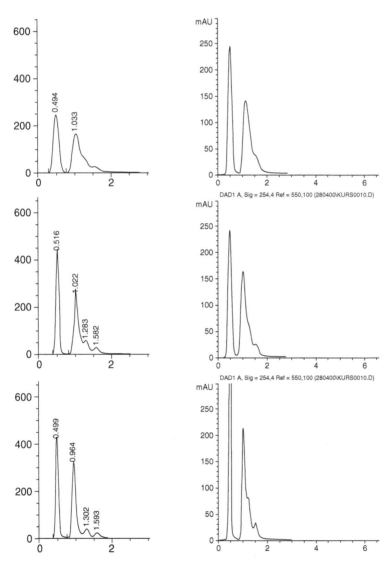

Figure 1.9.3 By reducing the time constant from 1 seconds to 50 ms, lower chromatograms, the peaks become narrower, the resolution is improved.

- Change of settings: A reduction of the rise time (time constant), e.g. to 50 ms (see Figure 1.9.3) and/or an increase of the sample rate to more than 20 Hz (for UHPLC separations best far above 50 Hz) improves the peak shape of early eluting peaks.
- Slow kinetics due to unwanted adsorption on various surfaces in the device can lead to peak broadening or tailing. Adhesion but also physisorption is observed on steel surfaces for certain analytes with partially negatively charged oxygen atoms on the molecule. If not elimination, at least an alleviation of the problem can be achieved by replacing steel capillaries with peek capillaries.
- Manipulation of the sample solution: Such measures certainly affect the method, but I would like to mention them briefly: Inject less, dilute with water and inject more, change the pH of the sample solution, add neutral salt, e.g. NaCl, to the sample solution, increase the density of the diluent by injecting guanidine, which elutes at dead time. Such manipulations prove to be particularly helpful for early eluting peaks.

1.9.4 Peak-to-Noise Ratio

All the options described above for improving the peak shape and thus the plate number logically lead to a better peak-to-noise ratio. This is also achieved by improved cell technology (e.g. "LightPipe") or by using a cell with a longer light path. In this case, it is worth consulting the manufacturer.

1.9.4.1 Noise Reduction

Here, too, suitable setting values are helpful: Large slit, e.g. 16 nm, and an equally large bandwidth, e.g. 12 nm for the diode-array detector. The following should also be considered when the focus is on noise reduction:

– In addition to the lamp, optical detectors should also be checked – at much longer intervals, of course – for circuit boards, mirrors, and lenses.
– Clean interfaces (LC–MS coupling and aerosol detectors), a clean UV cell, and finally a clean surface of the electrodes in the electrochemical detector also minimize noise.
– Keep an eye on the immediate surroundings: Possible heat generation from densely placed equipment? Frequency of the automatic dishwasher possibly noticeably different from that of the low-pressure pump and both devices are closed to each other? Rarely necessary, but it is nevertheless important to mention the use of an interference filter if required.

1.9.5 Coefficient of Variation, VC (Relative Standard Deviation, RSD)

One of the – certainly – many reasons for a large VC is a suboptimal peak shape or no baseline separation. It has been known since the 1960s that for not well-resolved

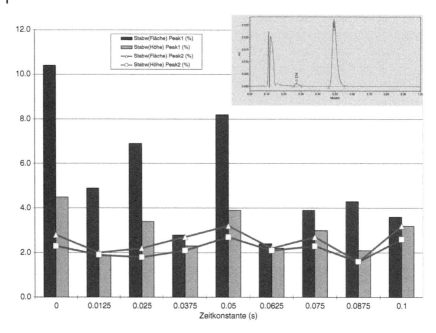

Figure 1.9.4 Coefficients of variation in the evaluation over the peak height and over the peak area in case of problematic peaks; for details, see 1.9.5. Source: Dr. Stavros Kromidas.

peaks, quantitative analysis by using the peak height is the lesser "evil." The clear positioning of the USP, a thoroughly cautious and conservative compendium, can be seen as an indication that this is indeed the case. It says: "Peak areas are generally used, but may be less accurate if peak interference occurs."Accurate "is the generic term for correct and precise. So if integration is wrong," the result is both incorrect (deviating peak areas) and less precise (large value for VC). In a series of experiments [Stavros Kromidas, unpublished results] we could confirm this, see Figure 1.9.4 A sample was injected ten times from a vial and the VC of the peak area and peak height was determined. For unproblematic peaks (in this example, the last peak), an evaluation via the peak height or via the peak area is practically equivalent, the VC value is also small. In the case of very small and/or tailing or nonbaseline separated peaks or a small peak-to-noise ratio, the VC of the peak height proves to be smaller than that of the peak area. And this is independent of whether the peak area evaluation is performed by drop, tangential, Gaussian, exponential skim, or valley to valley [4].

The decrease of VC by evaluation via the peak area is all the more significant the smaller the time constant is, the narrower the peaks are and the earlier they elute, see Figures 1.9.5–1.9.7. Ultimately, this means that what is discussed here is particularly important for fast separations: UHPLC separations or elution of peaks at a retention time smaller than about two to four times the dead time.

Let us return to the practice in the laboratories and note that, with a few exceptions, quantitative analysis is always carried out using the peak area, even in

1.9.5 Coefficient of Variation, VC (Relative Standard Deviation, RSD)

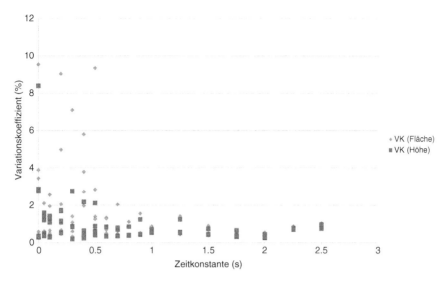

Figure 1.9.5 Coefficient of variation of peak height and peak area as a function of the time constant. Source: Dr. Stavros Kromidas.

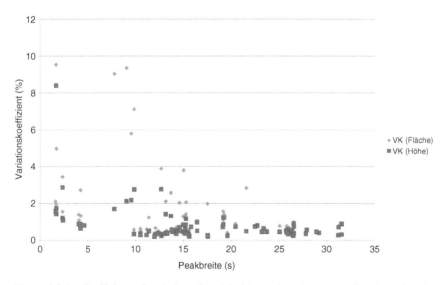

Figure 1.9.6 Coefficient of variation of peak height and peak area as a function of peak width. Source: Dr. Stavros Kromidas.

problematic cases. The above-mentioned reference in the USP and published results from numerous measurements thus remain without consequences in practice. Therefore, de facto only the following possibilities come into question: Sample rate recording between 5 and 10 Hz with time constants less than 0.05 seconds and 5 Hz with a time constant of 0.10 seconds: These seem to be reasonable values for a good reproducibility in area determination. At quite small (1–2 Hz) and large (50–70 Hz)

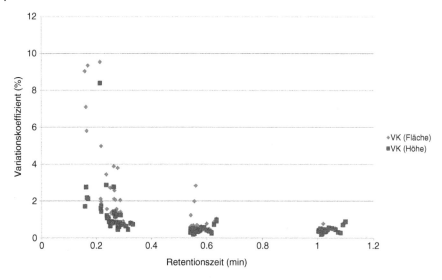

Figure 1.9.7 Coefficient of variation of peak height and peak area as a function of retention time. Source: Dr. Stavros Kromidas.

values, it is obviously difficult for a software to fix reproducibly peak start and end, the integration is not reproducible, see Figure 1.9.8

- Increase in number of measured values (...)
- Reduce aspiration speed ("Aspiration Time," "Aspiration Period"): The more viscous the sample solution (note: water is quite viscous), the lower the aspiration speed should be, the more precise repeat injections should be, the lower the VC.
- Avoid memory effects if possible: If a biocompatible system is out of the question for financial reasons, the following simple, yet effective measures should be considered: frequent and thorough rinsing of the system – remember to rinse periodically with EDTA or oxalic acid by presence of metal ions, use peek capillaries and vials with an inert glass surface, check the efficiency of the rinsing procedure and the rinsing liquid for the needle wash, change the steel injection needle for a ceramic one, for example, and finally consider changing the septa (material and manufacturer).
- If the sample chamber in the autosampler is not thermostatted, the use of preslit septa in the case of volatile solvents such as acetonitrile should be viewed critically when multiple injections are made from one or more vials: A (creeping) evaporation of solvent during the injection can lead to a concentration of the sample in the vial and thus to a drifting value for the VC.

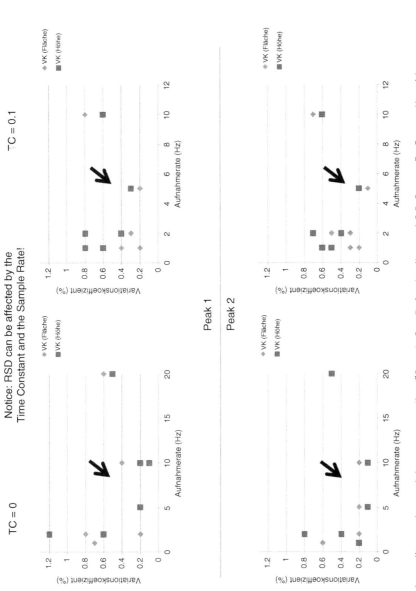

Figure 1.9.8 VC depending on the used data rate recording ("Sample Rate"); for details, see 1.9.5. Source: Dr. Stavros Kromidas.

References

1 U.S. Pharmacopeial Convention. http://www.usp.org/(accessed April 2018).
2 Park, H.H., De Pra, M., Haddad, P.R. et al. (2019). Localised quantitative structure-retention relationship modelling for rapid method development in reversed-phase high performance liquid chromatography (2020). *Journal of Chromatography A* 1609: 460508.
3 Stavros Kromidas. Colona – comparison and selection of HPLC-RP columns. www.colona.kromidas.de.
4 Kuss, H.J. and Kromidas, S. (eds.) (2009). *Quantification in LC and GC – A practical guide to Good Chromatographic Data*. Weinheim, Germany: Wiley-VCH.

Part II

Computer-aided Optimization

2.1

Strategy for Automated Development of Reversed-Phase HPLC Methods for Domain-Specific Characterization of Monoclonal Antibodies

Jennifer La[1], Mark Condina[1], Leexin Chong[1], Craig Kyngdon[1], Matthias Zimmermann[1], and Sergey Galushko[2]

[1]CSL Ltd., 45 Poplar Rd., 3059, Parkville, Australia
[2]ChromSword, Research and Development, Im Wiesengrund 49-b, 64367 Muehltal, Germany

2.1.1 Introduction

Recombinant monoclonal antibodies (mAbs) are a growing class of biopharmaceutical drugs targeting a number of diseases, including cardiovascular disease, autoimmune disorders, metabolic disorders, and cancer. The most common therapeutic class of mAb produced is immunoglobulin G (IgG). Recombinant mAbs are biologics and are therefore are subject to post-translational modifications (PTMs) that often occur during the manufacturing process. Oxidation is one such PTM, which can compromise product safety and efficacy. It requires little energy and can occur during cell culture, purification, or formulation in response to reactive oxygen species, light exposure or storage, and can be the leading cause of degradation during refrigeration [1]. Structurally, IgGs consists of four polypeptide chains – two identical heavy chains (HC) and two identical light chains (LC) connected by disulfide bonds. Each chain is composed of structural domains. Functionally, IgGs consist of one crystallizable fragment (Fc) and two antigen-binding fragments (Fabs) connected by a flexible hinge region. Each Fab consists of one LC and the N-terminal half of one heavy chain HC-Fab (Fd). Enzymatic cleavage of IgG molecules using IdeS generates Fc, Fab, and F(ab)' fragments. The reduction of disulfide bonds of these fragments leads to three antibody domains of ~25 KD, including LC, monomeric Fc (Fc/2), and Fd. Limited proteolysis of IgG molecules is a practically promising approach for the characterization of domain-specific modifications. The enzyme IdeS is used in this way to characterize domain-specific modifications [2]. Reversed-phase HPLC (RP–HPLC) is a technique to separate reduced domains and their oxidized products; however, only partial separation of the oxidized products has been achieved despite the use of long columns and substantial run times [2–4]. From a product development perspective, such challenges are not uncommon, and because different conditions of RP–HPLC are required for different mAbs, a straightforward strategy for method development of these samples is highly desired.

Optimization in HPLC: Concepts and Strategies, First Edition. Edited by Stavros Kromidas.
© 2021 WILEY-VCH GmbH. Published 2021 by WILEY-VCH GmbH.

Computer-assisted method development (CAMD) is an effective approach for optimization of separations; however, only a few applications of this technique have been published for mAbs [5, 6]. However, it should be noted that CAMD can be a time-consuming process when many columns and effects of different method variables are tested. A promising approach to increase productivity is automated method development. One of the main advantages of the automatic optimization is that a chromatographer can avoid complex tasks of the offline computer-assisted optimization. These include peak tracking, data input, method and sequence specifications, and other routine and nonroutine operations. In this approach, an analyst is responsible for defining a strategy and an intelligent chromatography method development data system plans and performs many routine and optimization experiments autonomously. A method development strategy can combine automated screening experiments with unattended optimization and followed by robustness studies using design of experiments (DOE). The results of some runs can also be used for offline simulation and optimization in CAMD. Automated HPLC method development and optimization were used successfully for mixtures of different small-molecule pharmaceutical samples [7–9]; however, no application has been described for protein samples. This chapter describes an effective strategy for automated HPLC method development, as applied to separate mixtures of mAb domains and related products. For the case study, the ChromSwordAuto method development chromatography data system was used.

Automated method development and software tools

Most automated HPLC method development approaches fall into one of three classes:

(1) Screening or running many column/solvent/method combinations to identify those with a reasonable separation
(2) Statistic or direct process optimization
(3) Mechanistic or model-based optimization

The simplest approach for automation of a method development strategy is to run high-throughput screening to test combinations of method variables and factors. This may include various columns, solvents, buffers, gradients, etc. Neither a mathematical process model nor a statistical DoE is required for the scouting approach. A chromatographer needs only to create a large sequence and then run it for new samples to rely on that some combinations of method variables and factors will provide practically reasonable separations. The scouting approach is used frequently for samples when specific optimization is not necessary (like chiral separations, for example). Specialized software applications for automated methods scouting are useful to rapidly create and edit long sequences and run them automatically. Agilent Scouting Wizard [10] can be used with Agilent LC instruments and ChemStation chromatography data system (CDS), Shimadzu Scouting Manager is used with LabSolution CDS and Shimadzu instruments [11]. ACD/Lab AutoChrom supports screening experiments with Agilent LC and Waters LC instruments via ChemStation and Empower CDS [12]. ChromSwordAuto Scout module [13] creates automatically and runs screening sequences with Waters, Agilent, Hitachi, and ThermoScientific instruments.

An alternative to the screening-based approach is to directly identify process optima based on the results of experiments that are planned by statistical software (such as repeated DoE). Unfortunately, both screening and statistical DoE are often not the most efficient strategies for mixtures when retention models have crossover in different regions of method variables. For such irregular samples, the direct and screening approaches can find the optimum only serendipitously. Nevertheless, for simple mixtures, experimental results from the direct approach can be useful. This approach can identify a local optimal separation region and estimate the sensitivity of method quality to specific parameter changes within the design space. Special software tools that include both features to create DoE and control of LC instruments to execute the DoE have substantial advantages against statistical software, which have only options to plan DoE. S-Matrix Fusion [14] and ChromSword AutoRobust [13] can create and run statistical DOE automatically with the difference that AutoRobust also contains functionality to build retention models and simulate chromatograms at different experimental conditions for 2D and 3D optimizations.

The most efficient, albeit complex approach for automated method development is the model-based optimization. In this case, mathematical models are utilized to reduce the number of experiments. In the recent final guidance for industry for the analytical method development, the U.S. Food and Drug Administration (FDA) recommends the submission of data to indicate a mechanistic understanding of the basic methodology [15]. However, the development of mechanistic models requires good chromatography understanding and reliable tests for parameter estimations. Limiting factors are computational time and reliability of the models that are applied for simulation and optimum search. The determination of mechanistic model parameters can be complicated for computer-assisted (offline) method development and requires time and operator qualification for optimization of multicomponent mixtures. Automatic optimization with mechanistic DoE incorporates engineering knowledge in the form of constrains, expert rules, and known fundamental relationships of liquid chromatography; therefore, this technology can find optimal conditions faster than the offline approach. Automatic optimization brings considerable advantages, as laid out above. ChromSwordAuto Developer application module [16] is a tool that unites the mechanistic or model-based optimization algorithms for unattended LC method development of small and large molecules.

2.1.2 Interaction with Instruments

Software for automated method development can operate in two modes – the primary/secondary and the primary mode. In the primary/secondary mode, the method development or scouting software operates as a primary process

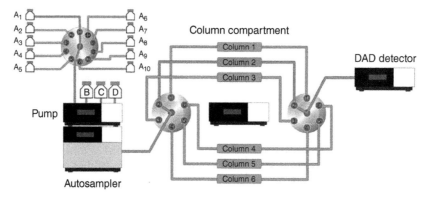

Figure 2.1.1 Schematic overview of the HPLC method development system.

planning next runs. A chromatography data system (CDS) operates as a secondary process to transfer commands from the method development software to modules, acquire and save data and methods. In the primary mode, a computer program operates as a CDS and acquires data, controls the HPLC instrument, and runs method development algorithms. A substantial advantage of the primary mode is that another CDS is not necessary and a user can operate with only one software. Scouting Wizard [10], AutoChrom [12], and Fusion [14] can operate only in the primary/secondary mode. ChromSwordAuto [13, 16] and AutoRobust [17] can operate both in primary/secondary and primary modes.

To apply different strategies for automated method development, it is preferable to use instruments equipped with columns and solvent-switching valves. Such valves enable the testing of different stationary- and mobile-phase combinations. To use buffers or other additives with absorbance in the UV range, two solvent switching valves are normally used – in channels A and B. In this case, both weak and strong solvents contain the same concentration of an additive to compensate for a baseline drift in gradient elution. Six- and eight-position column-switching valves are used to test different columns installed in one or several column compartments. An example of a method development system is shown in Figure 2.1.1. For higher productivity, two or more instruments can be used to run automated method development at the same time.

2.1.3 Columns

For separation of large molecules in RP–HPLC, wide-pore columns are used. It should be noted that column efficiency and selectivity can be different for different types of proteins and it is recommended to test different columns initially for better consistency for a particular project. In our case study, we tested columns listed in Table 2.1.1.

Table 2.1.1 Reversed-phase columns.

No.	Column	No.	Column
1	RHD Zorbax 300 C3[a]	12	Advance BioRP mAb SB C8[a]
2	RRHD Zorbax 300 C8[a]	13	Advance BioRP mAb Diphenyl[a]
3	RRHD Zorbax 300 C18[a]	14	Proswift RP-1S[b]
4	RRHD Zorbax 300 Diphenyl[a]	15	Proswift RP-4H[b]
5	Poroshell 300 SB-C3[a]	16	MAbPac RP[b]
6	Poroshell StableBond C8[a]	17	Aeris Widepore C4[c]
7	Poroshell StableBond C18[a]	18	Aeris Widepore XB-C8[c]
8	PLRP-S 5 µm 300 Å[a]	19	Aeris Widepore XB-C18[c]
9	PLRP-S 5 µm 1000 Å[a]	20	Acquity UPLC BEH300 C4[d]
10	PLRP-S 5 µm 300 Å[a]	21	Acquity UPLC BEH300 C18[d]
11	Advance BioRP mAb C4[a]	22	BIOshell A400 Protein[e]

a) Manufacturer: Agilent technologies.
b) Thermo Fisher Scientific.
c) Phenominex.
d) Waters.
e) Sigma Aldrich Fine Chemicals Biosciences.

2.1.4 Sample Preparation and HPLC Analysis

The native mAbs are oxidized with hydrogen peroxide or *t*-butyl hydroperoxide, then digested with IdeS, reduced by dithiothreitol (DTT) (Figure 2.1.2), and analyzed by RP–HPLC or RP LC–MS [2–4].

Chromatograms of samples normally contain three main peaks of LC, HC-Fab (Fd), and Fc/2 and oxidation products [2–4]. Protein oxidation can occur at cysteine, tryptophan, lysine, and other amino acids; however, methionine is often the most susceptible residue to oxidation of mAbs [18, 19]. Susceptible methionines are typically located on the surface of the protein and exposed to the solvent [20]. The most common product of methionine oxidation is methionine sulfoxide [21], which is more polar, less hydrophobic than methionine, and less retained in RP–LC than

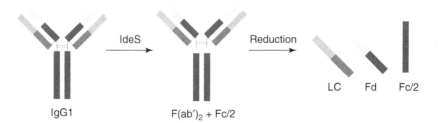

Figure 2.1.2 Limited proteolysis of IgG1 by IdeS.

their nonoxidized forms. The mass addition of 16 Da in the methionine-containing antibody domain is characteristic of methionine sulfoxide. Chemically stressed IgG molecules after digestion can be analyzed with online MS. In our experiments after optimization of separation, we observed four resolved peaks separated from the Fc/2 peak: one with Fc/2 + 16, two with Fc/2 + 32, and one with two Fc/2 + 48 masses. It indicates singly, doubly, and triply oxidized Fc/2. It was reported earlier that only two partially resolved peaks in RP–HPLC were observed for IgG1- and IgG2-oxidized fragments, which were identified as singly and doubly oxidized Fc/2 + 16 and Fc/2 + 32 correspondently [2–4]. Oxidation of more methionine residuals in our experiments can be explained by other oxidizing reagent and longer oxidation time. We used H_2O_2, which is substantially less in size than hydrophobic *tert*-butyl hydroperoxide and can access buried methionine residuals not accessible for *tert*-butyl hydroperoxide.

2.1.5 Automated Method Development

The automated method development strategies can involve the following steps:

(1) Automated screening of different RP protein columns with different temperature and additive concentrations.
   The goals of this step are a selection of the column method development set, defining additive concentrations and temperature range to provide peaks efficiency.
(2) Rapid gradient optimization for the column set with different combinations of solvent, temperature, flow rates.
   The goal of this step is to rapidly find practically reasonable methods and promising alternative combinations.
(3) Detailed sample profiling and gradient optimization with the best combinations.
   The goal of this step is to study a sample for possible coeluting peaks and to find the best and alternative gradient methods.
(4) Robustness tests of the final method and improving method performance.
   The goal of this step is to build the design space of the final method and determine method factor values, which provide the most robust method.

Special algorithms are built into ChromSwordAuto to support every step of automated method development. A combination of different steps can define different strategies of method development to meet requirements for a particular stage of a drug development project. For early drug development stages – when many drug candidates should be tested – the rapid gradient optimization mode with a standard column set can develop a practically reasonable method in a short time without the necessity to perform detailed optimization and robustness tests.

For early stages of drug development, the rapid optimization algorithm rapidly finds promising alternative combinations of stationary and mobile phases. The fine

2.1.5 Automated Method Development

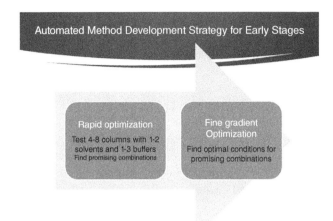

Figure 2.1.3 A method development workflow for early stages of drug development.

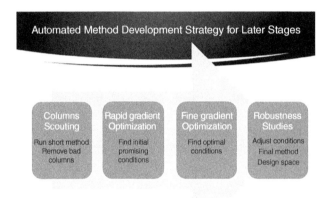

Figure 2.1.4 A method development workflow for late stages of drug development.

optimization algorithm can then be applied to find the best and alternative gradient methods (Figure 2.1.3).

For the development of the late-stage methods, all steps can be combined to provide robust and fast methods that can be transferred to other laboratories (Figure 2.1.4).

2.1.5.1 Columns Screening

Goals of columns screening are to select columns and critical method variables, which provide good efficiency and different selectivity to separate components of the mixture. The built-in algorithms of the CMDS include automated creation of different screening projects with a corresponding instrument and column conditioning. For this case study, 22 reversed-phase wide-pore columns (Table 2.1.1) were tested running a generic gradient 20–70% B/10 minutes with flow rate 0.15–0.5 ml/min, temperature 40–70 °C, different concentration of trifluoroacetic acid (TFA) (0.05–0.2%), and n-butanol (1–6%) in a mobile phase. For complex samples, rapid gradients cannot separate all target compounds; however, they identify column/mobile-phase combinations that generate broad and asymmetrical peaks, which can be excluded from the following optimization steps. An example of chromatograms for a promising and unpromising combination is shown in Figure 2.1.5. Six columns that exhibited the highest efficiency and difference in selectivity for separation of target peaks were selected as the columns method development set for the further rapid gradient optimization at temperature 70 °C and flow rate 0.3 and 0.15 ml/min. As was mentioned, the column set can be different for a different type of proteins and it is recommended to test more columns initially for a better column development set for a particular protein type.

It should be noted that method screening is the robotic process automation rather than intelligent automation. The screening approach cannot guarantee that the best column is found. Sometimes, an instrument method that delivered the best result after the screening is just more optimal for the worse column than for the potentially best one. Therefore, we consider that the main goal of the screening is to identify column/mobile-phase combination, which generates broad and asymmetrical peaks to exclude them from the following optimization steps when the intelligent automation mode is applied.

2.1.5.2 Rapid Optimization

The rapid optimization mode represents intelligent automation. This automation algorithm automates nonroutine optimization tasks and involves complex data processing and reasoning. During rapid optimization, the system performs an overview scouting and optimization gradient runs – typically 3–5 for every column/mobile phase/temperature/flow rate combination (Figure 2.1.6), and this enables it to obtain information on the peaks and the distribution of analyte retention times. Using these data, the software calculates optimal linear and multistep gradients. Results of a new run must be used for automated fine-tuning retention models and Monte Carlo optimization procedure. Using this function, a suitable method can be developed within a short period. Other important goals also can be achieved – to select the most promising column/mobile-phase combinations for further fine optimization procedure and find alternative column(s). In our experiments, the 50 × 2.1 mm AdvanceBio BioRP mAb C8 column with the solvent A: 0.1% TFA, 10% ACN, 87% H_2O, 3% butanol and the solvent B: 90% ACN, 0.1% TFA, 7% H_2O, 3% butanol with flow rate 0.3 and 0.15 ml/min was selected for the further fine optimization procedure.

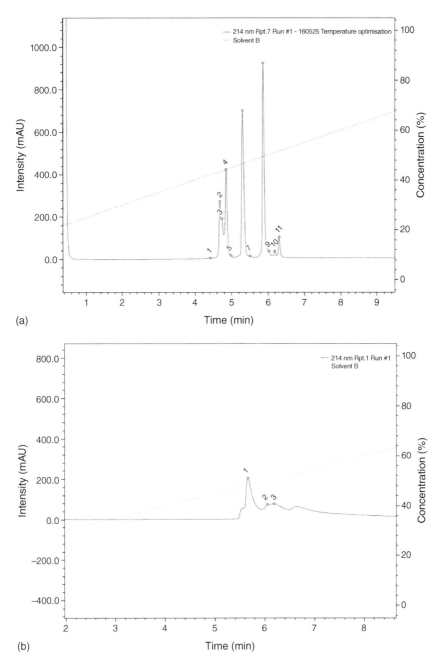

Figure 2.1.5 (a) Chromatogram after screening. Column Agilent Advanced Bio RP mAb SB C8. 100 mm × 2.1 mm. $T = 70\,°C$. 0.1% TFA, flow = 0.3 ml/min; (b) Chromatogram after the screening. Phenomenex Aeris Silica/C4. 50 mm × 2.1 mm 00B-4486-AN. $T = 70\,°C$. 0.1% TFA, flow = 0.3 ml/min.

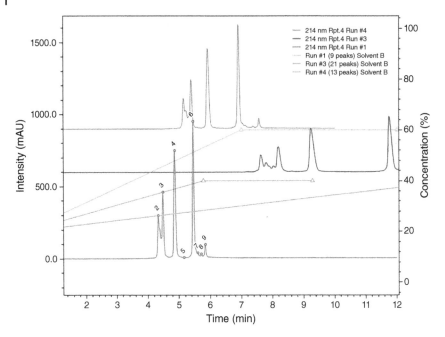

Figure 2.1.6 Results of the rapid optimization.

2.1.5.3 Fine Optimization and Sample Profiling

The goals of the fine optimization procedure are to find the most optimal conditions – analysis time, simplicity of the method, and separate maximal number of target components with practically acceptable resolution. For the fine optimization procedure, the most promising column/mobile phase/flow rate combinations selected from the previous stage are used. In this mode, the built-in algorithm of the chromatography method development data system performs more runs with different gradient profiles to study a sample and calculates, as closely as possible, appropriate retention models (dependence of analyte retention time on the optimization parameter). Using this information, the Monte Carlo and other optimization algorithms are used to find optimal conditions and several alternative linear and multistep gradient profiles are calculated and performed fully automatically (Figure 2.1.7).

2.1.6 Robustness Tests

The robustness of a method is extremely important for successful method transfer to other laboratories and instruments. Robustness can be described as the ability to reproduce the method in different laboratories or with different hardware without unexpected differences in the obtained results. Method values can be different at different laboratories and, in this case, preliminary robustness tests are necessary to study the effect of these variables on a method quality. Different critical quality

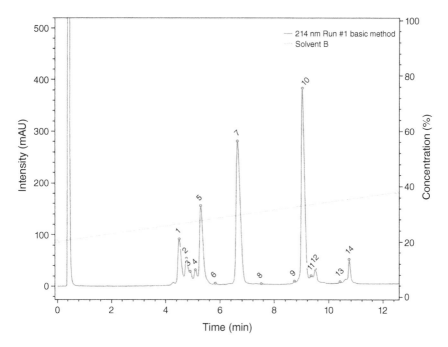

Figure 2.1.7 Chromatogram after the fine optimization. Column temperature = 70 °C; flow rate = 0.3 ml/min; gradient time = 12.6 minutes.

attributes (CQA) of a method can be tested – area, area%, retention time, resolution, and other CQAs. One of the most critical quality attributes for the HPLC method is the resolution between peaks of target compounds. The resolution values of a method should be within appropriate limits to ensure drug product quality.

Robustness tests projects can include the following steps:

(1) Selection of the method factors to be tested.
(2) Selection of the experimental design (DOE).
(3) Definition of the different levels for the variables.
(4) Creation of the experimental set-up.
(5) Performing the experiments.
(6) Calculation of effects.
(7) Statistical and graphical analysis of the effects.
(8) Doing conclusions from the analysis.
(9) Improving the performance of the method if necessary.

These different steps are considered in more detail below.

2.1.6.1 Selection of the Variables

For robustness tests, some operation factors should be considered. The selected variables can be quantitative like the gradient time or the flow rate and qualitative like the column batch. The selected factors should represent those that are most likely to be changed when a method is transferred between laboratories or instruments and that potentially could influence the method quality.

The following factors can affect the response of a method and should be included in the robustness tests:

- gradient time of linear gradients
- initial and final concentration of linear gradients
- time and concentration of each gradient node (step) for multistep gradients
- pH if a buffer is used
- flow rate
- column compartment temperature
- method equilibration time
- injection volume
- column batch

In reversed-phase HPLC of mAbs, acidic mobile-phase additives such as TFA and formic acid (FA) are typically used. A small amount of n-butanol in a mobile phase can increase efficiency for some columns [5]. The effect of concentration of TFA, n-butanol, and other additives in an MP can be easily tested in preliminary screening stages. If these parameters have no remarkable effect on the resolution in the range of 0.05–0.2% of TFA and 1–5% of n-butanol, then these factors can be excluded from robustness tests. The effect of wavelength for the UV detector is not usually tested because moderate differences in wavelengths (±3 nm) have practically no impact on chromatograms due to very similar spectra parameters of mAbs and related products.

2.1.6.2 Selection of the experimental design

The one-variable-at-a-time (OVAT), full factorial (FFD), and the Plackett–Burman partial factorial design (PBD) can be used for robustness tests. The OVAT is the fastest design; however, it cannot estimate interactions of different variables without preliminary studies. The FFD is the most comprehensive design to determine interactions of factors and describe the response surface for finding optimum factor-values; it however requires substantially more experiments. The PBD can be used as an alternative of the FFD; however, typically, a limited array of data points after the PBD cannot be used to determine chromatographic retention model parameters. In this case, a less reliable simplified model is usually used to calculate response; however, deviations between the predicted and experimental value of a CQA can be too high. Another problem is possible confounding of effects due to reducing the number of runs in PBD. In this case, the effects of different factors or interaction factors cannot be evaluated individually, and interpreting the results becomes difficult and even not correct. A practically reasonable and quite strict strategy for robustness projects can include two designs:

(1) The OVAT design, which can rapidly identify which of the tested variables have a significant effect on the response.

(2) The FFD of the critical variables, which were identified in step 1.

Both steps can be executed fully automatically with a reasonable number of experiments. When the number of runs is too high and not practically reasonable for running the FFD designs, then the PBD can be planned.

2.1.6.3 Definition of the Different Levels for the Factors

The factor levels of variables to be tested should be set around the nominal values specified in the operating (basic) method. The interval between the extreme values should represent the limits between which the factors are expected to vary when a method is transferred. The levels are defined by the analyst according to the results of a preliminary study of chromatographic retention behavior of compounds and instrument specifications taking into account the precision and the uncertainty with which a factor can be set and reset. To define the factor levels for the temperature, concentration, and time of gradient steps, it is recommended to study the effect of these variables in more detail.

The following values (+/−) for robustness tests of analytical HPLC methods of mAbs can be set:

Temperature	2–6 °C
Concentration of an organic solvent	1–2%
Gradient time	0.1–0.5 minutes
Flow rate	0.05–0.1 ml/min
Injection volume	0.5–3 µl

2.1.6.4 Creation of the Experimental Set-up

The variables are studied in DOE, which is selected as a function of the number of factors and levels to investigate. The simplest two-level screening designs allow testing a relatively large number of factors in a relatively small number of experiments. It is reasonable to use two-level design with center points for effects of concentration and gradient time and four-level design with center points for effects of flow rate and temperature. Such designs allow to establish the absence or presence of curvature and apply more complex models than the linear model. The creation of experimental designs manually takes substantial time even for OVAT. For planning FFD and PBD normally, special statistical computer programs are used and then the design plan should be transferred into a sequence of runs of a chromatography data system. This is also a time-consuming process and the best solution is that robustness tests software can create DOE and transfer it into a sequence of runs automatically. AutoRobust [17] software module provides a simple and rapid setup up to eight variables with 2–7 levels for OVAT, FFD, and PBD. The unlimited number of qualitative factors (column, solvent batches) can also be included in the DOE.

2.1.6.5 Execution of Experiments

The planned DOE is executed automatically. The method development system performs these runs interacting with a chromatography data system or directly with modules. For estimation of time effects and stability of the instrument and the column, several additional experiments at nominal levels should be added to the experimental design experiments. These replicate experiments are performed before, at regular time intervals between, and after the robustness test experiments. These experiments allow estimating for drift and column/instrument stability. Reproducible robustness experiments need to provide constant parameters both for injection and conditioning runs. Column and instrument wash-out, purging, and conditioning runs should be specified according to instrument and column specifications. Adequate time for column equilibration, not less than 10-column volume, has paramount importance for large proteins to obtain reproducible results. For more confidence, it is recommended to include the column equilibration time as a variable in the robustness tests DoE.

2.1.6.6 Calculation of Effects and Response and Numerical and Graphical Analysis of the Effects

Several responses can be determined from the performed experiments. The responses determined in robustness tests can be the resolution between each pair of neighboring peaks, the retention time, the area, the area% of compound peaks, peak width, and asymmetry coefficients. These parameters allow us to evaluate the quality of a method and the effects of variables and factors. One of the most critical quality attributes for HPLC methods is the resolution between peaks of target compounds. The resolution characteristic of a method should be within appropriate limits to ensure drug product quality. Two approaches can be used to evaluate the effect of method variables on resolution – descriptive and mechanistic. Traditionally, statistically based software use a descriptive approach and model response-surfaces with quadratic polynomials [14]. The main advantage of this approach is a simple and easy data-processing procedure. This approach does not use physical models of the separation process and peak tracking from run to run. From the theory and practice of computer-assisted HPLC method development, it is well known that quadratic dependence between resolution and method variables (concentration of organic modifier, gradient profile, temperature, pH) is often the exception rather than a rule for complex mixtures with irregular retention modes [22]. Retention models of compounds can overlap and dependencies $Rs = f(T, C, \text{gradient}, pH)$ can have one or several maxima and minima. Figures 2.1.8 and 2.1.9 show the resolution plots for a limited pair of oxidized fragments of mAb as a function of the change in temperature and flow rate. It is obvious that modeling of the resolution response with polynomial models, in this case, will lead to wrong conclusions regarding optimal conditions and robustness of the method. The mechanistic approach uses parameters of the chromatographic process responsible for the response; however, it requires studying the retention behavior of compounds to describe the effect of variables on the resolution. Those include peak

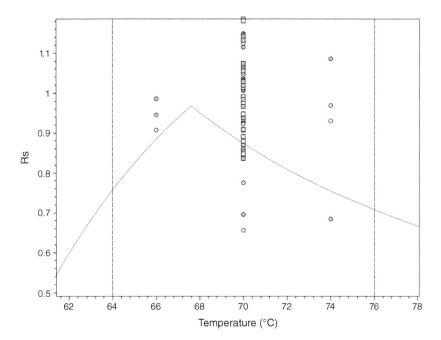

Figure 2.1.8 Effect of temperature on the resolution of the critical pair.

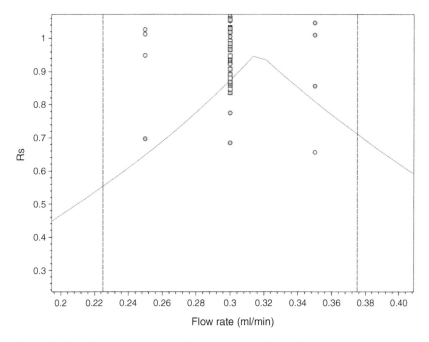

Figure 2.1.9 Effect of flow rate on the resolution of the critical pair.

tracking from run to run, evaluation of parameters of retention modes in gradient elution and under different temperatures, building system of equation, and solving them.

The mechanistic approach, which applies relations from the theory of liquid chromatography, is supported in the software for automated HPLC method development and robustness tests, which we used in this work [17]. After creation and performing the DoEs in an automated mode, data are processed for statistical and graphical analysis of responses. The effects of method variables on the resolution of critical pairs are shown in Figures 2.1.8 and 2.1.9. These variables have a substantial effect on resolution and the combination of these variables is necessary. The effect of two variables with a fixed nominal value for the other two variables is shown in Figures 2.1.10–2.1.12.

2.1.6.7 Improving the Performance of the Method

Analysis of the resolution maps for a combination of three different variables enables a chromatographer to determine areas where resolution can be increased or decreased. In our case study, small changes in temperature, flow rate, and gradient time enable us to improve the resolution and robustness of the method (Figure. 2.1.13) to provide a more robust method than that was used after optimization (Figure 2.1.7). Thus, the robustness studies can be considered also as an additional tool to improve the performance of the method.

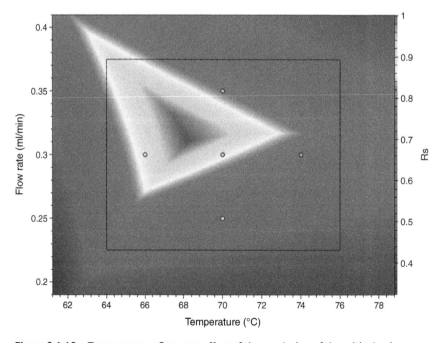

Figure 2.1.10 Temperature – flow rate effect of the resolution of the critical pair.

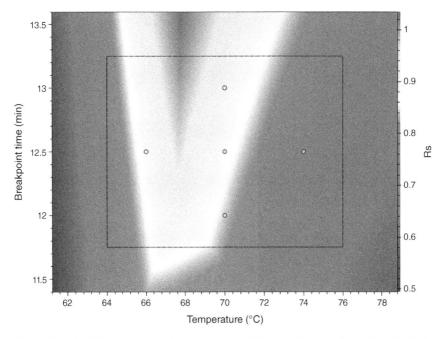

Figure 2.1.11 Effect of temperature and gradient time on the resolution of the limited pair.

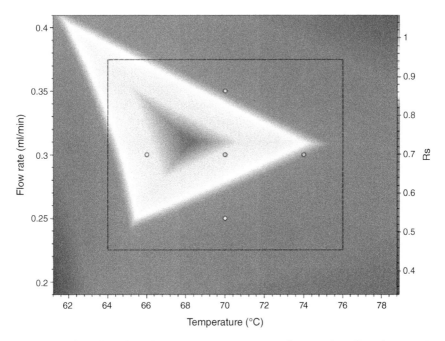

Figure 2.1.12 Effect of the flow rate and temperature at increased gradient time and the nominal concentration of ACN for gradient profile points.

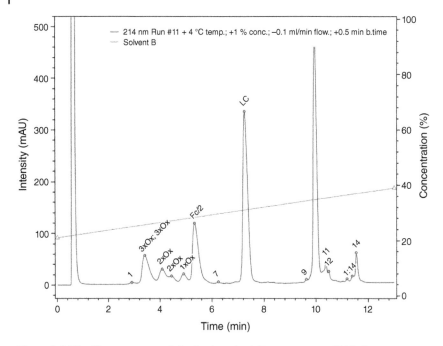

Figure 2.1.13 Chromatogram of the final method. Temperature = 68 °C, flow rate = 0.32 ml/min; gradient time = 13.5 minutes.

2.1.7 Conclusions

The use of an automated method development tool for stationary- and mobile-phase system screening, rapid and fine optimization, and robustness tests is a powerful and efficient approach for the development of a new HPLC method for separation of proteins. Appropriate strategy can significantly save the time of analytical scientists and would also increase the probability of developing an optimum method for the intended purpose.

References

1 Bertolotti-Ciarlet, A., Wang, W., Lownes, R. et al. (2009). Impact of methionine oxidation on the binding of human IgG1 to Fc Rn and Fc gamma receptors. *Molecular Immunology* 46: 1878–1882.
2 Yan, A., Ying, Z., Hans-Martin, M. et al. (2014). A new tool for monoclonal antibody analysis. Application of IdeS proteolysis in IgG domain-specific characterization. *mAbs* 6 (4): 1–15.
3 Pavon, J.A., Li, X., Chico, S. et al. (2016). Analysis of monoclonal antibody oxidation by simple mixed mode chromatography. *Journal of Chromatography A* 1431: 154–165.

4 Regl, C., Wohlschlager, T., Holzmann, J. et al. (2017). Generic HPLC method for absolute quantification of oxidation in monoclonal antibodies and Fc-fusion proteins using UV and MS detection. *Analytical Chemistry* 89 (16): 8391–8398.

5 Bobály, B., D'Atri, V., Beck, A. et al. (2012). Analysis of recombinant monoclonal antibodies by RPLC: toward a generic method development approach. *Journal of Pharmaceutical and Biomedical Analysis* 70: 158–168.

6 Fekete, S., Beck, A., Fekete, J., and Guillarme, D. (2015). Method development for the separation of monoclonal antibody charge variants in cation exchange chromatography, Part I: Salt gradient approach. *Journal of Pharmaceutical and Biomedical Analysis* 102: 33–44.

7 Xiao, K.P., Xiong, Y., and Rustum, A.M. (2008). Quantitation of trace betamethasone or dexamethasone in dexamethasone or betamethasone active pharmaceutical ingredients by reversed-phase high-performance liquid chromatography. *Journal of Chromatographic Science* 46: 15–22.

8 Xiao, K.P., Xiong, Y., Liu, F.Z., and Rustum, A.M. (2007). Efficient method development strategy for challenging separation of pharmaceutical molecules using advanced chromatographic technologies. *Journal of Chromatography A* 1163: 145–156.

9 Hewitt, E.F., Lukulay, P., and Galushko, S. (2006). Implementation of a rapid and automated high performance liquid chromatography method development strategy for pharmaceutical drug candidates. *Journal of Chromatography A* 1107: 79–87.

10 Edgar Naegele. New features of the agilent method scouting wizard for automated method development of complex samples. Agilent Technologies Technical Overviews. *Publication Part Number: 5991-6938EN*. http://www.agilent.com/cs/library/technicaloverviews/public/5991-6938EN.pdf.

11 Akihiro Kunisawa, Yusuke Osaka, Daiki Fujimura, Satoru Watanabe, Shinichi Kawano. Schimadzu C190-E211 Technical Report. https://www.shimadzu.com/an/sites/shimadzu.com.an/files/pim/pim_document_file/technical/technical_reports/10916/c190e211.pdf.

12 J. Hogbin, C. Lam. How a chromatographic method is more than a separation. International Labmate. (2020), 10–12. https://www.acdlabs.com/products/com_iden/meth_dev/autochrom/.

13 Galushko, S., Shishkina, I., Urtans, E., and Rotkaja, O. (2019). ChromSword software for method development in liquid chromatography. In: *Softwar-Assisted Method Development in High Performance Liquid Chromatography* (eds. S. Fekete and I. Molnar), 59. World Scientific http://www.chromsword.com/scout/.

14 Verseput, R. and Turpin, J. (2015). *Fusion QbD and the 'Perfect Storm' of Technologies Driving QbD-aligned LC Method Development*, 64–66. Chromatography Today http://www.smatrix.com/.

15 Publication of US Food and Drug Administration. Industry Analytical Procedures and Methods Validation for Drugs and Biologics. https://www.fda.gov/downloads/drugs/guidances/ucm386366.pdf

16 Galushko, S., Shishkina, I., Urtans, E., and Rotkaja, O. (2019). ChromSword software for method development in liquid chromatography. In: *Softwar-Assisted*

Method Development in High Performance Liquid Chromatography (eds. S. Fekete and I. Molnar), 59–60. World Scientific http://www.chromsword.com/developer/.

17 Galushko, S., Shishkina, I., Urtans, E., and Rotkaja, O. (2019). ChromSword Software for method development in liquid chromatography. In: *Softwar-Assisted Method Development in High Performance Liquid Chromatography* (eds. S. Fekete and I. Molnar), 64–71. World Scientific http://www.chromsword.com/autorobust/.

18 Davies, M.J. (2016). Protein oxidation and peroxidation. *Biochemical Journal* 473 (7): 805–825.

19 Sokolowska, I., Mo, J., Dong, J. et al. (2017). Subunit mass analysis for monitoring antibody oxidation. *mAbs* 9 (3): 498–505.

20 Pan, H., Chen, K., Chu, L. et al. (2009). Methionine oxidation in human IgG2 Fc decreases binding affinities to protein A and FcRn. *Protein Science* 18 (2): 424–433.

21 Xu, K., Uversky, V.N., and Xue, B. (2012). Local flexibility facilitates oxidization of buried methionine residues. *Protein & Peptide Letters* 19: 688–697.

22 Shyder, L.W. and Dolan, J.W. (2007). *High Performance Gradient Elution*, 228. Wiley.

2.2

Fusion QbD® Software Implementation of APLM Best Practices for Analytical Method Development, Validation, and Transfer

Richard Verseput[1,2]

[1] S-Matrix Corporation, Eureka, USA
[2] Ingo Green, Cromingo e.K., Klosterufer 2, Bordesholm 24582, Germany

2.2.1 Introduction

The S-Matrix Fusion QbD® Software Platform (Fusion QbD) is a comprehensive integration of chromatography-centric and advanced statistical tools in an automated experimentation platform. Fusion QbD provides all the tools and automation capabilities needed to successfully execute the Stage 1 – Analytical Procedure Design and Development activities associated with the modern analytical procedure lifecycle management (APLM) approach to analytical development [1]. This includes ability to fully characterize the robustness of an entire multiparameter experimental region to establish a true robust *Design Space* using industry- and regulatory-accepted robustness metrics, which can be applied to all critical method performance characteristics [2, 3]. Note that the design space is also referred to in analytical QbD publications and guidances as the method operable design region (MODR).

Fusion QbD also contains a full method validation experiment suite to support APLM Stage 2 – Procedure Performance Qualification. The validation experiment suite includes an Analytical Capability experiment protocol, which fully characterizes the sample preparation and sample injection components of overall method variation. This enables the user to select the most efficient *Replication Strategy* needed to meet final method performance specifications. A powerful new addition to the validation experiment suite is the integration of USP <1210> Tolerance Interval (TI) calculations and reporting integrated within Accuracy and Repeatability experiment analysis and reporting, which can be used to support *Method Transfer*.

Developed over a 20-year period in close cooperation with international pharmaceutical company customers, regulatory agencies, and international instrument manufacturer partners, Fusion QbD additionally includes full closed-loop method development and validation experiment automation with multiple Chromatography Data Software (CDS) programs for liquid chromatography (LC) methods and hyphenated LC methods utilizing mass spectrometry (LC–MS). Of critical importance to APLM best practices, Fusion QbD also offers full 21 CFR 11 compliance support, including bidirectional auditing of all CDS data exchanges, to support cross-platform data integrity (see also Chapter 4.5).

2.2.1.1 Application to Chromatographic Separation Modes

Fusion QbD can be used universally for any type of chromatography, with built-in templates tailored to the following chromatographic separation modes:

- Reversed phase
- Normal phase
- Chiral
- Ion exchange
- Hydrophilic and hydrophobic interaction
- Size exclusion
- Supercritical fluid

2.2.1.2 Small- and Large-Molecule Applications

Fusion QbD fully supports the development and validation of methods for small and large molecules, including applications such as peptide mapping and optimization of monoclonal antibody (mAb) methods. Fusion QbD was the method development software platform used in the work carried out in the development of the NIST book chapter on *State-of-the-Art and Emerging Technologies for Therapeutic Monoclonal Antibody Characterization*, Volume 2 [4]. And only Fusion QbD has a large-molecule-specific validation analysis and reporting suite developed under sponsorship of a major international pharmaceutical company, and used successfully in many regulatory filings. In fact, for large-molecule method development where no baseline chromatographic separation is expected, only Fusion QbD can model strategic large-molecule separation metrics such as peak height-to-valley ratios and retention time differences, such as those between the main peak and the acidic and basic regions.

2.2.1.3 Use for Non-LC Method Development Procedures

Fusion QbD also contains a generalized DOE module, which enables input of user-specified study variables. In addition to its use in formulation and process development, this module is used for the development of Non-LC methods such as gas chromatography (GC), capillary electrophoresis (CE), and dissolution. It is also used successfully in other analytical development application areas such as optimization of sample preparation procedures and optimization of mass spectrometer detection parameters.

Fusion QbD's generalized DOE module also provides CDS testing and data capture automation support for studies such as GC, CE, dissolution, and sample preparation, which involve analysis of chromatographic results data. Again, the integrated bidirectional data exchange with the CDS reduces most of the work, which must otherwise be carried out manually by the user, and eliminates the transcription errors associated with manual data transfers.

2.2.2 Overview – Experimental Design and Data Modeling in Fusion QbD

Fusion QbD has underlying math engines for statistical experimental design, often referred to as Design of Experiments (DoE or DOE). DOE is specifically mentioned in multiple guidances, including ICH Q8(R2) [5] and the recent European Compliance Agency guidance on Analytical Procedure Lifecycle Management [6]. The reason for this is that statistical experimental design is an extremely efficient approach for identifying critical method development parameters and gaining required knowledge of all their important effects on critical method performance characteristics. Because Fusion QbD was designed for the working scientist, the software has an automated mode for DOE, which automatically generates the most efficient and defensible design for the user's specified study parameters and stage of work (e.g. Screening or Optimization).

An important reason for the emphasis on DOE in the regulatory guidances is that the methodology is fundamentally a multifactor approach – i.e. combining two or more study factors in the same experiment, and DOE optimization experiments can generate data, which expresses all important study factor effects, including nonlinear and combined (interaction) effects. However, it should be noted that a companion flexible data modeling capability is also required to generate equations from the data which correctly characterize and quantify these effects. Many traditional linear chromatographic models are not capable of characterizing nonlinear effects or multifactor interaction effects. However, they are still used to predict the magnitudes of these effects in a limited approach to in silico robustness assessments of a candidate method. Since these effects are not part of the linear chromatographic models, predictions of these effects are not valid, and will result in false-negative assessments of their impacts on method robustness. Therefore, Fusion QbD also has an Automated Mode for one-click data modeling using the most advanced automated linear regression (ALR) modeling capability available in software today. This enables Fusion QbD to fully characterize all important study parameter effects on all critical performance characteristics.

2.2.3 Analytical Target Profile

The analytical target profile (ATP) defines the quality of the output product of the analytical procedure – i.e. the reportable result [7]. Defining the ATP in terms of the reportable result makes it agnostic to the analytical procedure. Therefore, the ATP can guide the selection of the analytical procedure, since the procedure must be able to meet the quality specification for the reportable result defined in the ATP.

The traditional approach to setting reportable result requirements, as described in ICH Q2(R1) [8], defines separate requirements for accuracy (bias) and repeatability (precision). For example, a tablet drug product with a target of 100% and specification range of 95–105% may have the commonly applied acceptance criteria of ≤3.0% for accuracy studies and ≤2.0% RSD for repeatability studies. However, an assay may meet both individual specifications and still produce out-of-specification (OOS) results depending on the magnitudes of the bias and the % RSD.

In the APLM approach, bias and precision limits are combined into a target measurement uncertainty (TMU), which can be represented by a single Tolerance Interval (TI) – a "statistical concept that describes the proportion or fraction of future results that will fall within a given range with defined level of confidence." The TI incorporates the user's acceptable level of risk, which can be defined as the number of failures per thousand reportable results, and can be used as a guide in the procedure validation and transfer work done in APLM Stage 2.

Fusion QbD's Analytical Method Validation module contains two strategic implementations of the TI to support critical elements of APLM Stage 2. The first is within its Analytical Capability experiment to support Replication Strategy characterization and optimization. The second implementation is within its Accuracy and Repeatability analytics to support Method Transfer. These features are described in detail within Sections 2.2.7 and 2.2.8 of this chapter.

2.2.4 APLM Stage 1 – Procedure Design and Development

2.2.4.1 Initial Sample Workup

It is never too early in method development to start using multifactor DOE techniques. The only requirement for an initial chemistry system screening study is a starting-point method which can generate an integrable chromatogram – this means that the peaks (i) are initially retained, (ii) elute before the end of the run, and (iii) are on scale using an appropriate 2D wavelength. For screening studies, Fusion QbD uses its patented Trend Responses™, which obtain analyzable data from only basic integration of the experiment chromatograms – no peak tracking is required for screening studies. Therefore, the peaks are not expected to be well separated and with good shape, only that the starting-point method will yield a chromatogram with peaks which can be integrated. This will be discussed in the next section.

Figure 2.2.1 presents a typical small molecule starting-point chromatogram for an initial chemistry system screening study. Note that most peaks are at least partly coeluted, and the peak shapes are suboptimal. However, the peaks are on scale using the selected 2D wavelength, they are all retained, and they elute before the end of the run.

2.2.4 APLM Stage 1 – Procedure Design and Development

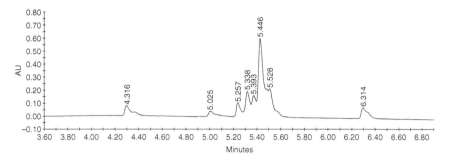

Figure 2.2.1 Example chromatogram at start of chemistry screening study.

When in early method development you do not have a starting-point method for your sample capable of generating a chromatogram with integrable peaks, then you can carry out a quick Initial Sample Workup study. Continuing with a reversed-phase method example, the Initial Sample Workup study can be quickly run using the following approach:

Experiment variables:

- *pH – 2.70 and 3.70*:
  For LC systems with quaternary pumps Fusion QbD's built-in automated Buffer Selector supports online preparation of target pH levels. For example, selecting the Formate Buffer System in the Buffer Selector would enable the software to automatically achieve the two target pH levels specified above. Note that the automated Buffer Selector also minimizes manual buffer preparations for broad pH screens.
- *Gradient time*: use the range in table below appropriate to your column and LC system (Table 2.2.1).

Experiment constants:

- *Column type*: alkyl-bonded phase: e.g. BEH C18
- *Sample concentrations (UV-absorbing compounds)*:
  o API(s): 0.10–0.15 mg/ml.
  o Related compounds and impurities: 5.0–10.0% of API concentrations.
- *Detection wavelength*:
  Use a PDA detector. Acquire a 3-D spectrum. You will use the spectrum data to determine a 2-D channel wavelength for your subsequent method development work, as described in the "Analysis of Experiment Results" section below.
- *Strong solvent*: acetonitrile
- *Initial% and Final% strong solvent (Constant – consider solubility)*:
  Single step: 5.0–95.0%
  [Initial% strong solvent should be as low as possible to support initial retention. Final% strong solvent should be as high as possible to support elution.]

Table 2.2.1 Gradient time ranges for sample workup and chemistry screening.

HPLC			5.0–95.0%	Slope equivalent gradient time range (minutes)
Column dimensions – W × L (mm)	Particle size (μm)	Pump flow rate (ml/min)	Gradient slope range (%/min)	
4.6 × 50	3.5	1.00	6.0–18.0	5.0–15.0
4.6 × 100	3.5	1.00	3.6–9.0	10.0–25.0
4.6 × 150	3.5	1.00	3.0–9.0	10.0–30.0
UHPLC			5.0–95.0%	Slope equivalent gradient time range (minutes)
Column dimensions –W × L (mm)	Particle size (μm)	Pump flow rate (ml/min)	Gradient slope range (%/min)	
2.1 × 50	1.7	0.50	6.0–18.0	5.0–15.0
2.1 × 100	1.7	0.40	3.6–9.0	10.0–25.0
2.1 × 150	1.7	0.30	3.0–9.0	10.0–30.0

2.2.5 Chemistry System Screening

A chemistry screening study is intended to investigate the chemistry of the method. Therefore, the understood or anticipated major effectors of selectivity should be included in the chemistry screening study. For example, for a reversed-phase method, the user can aggregate major shape and selectivity factors such as multiple stationary phases (columns), a broad range of pH, strong solvent types, and gradient slope into a comprehensive screening experiment. The combination of DOE, CDS automation, and advanced data modeling within Fusion QbD enables the user to efficiently and quickly screen multiple chemistry systems. Given the modern UHPLC instrumentation commonly used in early method development, which includes quaternary pump modules and temperature-controlled column oven compartments with multiposition switching valves, in most cases the software's full CDS automation support enables the entire experiment to be completed overnight in walk-away mode. It is noteworthy that some modern texts on liquid chromatography explicitly state the practical advantages of chemistry system screening, for example in the third edition of the text by Snyder, Kirkland, and Dolan the authors state:

> For methods involving a large number of samples, and where adequate resolution must be combined with run times that are as short as possible, it can be

profitable to spend more time initially on "scouting" experiments. The experimentation may be with different columns, different B-solvents, and variations in mobile-phase pH and temperature. Use of Gradient elution during the experiments can help avoid the need to separately optimize values of %B for each variable studied [9].

2.2.5.1 Starting Points Based on Molecular Structure and Chemistry Considerations

A starting-point pH and stationary phase (column) are sometimes obtained from structure- and chemistry-based considerations. For example, a starting-point pH and stationary phase may be selected based on the pKa and LogD values of one or more compounds, which were obtained either experimentally or from a literature search. In this case, these factors are set to a constant, and method development trials are initiated using other factors such as column oven temperature and gradient slope – usually by holding the initial and final percent strong solvent constant and varying gradient time (t_G). This approach has a high risk, since as also stated in the third edition of the text by Snyder, Kirkland, and Dolan:

> Still another approach is to search the literature for separation of the same or similar sample. Trial-and-error modifications of conditions are then followed until an acceptable separation is achieved. *We do not recommend this approach* [emphasis added by original authors] because possible deficiencies in literature methods can delay subsequent attempts at achieving a final, acceptable separation [9].

When an initial pH and column are selected based on structure- and/or chemistry-based considerations, it is critical to reduce the risk in this approach by carrying out a comprehensive chemistry system screen, which incorporates the initially chosen pH and column within broad combined ranges of pH and t_G, and a variety of stationary-phase chemistries. Such a screening approach would either confirm the correctness of these initial selections, in which case there would now be data-based evidence to support them, or it may identify a different combination of pH and column, which are demonstrated to be a much more advantageous chemistry system. This latter outcome is often the case, since these factors *interactively* affect selectivity and shape. A multifactor DOE study construct enables these effects to be expressed in the data and characterized in the analysis [10, 11].

2.2.5.2 Trend Responses and Data Modeling

Chemistry system screening experiments typically generate many experiment chromatograms in which both the peak shapes and the peak elution order can differ greatly from chromatogram to chromatogram. Coelution of at least some peaks is also expected, and complicated by the fact that different peak pairs can be separated or coeluted in the different chromatographic conditions. These circumstances make accurate peak tracking extremely difficult at best, and most often impossible.

S-Matrix therefore developed the Trend Response automation capability within Fusion QbD in response to these challenges. This capability enables the user to define any number of logical and analyzable metrics of shape and selectivity, which the software will then automatically extract from each experiment chromatogram. As mentioned previously, the only requirement is basic integration of the experiment chromatograms, meaning that (i) all peaks of interest in a given chromatogram are integrated, with the exception of course of coeluting peaks, and (ii) baseline noise "peaks" are not integrated.

The overarching goal of a chemistry system screening study is identification of the most advantageous chemistry system – the combination of Column Type, pH, Strong Solvent, and Slope (t_G) – to promote to an optimization experiment. This qualitative goal is normally represented by individual screening-level performance requirements such as:

- The total number of integrable peaks.
- The number of peaks, which are baseline resolved.
- The number of peaks with acceptable tailing.
- The number of peaks with required S/N ratio.
- The resolution of the API from its immediately earlier and later eluting peaks.
- API tailing.

Trend Responses are therefore configured by the user to automatically extract these performance metrics, and/or any other metrics important to the user, which Fusion QbD can then model to identify the chemistry system which simultaneously meets all specified screening goals. The best combination of the study factors can then be promoted to a multifactor optimization experiment.

Figure 2.2.2 shows a "trellis" series of three graphs generated from models of Trend Response data obtained from a typical chemistry system-screening study. The three graphs have the X-axis and Y-axis variables in common, and differ only in the stationary phase. In the graphs, each Trend Response is assigned a color by Fusion QbD (shown here as a grayscale – see Table 2.2.2), and the experimental region where the

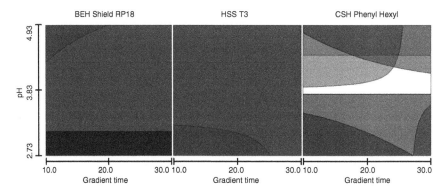

Figure 2.2.2 Screening Trellis graph series – three stationary phases.

Table 2.2.2 Trend response goals for the chemistry screening study.

Name	Units	Goal	Color	Lower bound	Upper bound
No. of peaks	*	Target	10% gray	7.0	9.0
No. of peaks >= 1.50 – USP resolution		Target	20% gray	6.0	8.0
No. of peaks >= 2.00 – USP resolution		Target	30% gray	6.0	8.0
No. of peaks <= 1.50 – USP tailing		Target	40% gray	7.0	9.0
First peak – Retention time	min	Maximize	50% gray	0.90	

methods ***do not meet*** the goal for that response is shaded with the color. Therefore, the remaining ***unshaded*** region in a given graph corresponds to method conditions which simultaneously meet all response goals.

A comparison of the graphs in Figure 2.2.2 shows that only the graph associated with the CSH Phenyl Hexyl column contains an unshaded region. This therefore defines that the CSH Phenyl-Hexyl column should be promoted to an optimization study. In addition, the narrower pH range of 3.60–4.20 should also be promoted to the optimization study along with the narrower t_G range of 20–30 minutes. These ranges of pH and t_G should be included in an optimization study for two reasons:

1) Given careful integration of the experiment chromatograms, the models obtained from screening studies have the predictive accuracy needed to identify the most advantageous chemistry system to promote to method optimization. However, these models do not have the predictive accuracy needed to identify the optimum performing and most robust methods. Therefore, the unshaded region in the graph associated with the CSH Phenyl-Hexyl column in Figure 2.2.2 should not be considered a design space, but rather a region in which the methods consistently meet the specified screening performance goals.
2) These factors must be included as study variables in the optimization experiment to obtain models with the precision required to accurately characterize the mean performance and robustness of all possible methods within the optimization experimental region with respect to the expected variation in these parameters on transfer and normal use over time.

2.2.6 Method Optimization

2.2.6.1 Optimizing Mean Performance

In the optimization phase, the best conditions of categorical study factors such as stationary phase and strong solvent type are set to a constant, and the narrower workable ranges of the numerical parameters such as pH and t_G are then examined more granularly. Factors which can affect mean performance and/or robustness, but for which the magnitudes of the effects are expected to be less than those of the major effectors included in the screening studies, can now also be included. For example, for a reversed-phase method, the optimization study may include parameters such as column oven temperature, buffer concentration, and pump flow rate, in addition to the narrower workable ranges of pH and t_G.

Holding categorical study factors such as stationary phase and strong solvent-type constant in optimization experiments reduces the frequency of changes in compound elution order across the experiment run chromatograms. This aids the peak tracking needed to derive equations (models) which represent all study parameter effects on all included critical method performance characteristics, and which have the predictive precision required to fully characterize the MODR.

Fusion QbD has an automated ***PeakTracker*™** feature, which utilizes traditional peak results data, along with UV and MS spectra data when available, to identify and track individual peaks in the experiment chromatograms. ***PeakTracker*** operation requires only basic integration of the experiment chromatograms within the CDS. Fusion QbD then automatically imports all chromatogram results needed for displaying the experiment data chromatograms and tracking the peaks. Once tracking is complete, ***PeakTracker*** automatically assigns the identified peak names to the results data imported for each peak in each chromatogram for instant data modeling. Noteworthy capabilities of ***PeakTracker*** include:

- Auto-deconvolution of coeluted peaks.
- Identification of two or more peaks with the same mass.
- Identification of nonabsorbing and nonionizing compounds.

PeakTracker's peak deconvolution capabilities, along with its ability to autogenerate retention time data for identified coeluted peaks, mean that Fusion QbD will have data, which would normally be missing for coeluted peaks in the CDS data. This means that runs, which would otherwise have missing data, now have correct representative data. This enables Fusion QbD to generate "hyper-accurate" models, since it has data that would otherwise be missing.

Fusion QbD also contains the additional, extremely powerful capability of full automation support for forced degradation studies. This capability enables users to set up an experiment execution protocol in which each experiment run will be repeated for a user-specified number of sample preparations, with each preparation corresponding to an included degradation path, and for which Fusion QbD assigns a different injection vial in the sequence it builds for auto-execution of the experiment in the CDS. Once the peaks are tracked in the experiment chromatograms, either by ***PeakTracker*** or manually within the CDS, Fusion QbD aggregates the peak results data from the Sample Preparation (degradation path) replicates for each run into a "composite chromatogram" data set for the run to be used in data analysis, robust method optimization, and prediction chromatogram visualization.

Figure 2.2.3 presents two "Resolution Map" graphs generated by Fusion QbD as part of modeled data visualization. The left graph is a traditional 2D graph, which visualizes the resolution of the least-resolved peak pair across the experiment conditions. The right graph is a 3D response surface visualization of the same 2D graph. Note that in many cases the different regions of this type of graph will represent

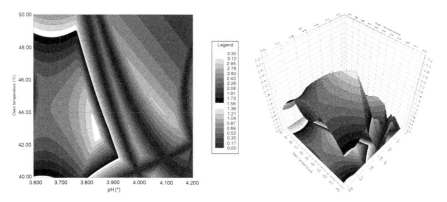

Figure 2.2.3 2D and 3D resolution map graphs.

Table 2.2.3 Mean performance goals for the optimization study.

Name	Units	Goal	Color	Lower bound	Upper bound	Crosshair prediction
G – USP resolution	*	Maximize	10% gray	2.000		30.793
F – USP resolution	*	Maximize	60% gray	2.000		6.102
A – USP resolution	*	Maximize	50% gray	2.000		10.212
API – USP resolution	*	Maximize	40% gray	2.000		5.546
DDeg – USP resolution	*	Maximize	30% gray	2.000		4.307
E – USP resolution	*	Maximize	20% gray	2.000		2.489
B – Retention time		Maximize	10% gray	1.00		1.19
API – USP tailing		Minimize	20% gray		1.40	1.35

different peak pairs for which the resolution decreases and increases as the peaks move closer or farther apart, and in some cases goes to zero as an individual peak pair changes elution order.

Figure 2.2.4 presents the corresponding 2D Overlay graph, which enables the user to focus on critical peak pair resolutions and reduce the ranking of, or eliminate from consideration, peaks that are less important to resolve. As Table 2.2.3 shows, goals for other responses can also be included in this graph. For example, the graph in Figure 2.2.4 also includes the retention time of the early eluting first peak and the USP Tailing of the API.

It is important to understand that all models either updated with or directly derived from experimental results are only able to predict the mean result for a given response associated with the input level settings of the study parameters. In other words, for each possible method in the multifactor experimental region, the model output is a prediction of how the method will perform *on average* for the response. For example, if 1000 sample injections are carried out using the method, the model prediction of the resolution of a critical peak pair is an estimate of the

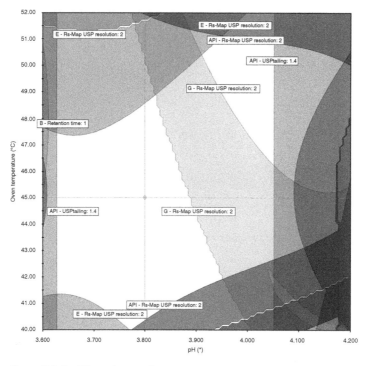

Figure 2.2.4 2D overlay graph.

mean (arithmetic average) of the 1000 individual resolution values that would be obtained. The model does not directly predict the magnitude of the variation in the 1000 injections, and so cannot directly provide any knowledge as to the relative robustness of the method. This is true no matter what software is used to obtain the model. This means that even a well-designed and executed DOE experiment can only *directly* provide knowledge of method mean performance. As a result, poor robustness performance in a selected method may not be identified until a robustness experiment is conducted as part of method validation prior to transfer. This is clearly too late in the process. What is therefore needed is the ability to also predict the robustness of any given method within the experimental region. The next section of this chapter will describe in detail how Fusion QbD can accomplish this from the optimization experiment data using *in silico* robustness simulation modeling without the need to conduct additional experiments.

2.2.6.2 Optimizing Robustness In Silico – Monte Carlo Simulation

Monte Carlo simulation is a computational technique by which a mean performance model for a given response is used to obtain predictions of performance variation for any given method within the experimental region [12]. This technique correctly represents the expected total setpoint variation in each study factor on transfer and normal use as a random, normally distributed setpoint error. It then generates an estimate of the given method's cumulative total variation for a given

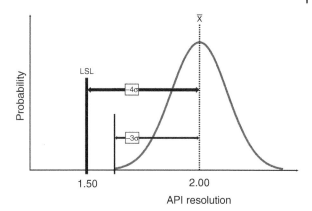

Figure 2.2.5 Example of method with mean response = 2.00 and $C_{pk} = 1.33$.

response resulting from the combined setpoint error distributions of all the included study factors.

For a given candidate method, the Monte Carlo simulation provides the quantitative estimate of the variation in a given modeled response in the form of an error distribution from which several metrics can be obtained. Familiar statistical metrics of such a distribution which Fusion QbD provides include the variance, the 1σ–6σ standard deviation values, the relative standard deviation (RSD), and the percent relative standard deviation (% RSD).

Fusion QbD reflects the estimated variation in a given response obtained from the Monte Carlo simulation against the user's specified tolerance limit or limits for the response to obtain a *Process Capability* (CP) index. CP indices are industry- and regulatory-accepted statistical metrics of robustness, which have been presented by the FDA for many years in public forums, and are also presented in several regulatory guidances [3, 6, 13, 14]. Several CP indices can be calculated from the estimated variation in a given response depending on the response's defined failure mode: 1-sided or 2-sided, with or without a target, and for 2-sided failure modes with a target, whether the specification limits are symmetrically located around the target.

Figure 2.2.5 illustrates the C_{pk} index used for the resolution response, which in this example has a 1-sided specification limit (lower bound only) of ≥1.50. C_{pk} is calculated as

$$\hat{C}_{pk} = \min(C_{pu}, C_{pl}), \text{ where } \hat{C}_{pu} = (USL-\bar{x})/3\sigma, \ \hat{C}_{pl} = (\bar{x} - LSL)/3\sigma$$

where USL and LSL are the upper and lower specification limits, respectively, and σ is the standard deviation of the response variation obtained from the from the Monte Carlo simulation.

C_{px} indices are scaled measures of variation relative to specification limit(s). In traditional statistical process control (SPC), a process is deemed capable when its measured C_{px} value is ≥1.33. The value of 1.33 means that the inherent variation, as defined by the ±3σ interval limits, is equal to 75% of the specification limits (4/3 = 1.33). Conversely, a process is deemed not capable when its measured Cp is ≤1.00, as the value of 1.00 means that the ±3σ interval limit(s) are coincident with

Table 2.2.4 Maximum expected variations in study parameters used in the simulation.

Variable name	Units	Maximum expected variation ($\pm 3\sigma$)
Pump flow rate	ml/min	0.050
Gradient time	min	2.0
Oven temperature	°C	3.0
pH	*	0.15

the specification limit(s). The appropriateness of applying C_{px} indices to analytical method performance requirements is evidenced by the statement:

> The lifecycle concept described in ICH Q8 is adaptable to analytical procedures if we consider an analytical procedure as a process and the output of this process as the reportable result, that is, the value that will be compared to the acceptance criterion [1].

Table 2.2.4 presents the maximum expected variations defined for the optimization experiment study variables in this example. Figure 2.2.6a represents a zoomed-in view of a subregion within the overall experimental region where the methods meet both the mean performance requirements previously presented in Table 2.2.3 and the coordinated method robustness metrics defined in Table 2.2.5. The responses in Figure 2.2.6a are those which have edges of failure for mean performance and/or robustness within the graphed region. Figure 2.2.6b presents the chromatogram associated with the final robust method conditions.

As Figure 2.2.6a shows, a much-reduced proportion of the overall experimental region contains methods which meet both mean performance and robustness requirements for all critical performance characteristics (remaining unshaded region). Comparing Figure 2.2.6a with Figure 2.2.4 demonstrates the power and clarity of Fusion QbD's integrated Monte Carlo simulation and Process Capability

Table 2.2.5 Coordinated robustness goals for the optimization study.

Name	Units	Goal	Color	Lower bound
Mean performance – all goals			10% gray	
B – retention time – Cpk		Maximize	10% gray	1.33
API – USP tailing – Cpk		Maximize	30% gray	1.33
G – USP resolution – Cpk		Maximize	50% gray	1.33
F – USP resolution – Cpk		Maximize	10% gray	1.33
A – USP resolution – Cpk		Maximize	30% gray	1.33
E – USP resolution – Cpk		Maximize	50% gray	1.33

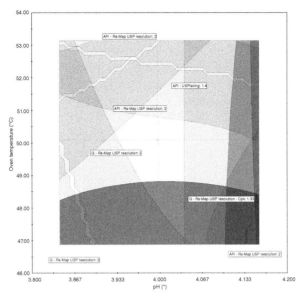

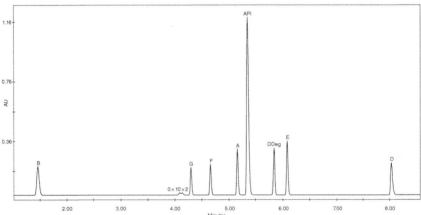

Figure 2.2.6 (a) Example of MODR established for mean performance and robustness, (b) prediction chromatogram of optimized method.

metrics capabilities in terms of quantitative identification and visualization of the most robust methods.

2.2.6.3 A Few Words About Segmented (Multistep) Gradients and Robustness

A common goal in chromatographic method development, which is well understood, but not normally quantitatively specified, is keeping the early eluting peaks away from the start of the gradient and the late eluting peaks away from the end of the gradient. The reason is that mobile-phase composition is perturbed at these inflection points, and so there is an associated reduction in robustness, which can affect reproducibility. Consider the following statement, again from 3rd edition of the text by Snyder, Kirkland, and Dolan:

> Increasing resolution by adjusting selectivity for different parts of the chromatogram can sometimes be achieved with a segmented gradient; ... Segmented gradients are not often used for improving resolution ... because their ability to enhance resolution without increasing run time is usually limited... However, there are other – generally more useful – means of optimizing resolution by changing selectivity and relative retention. Also, separations that use segmented gradients to improve resolution are likely to be less reproducible when transferred to another piece of equipment [9].

2.2.7 APLM Stage 2 – Procedure Performance Verification

2.2.7.1 Replication Strategy

Fusion QbD's method validation suite includes an Analytical Capability experiment. This is a components-of-error study construct specifically adapted to LC methods, which quantitatively characterizes the independent magnitudes of sample preparation error and sample injection error, and their relative contributions to the overall method repeatability. The analysis defines the overall error associated with the method given any combination of preparation and injection repeats, and incorporates the USP <1210> Tolerance Interval calculations. This enables identification of the most efficient replication strategy (optimum combination of preparation and injection repeats) to reduce the variability in the final method when the minimum strategy of one preparation and one injection does not meet the specified risk-based method performance requirements in terms of number of failures per thousand uses of the method (fpt). Example analysis results are shown graphically in Figure 2.2.7a,b. Given an upper-limit fpt specification of ≤50, Figure 2.2.7a demonstrates that there will be an unacceptable number of fpt when the final method specifies one preparation and one injection (fpt ≥ 86). However, Figure 2.2.7b demonstrates that changing the replication strategy to two preparations, two injections per preparation, results in a reportable value, which is the mean of four injection results. Figure 2.2.7b demonstrates that this replication strategy will reduce the overall error in the method to acceptable limits (fpt ≤ 22).

2.2.8 The USP <1210> Tolerance Interval in Support of Method Transfer

Fusion QbD's method validation suite includes the new USP <1210> Tolerance Interval (TI) and Prediction Interval (PI) calculations and reporting [15] integrated within its Accuracy and Repeatability experiments, and within its combined

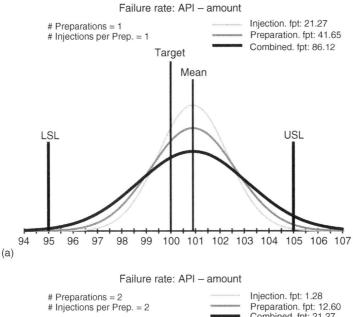

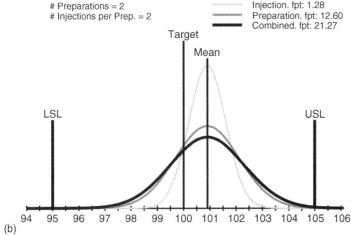

Figure 2.2.7 (a) and (b) Example replication strategies and associated T.I. results by component of variation (ftp = failures per thousand).

Accuracy_Linearity_Repeatability experiment. The TI calculation approach establishes a single criterion that can be used to simultaneously validate both accuracy and precision with more sensitivity to the likelihood of method failure. Given that the originating lab has used a risk-based approach to establish a TI which supports the performance requirements defined in the ATP, the TI would then be an appropriate performance metric to apply to any receiving lab as a gauge of the success or failure of the method transfer.

Fusion QbD's CDS automation capabilities enable a transferring lab to directly compare its accuracy, linearity, repeatability, and TI analysis results with those obtained from a receiving lab within the same validation study, all under full 21 CFR

Table 2.2.6 Tolerance interval analysis results – receiving lab.

Statistic name	Statistic value
Interval specification	95.00 < % Recovered < 105.00
Overall interval	97.61 < % Recovered <107.50
Result	Fail

11 compliance support and with full bidirectional audit trail. Table 2.2.6 presents the Tolerance Interval analysis results table which Fusion QbD automatically includes within its Accuracy and Repeatability Analysis reports when the user enters a required TI (or PI) specification and associated Confidence Probability in the simple analysis wizard. As the table shows, the reported results include a clear Pass/Fail assessment based on an automated comparison of the analysis results with the defined specification.

2.2.9 What is Coming – Expectations for 2021 and Beyond

Several important new guidance documents are either available now or will be available in the near future. These include:

- European Compliance Agency guidance on APLM
- Revision to ICH Q2(R1)
- ICH Q14: Analytical Procedure Development
- USP General Chapter <1220>

Below is a list of what the authors anticipate will be important consistent elements within these new guidance documents supporting analytical development and validation.

1) *Data integrity*
   Repositioning method robustness from a validation activity to an integrated element of analytical method development means regulatory expectations of data integrity will also move "upstream." This was highlighted in a statement made by an FDA representative at the 2018 USP Workshop on Enhanced Approaches for Lifecycle Management that *as long as the data integrity associated with the method development work matches what would be done in a formal Validation Robustness effort, then the results are acceptable.*
2) *Data and calculation transparency*
   The emphasis and expectation will be on data-based decisions. The relevant data should therefore be well documented, and based on unbiased, statistically defensible study constructs and data treatments. Therefore, there will be an expectation of a verifiable pedigree to the data, and also of the transparency of the computer

analytics and modeling which generated the results on which the decisions were based.

3) *Analytical target profile (ATP) and tolerance interval (TI)*

Selection of the analytical procedure will be expected to be guided by the ATP. And the ATP will, in turn, be expected to incorporate the concept of a TI, which combines bias and precision. There will also be an expectation that the TI is used to optimize and verify the Replication Strategy defined for the method by which the reportable results are generated, and that the TI will also be incorporated into Method Transfer assessments.

4) *Design of experiments (a.k.a.) statistical experimental design*

Design of Experiments (DoE or DOE) is a core element within many process and analytical development guidances, and its use by pharma worldwide in analytical development is growing. It is therefore anticipated that other study approaches used to generate decision-based data will need to be defended by demonstrating that the study protocol possesses the level of statistical quality required to support the decisions which were made based on the study results.

5) *Full characterization of a MODR*

It is well understood that there is significant risk associated with only characterizing the performance of a proposed final method. Such a limited characterization also does not characterize the operable ranges of the critical method parameters in terms of both mean performance and robustness, which is required for post approval operational flexibility. In fact, FDA has made many public presentations at conferences such as Pittcon and IFPAC regarding the value of a fully characterized design space, or MODR, in terms of both mean performance and robustness for all critical system suitability characteristics. The major benefit of a quantitatively characterized multi-dimensional MODR was initially represented as providing the opportunity for post approval regulatory flexibility. However, the most important benefits will likely be the translation of the MODR into (i) correctly specified precision limits of the critical method parameters required to support successful method transfer, and (ii) correct SQC and SPC control chart limits and associated corrective actions for use in APLM Phase 3. Therefore, there will be a growing expectation of a fully characterized robust design space as a representation of the method understanding underlying the mean performance and robustness claims in regulatory submittals, as well as the basis of any proposed requests for post approval operational flexibility.

References

1 Martin, G.P., Barnett, K.L., Burgess, C. et al. (2013). Stimuli to the revision process: lifecycle management of analytical procedures: method development, procedure performance qualification, and procedure performance verification. *Pharmacopeial Forum* 39 (5): 6–7.

2 Chatterjee, S., FDA. (2013). ONDQA/CDER/FDA, *QbD Considerations for Analytical Methods*, IFPAC Annual Meeting.

3 Kauffman, J.F., Mans, D.J., and FDA DPA (2015). *Experimental Design and Modeling to Improve HPLC Method Performance for Small Molecules*. Europe: CASSS CMC Strategy Forum.

4 Schiel, J., Davis, D., and Borisov, O. (eds.) (2015). *State-of-the-Art and Emerging Technologies for Therapeutic Monoclonal Antibody Characterization*, vol. 2. Biopharmaceutical Characterization: The NIST mAb Case Study, American Chemical Society.

5 ICH Q8(R2) (2009). *Guidance for Industry, Q8(R2) Pharmaceutical Development*. Geneva, Switzerland: International Council for Harmonisation of Technical Requirements for Pharmaceuticals for Human Use (ICH).

6 *Laboratory Data Management Guidance*. (2018). *Analytical Procedure Lifecycle Management (APLM)*, European Compliance Agency, Final r1.

7 Barnett, K.L., McGregor, P.L., Martin, G.P. et al. (2018). Analytical target profile: structure and application throughout the analytical lifecycle. *Pharmacopeial Forum* 42 (5): 3–7.

8 ICH Q2(R1). (2005). Validation of analytical procedures: text and methodology.

9 Snyder, L.R., Kirkland, J.J., and Dolan, J.W. (2010). *Introduction to Modern Liquid Chromatography*, 3e. Hoboken, New Jersey: John Wiley and Sons.

10 Montgomery, D.C. (2009). *Design and Analysis of Experiments*, 7e. New York: John Wiley and Sons.

11 Myers, R.H. and Montgomery, D.C. (1995). *Response Surface Methodology*. New York: John Wiley and Sons.

12 Christian, R.P. and Casella, G. (2004). *Monte Carlo Statistical Methods: Second Edition*. New York: Springer Science+Business Media Inc.

13 Chen, C., and Moore, C., FDA. (2006). *Role of Statistics in Pharmaceutical Development Using Quality-by-Design Approach – an FDA Perspective*, FDA/Industry Statistics Workshop, 2006.

14 Yu, L.X., FDA. (2012). *Quality by Design: Objectives, Benefits, and Challenges*, AAPS Annual Meeting.

15 USP. (2018). *<1210> Statistical Tools for Procedure Validation*, The United States Pharmacopeial Convention.

Part III

Current Challenges for HPLC Users in Industry

3.1

Modern HPLC Method Development

Stefan Lamotte

BASF SE, Competence Center Analytics, Carl-Bosch Str. 38, 67056 Ludwigshafen, Germany

If one looks at the resolution equation of chromatography, it becomes apparent that besides the number of theoretical plates and retention, selectivity plays a role. In capillary gas chromatography, the factor that is crucial to resolution is the number of theoretical plates. The high impact of the parameter results in the dominance of the theoretical plates for the separation in capillary gas chromatography. In liquid chromatography, the number of theoretical plates is usually a factor of 10 smaller than in gas chromatography. Therefore, the most important parameter of the resolution equation here is selectivity. This parameter is in the spotlight for every method development. In HPLC, selectivity can be controlled by selecting the separation column, the mobile phase, and the temperature.

In HPLC method development, the optimization of the mobile phase is often given a much greater importance than that of the stationary phase. Most computer-aided optimization programs are also aimed for optimizing the parameters of the mobile phase. Certainly, the type of solvent used (proton acceptor or donor, or protic or aprotic solvent, its solvent strength, and also its dipole moment), the pH value of the mobile phase, and the pH of buffers or only salts have a decisive influence on the selectivity. Nevertheless, the proper choice of the separation column must not be lost sight of. Above all, it is important to optimize the stationary phase before optimizing the mobile phase. If the optimal stationary phase is available right at the beginning of the optimization of the mobile phase, this usually results in simpler, more robust, and faster separation systems and methods. For example, you can run simple, binary, linear gradients, or it is even possible to work isocratic if the right stationary phase is used right in the beginning. In particular, an isocratic mode is good for robustness, and massively simplifies the transferability and validation of liquid chromatographic methods. It is therefore very worthwhile to invest time in the selection of the optimal stationary phases.

However, the ideal approach to optimizing the stationary phase depends crucially on the number of analytes to be separated and the complexity of the sample matrix.

In the rough, the following rule of thumb can be formulated depending on the analytes to be separated:

1-6 Analytes: Screen different separation columns and select the separation column that best separates your most critical peak pair (the least resolution peak pair).

7-20 Analytes: In most cases, we do not manage to achieve a sufficiently robust separation simply with only one standard separation column and simple mobile phases. However, a combination of different stationary phases helps here. The so-called POPLC (stationary-phase optimized liquid chromatography) is suitable. This method makes use of the additivity of the retention times. With the help of software, the ideal separation column composition can be determined from basis experiments, and these can subsequently be assembled and applied [1]. The following diagram explains how to do this (Figure 3.1.1).

More than 20 analytes: In this case, you need appropriate separation efficiency (peak capacity) and also mandatory gradient elution. This means that correspondingly long separation columns must be used, and these must be applied in combination with flat gradients (gradients between 10 and 20 column volumes). Chromatography times of hours are achieved quite quickly, which slows down the throughput very much.

Already with the separation of more than 20 components, one should consider whether it might not be more clever to use multidimensional separation methods. If you operate the HPLC in a multidimensional mode, the required peak capacity can be achieved much faster (see Chapters 1.1, 4.1, and 4.4). However, the method development and validation of such a method becomes more demanding. In view of this, it is necessary to consider whether the method development effort is worthwhile. The answer to this is certainly very individual and depends on the amount of samples in each case. With a few and isolated samples, it may well make sense to accept one-dimensional separations with long analysis times.

3.1.1 Robust Approaches to Practice

3.1.1.1 Generic Systems for all Tasks

When analytical tasks and separation systems change frequently in HPLC labs, high sample throughput is severely limited because the speed-determining step is the setting up and equilibrating of HPLC methods. In the search for a solution to this problem, other boundary requirements must also be met. The HPLC methods in an industrial laboratory must be simple, robust, and automatable.

As shown in previous work [2], very high-resolution methods with high peak capacity can be used. However, these have the limitation of long analysis times, and are therefore not useful for all questions and also not in the permanent application. But more on that later.

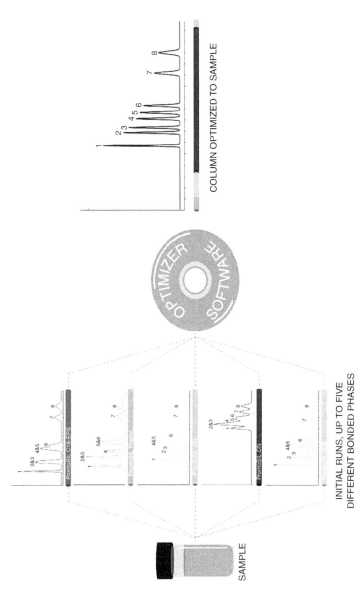

Figure 3.1.1 POPLC Scheme

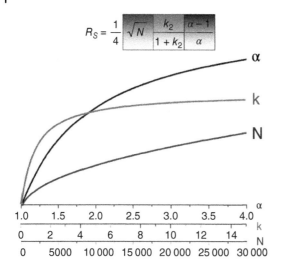

Figure 3.1.2 Resolution equation: RS, resolution, N, theoretical plate count, k, retention factor, α, selectivity.

Therefore, we investigated orthogonal selective separation systems in HPLC. With a handful of separation systems, it has been possible to solve the majority of all separation problems. In contrast to the classic method of working, where the HPLC method of the sample is adapted, the sample migrates here and is adapted to the HPLC method.

But from the beginning: How do we get to our orthogonal separation systems for the HPLC in the first place, and are they actually universally applicable for all questions?

In the beginning is the targeted experiment.

Before we could even start with our measurement series, it was necessary to consider the basic variables in the HPLC determining the resolution between two component-determining variables (see Figure 3.1.2).

The greatest influence on the resolution between two analytes, and thus their separation, has been the selectivity. In addition to the stationary phase, this is influenced by the mobile phase and temperature. In the HPLC, the mobile phase is classically optimized, with the disadvantages mentioned above: long equilibration and set-up times. But what if you keep the mobile phase constant and choose the separation columns as orthogonally as possible? To investigate this, we have selected a set of separation columns with the most orthogonal properties possible. For this purpose, we select from 3 groups each 6 or 8 separation columns, which primarily separate according to hydrophobic interactions, π–π interactions, cation-exchange and hydrogen bond interactions. These then feed with a set of the 30 most common analytes in our HPLC laboratory and record the retention data.

The evaluation criteria according to which we approached the results:

- Which separation column shows the highest retention for very polar analytes (marker: formic acid)?
- Which separation column shows the most symmetrical elution bands for the analytes?

And then for the groups among themselves:

- Which separation system separates the analytes, which cannot be separated on the separation system of the other groups?

The overview shows graphically how we get to the selection (Figure 3.1.3).

The green spots mean that the component is resolved on this separation system. The red spots mean there are coelutions with the analytes mentioned in the spot. Yellow means this combination was not measured, because the corresponding separation column was already disqualified beforehand (bad peak shape due to wide or asymmetric elution bands or almost no retention), or the separation task had already been solved on another separation system. The third separation system was measured with both methanol as a solvent and with acetonitrile (field on the far left). It is clearly visible that the number of green fields when using the protic solvent methanol is greater. This finding supports the previously found results that polar stationary phases have higher selectivity with protic solvents in the mobile phase, and thus is preferable to use methanol over acetonitrile as a mobile phase for polar stationary phases.

From the experimental data, the following 3 separation systems emerged:

3.1.2 The Classic Reverse-phase System

Column:	Triart C18 ExRS, 3 µm, 150 x 3.0 mm, YMC			
Temperature (°C):	25			
Injection (µL):	10			
Flow rate (mL/min):	0.7			
Cycle time (min):	20			
Typical back pressure (MPa):	40			
Gradient				
Mobile Phase A:	1000 ml H_2O + 1 ml H_3PO_4			
Mobile Phase B:	20 ml H_2O + 980 ml Acetonitrile + 1 ml H_3PO_4			
time (min):	0	12	15	16.5
A (%):	90	0	0	90
B (%):	10	100	100	10
UV-Detector				
Wavelength (nm):	205			

This separation system is characterized by a particularly strong hydrophobic retention and methylene group selectivity, and covers a wide polarity range. The marker for hydrophobic retention (formic acid) shows a stronger retention on this separation system despite 10% acetonitrile in the mobile phase than on reverse phases (AQ phases) with acidified water as a mobile phase, which is acceptable for purely aqueous media. Furthermore, all analytes with very symmetrical elution bands are eluted from the HPLC column.

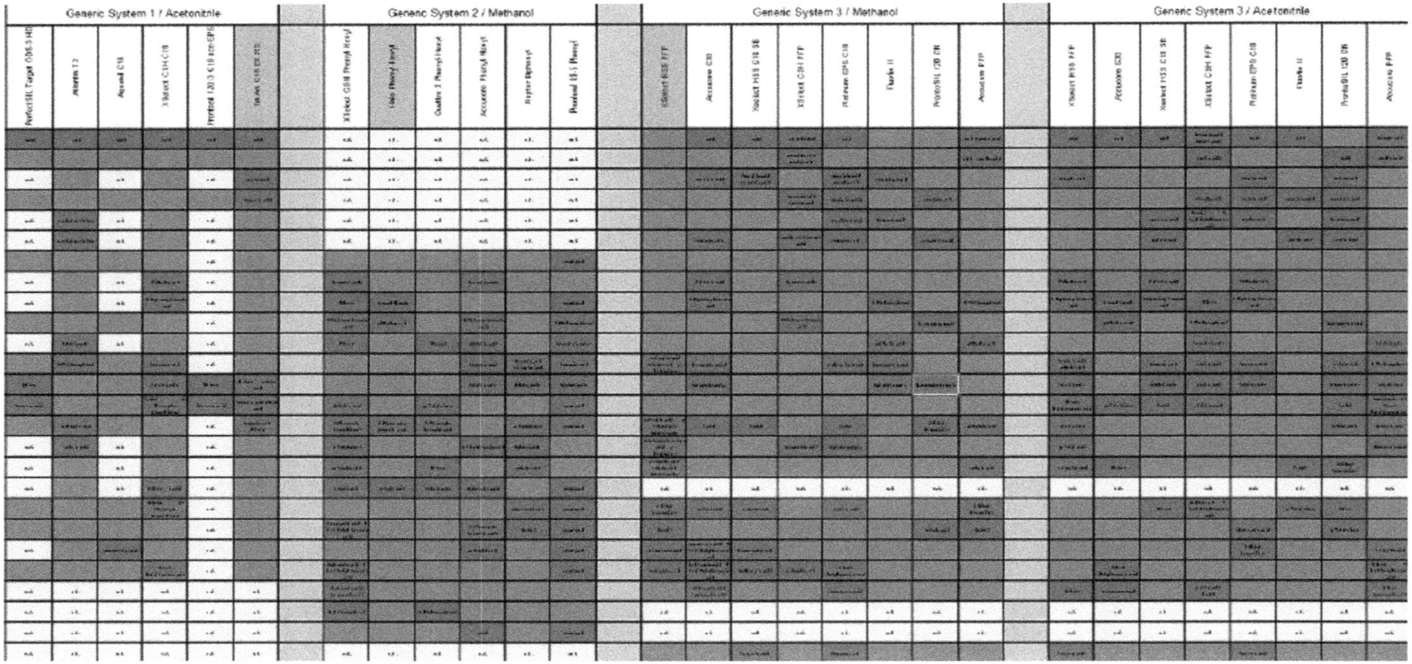

Figure 3.1.3 Overview of the selectivity of the individual separation systems.

3.1.3 A System that Primarily Separates According to π-π Interactions

Column:	Halo Phenyl-Hexyl, 2.7 µm, 150 x 3.0 mm, AMT Inc.
Temperature (°C):	25
Injection (µL):	10
Flow rate (mL/min):	0.7
Cycle time (min):	20
Typical back pressure (MPa):	31
Gradient	
Mobile Phase A:	1000 ml H_2O + 1 ml H_3PO_4
Mobile Phase B:	20 ml H_2O + 980 ml Methanol + 1 ml H_3PO_4
time (min):	0 12 15 16.5
A (%):	100 0 0 100
B (%):	0 100 100 0
UV-Detector	
Wavelength (nm):	205

The complementary selectivity to the first separation system is important in the selection of this separation system. In particular, the late eluting components, and those showing a coelution on the first separation system, can be separated here. The separation system, like separation system 1, is characterized by symmetrical elution bands.

3.1.4 A system that Primarily Separates According to Cation Exchange and Hydrogen Bridge Bonding Selectivity

Column:	XSelect HSS PFP, 3.5 µm, 150 x 3.0 mm, Waters Inc.
Temperature (°C):	25
Injection (µL):	10
Flow rate (mL/min)	0.7
Cycle time (min):	20
Back pressure (MPa):	23
Gradient	
Mobile Phase A:	1000 ml H_2O + 1 ml H_3PO_4
Mobile Phase B:	20 ml H_2O + 980 ml Methanol + 1 ml H_3PO_4
time (min):	0 12 15 16.5
A (%):	100 0 0 100
B (%):	0 100 100 0
UV-Detector	
Wavelength (nm):	205

The HPLC column of this separation system has no endcapping. Compared to the other two separation systems, the stationary phase has a high silanophilic activity (cation exchange capacity). Therefore, in particular, polar interactions between stationary phase and analytes dominate this separation system. We observe quite high retention for early eluting analytes in separation system 1. In particular, the coelutions in separation system 1 can be resolved. This separation system is complementary to the other two.

With these 3 complementary separation systems, approximately 80% of all HPLC separation tasks for small molecules (<1000 g/mol) can be solved.

The separation power is also impressive in these separation systems. The systems have a peak capacity of approximately 150.

Nevertheless, there are cases where these separation systems fail. If analytes are either so polar that they are not retained on the above systems, or are so nonpolar that they cannot be eluted from any of these 3 separation systems. For these two cases, we have developed the following alternatives:

3.1.5 System for Nonpolar Analytes

Column:	HALO C8, 2.7 µm, 150 x 3.0 mm, AMT Inc.			
Temperature (°C):	30			
Injection (µL):	20			
Flow rate (mL/min):	1.0			
Cycle time (min):	20			
Typical back pressure (MPa):	54			
Gradient				
Mobile Phase A:	H_2O / Acetonitrile (70/30) (v/v)			
Mobile Phase B:	Acetonitrile / MTBE (70/30) (v/v)			
time (min):	0	12	15	15.1
A (%):	100	0	0	100
B (%):	0	100	100	0
UV-Detector				
Wavelength (nm):	205			

Here, a separation column with a low carbon content (approximately 5%) in combination with mobile phases of high solvent strength is used. Thus, very hydrophobic analytes can also be eluted. This separation system delivers sharp and symmetrical peak shapes.

3.1.6 System for Polar Analytes

If the analytes are so polar that they show little or no retention on the other separation systems, a reverse-phase HPLC orthogonal separation system must be used. This is aqueous normal-phase chromatography, also known as HILIC (hydrophilic interaction liquid chromatography, see Chapter 1.2). This is where the separation occurs as partition between a more polar and nonpolar mobile phase. The separation process can be compared with a liquid–liquid extraction in a separating funnel. The figure below shows it schematically.

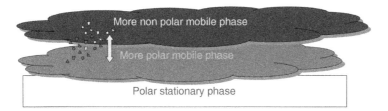

Often in these separation systems the strong polarity of the analytes is associated with a missing chromophore group, so that even rarely the UV detector can be applied as a detector and, instead, an aerosol detector (mostly ELSD) must be used.

Column:	Nucleodur HILIC, 1.8 µm, 100 x 3.0 mm, MN			
Temperature (°C):	25			
Injection (µL):	10			
Flow rate (mL/min):	1.0			
Cycle time (min):	20			
Typical back pressure (MPa):	35			
Gradient				
Mobile Phase A:	50 ml H_2O + 950 ml Acetonitrile			
Mobile Phase B:	300 ml H_2O + 700 ml Acetonitrile			
time (min):	0	12	15	15.1
A (%):	100	0	0	100
B (%):	0	100	100	0
UV-Detector/ELSD				
Wavelength (nm):	205 nm/T: 40 °C, N_2-Flow rate: 1.2 ml/min			

One disadvantage of HILIC is the slow equilibrium adjustment. Therefore, longer equilibration times (or volumes) and flatter gradient volumes compared to RPHPLC are usually required here. This disadvantage can be compensated by the use of low-viscosity mobile phases. Therefore, the use of supercritical chromatography, SFC, is a good option (see Chapter 1.6). Strictly speaking, the mobile phase is not in the supercritical stage, but the diffusion processes are significantly accelerated when acetonitrile is replaced by CO_2 as an organic solvent in the mobile phase.

Often, the practical application of HILIC chromatography is difficult and offers some pitfalls. In practice, it is mainly important to solve the sample in a solvent mixture, which comes as close as possible to the mobile phase A (start eluent of the gradient). If this is not done, the elution bands distort, resulting in peak fronting up to double peaks or complete elution of the analytes at void time. Remedy may in some cases reduce the volume of injection. Injecting less than 1% of the column volume, solvent strength effects are largely pushed back. If this is not possible, however, it becomes difficult, since the analytes are often not or very difficult to be soluble in the rather nonpolar solvent mixtures of a HILIC system. Because of this, only lower amounts of the analytes can be injected and consequently only higher detection limits compared to reverse-phase chromatography can be achieved.

Column:	Nucleodur HILIC, 1.8 µm, 100 x 3.0 mm, MN			
Temperature (°C):	50			
Injection (µL):	10			
Flow rate (mL/min):	0.5			
Cycle time (min):	11			
Pressure ABPR (psi):	1800			
Gradient				
Mobile Phase A:	CO_2			
Mobile Phase B:	H_2O/MeOH (90/10) (v/v)			
time (min):	0	4.8	7.8	7.9
A (%):	50	2	2	50
B (%):	50	98	98	50
UV-Detector/ELSD				
Wavelength (nm):	205 nm/T: 40 °C, N_2-Flow rate: 1.2 ml/min			

A very interesting aspect is the complete orthogonality of the separation systems 1 and 5. Analytes, which are very well retained on system 1, elute on system 5 at flow time and vice versa.

3.1.7 Conclusion

With these 5 separation systems, approximately 95% of all HPLC questions for small molecules (<1000 g/mol). The combination of the developed separation systems in multidimensional solutions also helps to solve separation problems and can be regarded as the ultimate approach.

3.1.8 The Maximum Peak Capacity

Multidimensional separation systems are first choice, and also show ways to maximize peak capacity [3]. By adding more and more dimensions, the peak capacity is in principle not limited; however, the increase in complexity leads to a considerable reduction in robustness and also to an increase in the measurement uncertainty of such systems. It is therefore questionable whether this can still cover the required accuracy and precision in a regulated environment.

Therefore, the question arises in how far one can build a simple, one-dimensional separation system, which can cover the needs of complex samples and at least show at which points of the chromatogram optimization is needed. To achieve the highest possible peak capacity, you need long separation columns filled with particles of larger diameter and operated with flat gradients [2]. The following separation system turns out to be ideal for practical application:

Column:	Halo C18, 5 μm, 7 x 250 x 3.0 mm (1750 x 3.0 mm), AMT Inc.			
Temperature (°C):	50			
Injection (μL):	50			
Flow rate (mL/min):	0.46 (≈ 2.5 mm/s)			
Cycling time (min):	420			
Typical back pressure (MPa):	55			
Gradient				
Mobile Phase A:	50 ml acetonitrile + 950 ml H_2O + 1 ml acidameisen (formic acid oder H_3PO_4)			
Mobile Phase B:	20 ml H_2O + 980 ml acetonitrile + 1 ml acidameisen (formic acid oder H_3PO_4)			
Time (min):	0	270	360	361
A (%):	100	0	0	100
B (%):	0	100	100	0
UV-detector				
Wavelength (nm):	205			

The gradient steepness is 15 column volumes. In this separation system, a peak capacity between 1000 and 1500 can be achieved. This is usually sufficient to detect coelutions and critical separation areas in the chromatogram. This separation system has proven to be very helpful in practice to show the heterogeneity of a sample before actually developing the method. This ensures that no unexpected component suddenly appears toward the end of a method development process that was not previously considered.

3.1.9 Outlook

Sometimes, however, it is not so much the separation, but the preseparation, i.e. the preparation of samples, that is decisive. The samples must first be separated from the matrix and brought into solution. The compatibility of these steps with the separation itself still requires a lot of research. Much remains to be done.

References

1 Nyiredy, S., Szűcs, Z., and Szepesy, L. (2006). Stationary-phase optimized selectivity LC (SOS-LC): separation examples and practical aspects. *Chmatographia* 63: 3–9. https://doi.org/10.1365/s10337-0006-0833-7.

2 Lamotte, S., Gruendling, T., Loeb, U. et al. (2017). Generic ultrahigh resolution HPLC methods: an efficient way to tackle singular analytical problems in industrial analytics. *Chromatographia* 80: 763–769. https://doi.org/10.1007/s10337-017-3300-8.

3 Stephan, S., Jakob, C., Hippler, J., and Schmitz, O.J. (2016). A novel four-dimensional analytical approach for analysis of complex samples. *Analytical and Bioanalytical Chemistry* 408: 3751–3759. https://doi.org/10.1007/s00216-016-9460-9.

3.2

Optimization Strategies in HPLC from the Perspective of an Industrial Service Provider

Juri Leonhardt and Michael Haustein

Currenta GmbH & Co. OHG, Production Analytics, CHEMPARK, Alte Heerstraße, 41538 Dormagen, Germany

3.2.1 Introduction

An analytical method developed for eternity – vision or reality? The answer is: vision, as the chemical industry changes its demands every day, due to different effects within the production chain (e.g. availability, costs, political situation, changes in the toxicological evaluation of raw materials) or due to changed requirements and regulations by law (e.g. necessity for lower detection limits). Therefore, the introduction of new starting or raw materials, the demand for lower or higher tons of production amounts, or change of manufacturing pathways may result. Consequently, these aspects often influence the analytical service provider by change of analytical demand (number and scope of analyses), the requirement to detect and determine new analytes, the need of new and/or additional analytical techniques, or the lack of robustness of the established analytical method. All these possible effects require an optimization of the established method or may even cause the development of a new one.

The analytical spectrum of laboratories working for chemical industry can be classified into three main fields. These are research and development, quality control, and process control analytics. All three fields cause different requirements for an analytical method and thus differ significantly with regard to method optimization. Therefore, the field of application of the new method must be defined in advance. However, it is not always easy to determine the final field of application, as this can change after the development process. For example, a method developed for research and development could be used for quality control at a later time. In this case, other requirements, such as robustness in continuous operation, may be more important. As a consequence, it is mandatory to know the field of application of the method prior to method optimization steps.

3.2.2 Research and Development

The main focus of a method developed for research is often not the robustness, but selectivity and, in some cases, sensitivity (limit of detection). In most cases, the aim

Optimization in HPLC: Concepts and Strategies, First Edition. Edited by Stavros Kromidas.
© 2021 WILEY-VCH GmbH. Published 2021 by WILEY-VCH GmbH.

is to obtain as much information as possible from just a few individual samples. Therefore, parameters such as selectivity and limit of detection are important for the development or optimization of an analytical method. As in most cases, there is no technical or chromatographic information on how the final method should look like or which kind of instrument should be used, the degree of freedom in the optimization is very high. The typical time frame provided for optimization can range from a few days to several months.

First of all, it is helpful to look at the physicochemical properties (e.g. solubility, polarity, etc.) and the structure of the target analyte. These could provide initial hints for the development or optimization of the method. Polarity, solubility, and molecular structure provide information about the selection of the stationary phase. Additionally, the structure (e.g. chromophores) and molecular mass can influence the type of detection, e.g. UV/Vis spectroscopy or mass spectrometry.

Based on this information, a generic method (e.g. RP-C18; H_2O/acetonitrile linear gradient 5–95% in 30 minutes) may be used as a starting point to obtain first results. In further steps, other parameters such as column temperature, gradient profile and time of elution, solvent composition, etc. can be varied until the required separation of the target component(s) is achieved. To reduce the number of individual measurements, the described process may be structured by commercially available optimization tools like DryLab™.

Using predefined basic measurements (e.g. separation at two different column temperatures), it is possible to create a two-dimensional resolution model. Based on this model, it is possible to simulate or calculate, for example, the retention time of targeted analytes within the model limits. However, the simple model does not contain all parameters. Consequently, basic measurements for additional parameters must be performed. However, in the field of research and development, this additional demand of work may be acceptable.

Methods from research laboratories are often used for the registration process of new products. Their subsequent use in quality control laboratories sometimes causes insufficient robustness as these methods were originally developed for a different purpose.

3.2.3 Quality Control

In the field of quality control, analytical methods are typically provided by the sponsor together with product specifications, as these data were already developed during the registration process. In most cases, quality control has to be performed according to these methods.

Therefore, the main characteristics of such methods are robustness and easy handling, as they may be used in different laboratories and in addition may be within continuous use. The main target is to use the given methods without modifications since any change requires sponsor approval and, in addition, revalidation activities.

To simplify the implementation of the registered method, it is helpful to use identical analytical equipment and separation column that was used to develop or optimize this method. This approach eliminates, e.g. retention time shifts related to a different gradient dwell volume.

It is recommended to perform a so-called method transfer between the developing and the applying laboratory, to verify the implementation of the method. It can be achieved by analyzing a set of identical samples using the method to be implemented in both laboratories followed by statistical evaluation of the resulting data pairs.

In some cases, however, it is necessary to change the given method. The reasons for this are different. Lack of robustness is the main reason, which results in unstable retention times, changing peak areas and/or shapes and rapid pressure increase. However, it is not recommended to use new mobile or stationary phases for the optimization. The degree of freedom in method optimization is, in general, minimal as time and costs have to be taken into account.

In principle, an in-house method development is possible, but usually not preferred by the sponsor. Moreover, due to the lack of capacities in the quality control laboratory analyzing hundreds of samples a day, it is often not applicable.

Nevertheless, this does not mean that registered methods per se cannot be optimized.

Within the life cycle of a product, there are regularly planned reviews of the specification and consequently of the registered methods. Thus, via direct and early communication between user and developer, it is possible to suggest improvements and optimizations or to develop them together for implementation in a new or optimized method.

3.2.4 Process Control Analytics

This kind of analytics monitors the individual reaction steps of the production process and therefore supports managing the plant. Over the entire production process, samples are taken and analyzed at essential synthesis steps to evaluate the composition of intermediates and the progress of the reaction. Critical reaction steps usually require a very short processing time of a few hours or less. This means that the analytical response time of such critical samples (including system calibration and evaluation) must be adjusted to the reaction time of the synthesis step. The main focus of method optimization is therefore on speed and robustness. Analytical systems in the field of production control have to be available 24/7 and should be set up at least partially in redundancy. In this respect, a standardized series of equipment is of significant importance, as it minimizes the number of training activities for laboratory employees and simplifies the storage and purchasing of replacement parts as well. Any smallest delay in sample preparation, system calibration, or even maintenance and service activities should be avoided, as this has a direct impact on the grade of utilization. A positive aspect in this field is the high degree of freedom, which allows a wide range of optimizations with regard to method design and equipment configuration, respectively.

The decisive factor in process-related analytics is time and thus the robustness of the analytical method. The more fragile the combination of technique and method is, the higher the maintenance effort (time and costs) might be. Therefore, it is recommended to avoid brand-new and fragile techniques and to focus on established and robust systems. For example, in the field of liquid chromatography, it is recommended to use an HPLC system (≤600 bar) with a so-called narrow (small) bore column (internal diameter: 2.1–4 mm) instead of a UHPLC system (600–1400 bar) and a microbore column (internal diameter: 1–2.1 mm).

Useful specifications for an HPLC system setup:

- *System*:
  - Pressure: <600 bar
  - Flow rate: 0.3–3 ml/min
  - Homogeneous system design; for example, unified instrument supplier
- *Mobile phase*:
  - Avoid highly concentrated and/or complex buffer systems
  - Prefer pure solvents
  - Use as few additives as possible
  - Use acidified solvents to avoid biological contamination
  - Example: H_2O/acetonitrile + 0.1 vol% formic or acetic acid
- *Stationary phase*:
  - Inner diameter: 2.1–4 mm
  - Length: 5–15 cm
  - Modification: variable, but stable material that is commercially available for at least a decade
  - Column temperature: approximately 30–40 °C
- *Detection*:
  - Stable and easy in use, avoid maintenance activities
  - Simplicity in data evaluation is the main focus
  - Where possible use, e.g. UV/VIS, fluorescence, refraction index, conductivity detection
  - Avoid mass spectrometers as far as possible, as they require a lot of maintenance work

For method development or optimization, there are several approaches. For example, registered methods from the field of quality control may be used as a basis and modified if required. The benefit of this approach is the certainty that the stationary and the mobile phase as well as general chromatographic conditions are suitable. This procedure is often preferred because it simply saves time. If this approach is not working, the previously described R&D approach (see Section 3.2.2) may be used. Software-based optimization is principal applicable, but is rarely used in practice due to limited time resources combined with insufficient knowledge of the tool.

A significant reduction of analytical run time may be achieved by using an alternative method of system calibration and evaluation. To keep the calibration effort as low as necessary and therefore the system availability as high as possible, the use of quality control samples is recommended. These should be analyzed periodically, e.g. several times a day, which provides valid information on the current calibration status of the system and reducing the calibration effort. Ideally, control samples should be as similar as possible to the analyzed samples to include the matrix effect. Therefore, it is obvious to use an authentic sample from the plant as a control sample. Such samples usually have a nearly identical matrix profile (impurities) and are in the correct calibration range of the method. However, it is mandatory to have the control sample available in an adequate quantity and to be stable over a sufficient period of time.

3.2.5 Decision Tree for the Optimization Strategy Depending on the Final Application Field

See Figure 3.2.1.

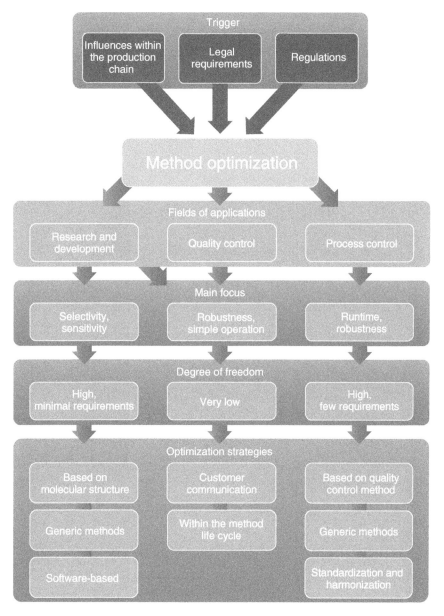

Figure 3.2.1 Pathway for the development of an optimized method.

3.3

Optimization Strategies in HPLC from the Perspective of a Service Provider – The UNTIE® Process of the CUP Laboratories

Dirk Freitag-Stechl and Melanie Janich

CUP Laboratorien Dr. Freitag GmbH, Carl-Eschebach-Strasse 7, D-01454, Radeberg, Germany

3.3.1 Common Challenges for a Service Provider

As a contract laboratory, CUP laboratories typically receive requests for method development and validation when the project is planned. This means that relatively little is known about the method at this stage and the customer's own experience with the method is little or nonexistent. Often, the method developed and validated by the API manufacturer is "simply" to be transferred to a finished product. In many cases, there are also no really "hard" specifications. However, according to the clients, the method urgently needs to be validated within three months at low cost and "GMP-compliant" because, for example, a clinical trial is already planned and patients are urgently waiting for the new drug.

Another, no less challenging, variant of the request is that the customer has already put a lot of work into method development and is now pressing for time, the sales department is pressing for delivery, and the stability program is due to start soon.

In both and many similar cases, the contract laboratory is expected to make promises and commitments before the project begins, regarding cost and time frames that are very difficult to meet without knowledge of the method. However, these are expected to be binding and often have a decisive influence on the award of the contract and later the overall success of a project and thus on large assets.

Thus, projects for method development and validation often give the impression of dealing with an unravelable tangle of questions. In the following, the UNTIE® process is used to present the approach of the CUP Laboratories to solve such knots (Figure 3.3.1).

3.3.2 A Typical, Lengthy Project – How it Usually Goes and How it Should not be Done!

To begin with, we will briefly describe a typical example of a lengthy project in which the CUP laboratories, despite a great deal of internal effort, did not live up to the claim "We benefit the customer" with the classic "straightforward" approach.

Optimization in HPLC: Concepts and Strategies, First Edition. Edited by Stavros Kromidas.
© 2021 WILEY-VCH GmbH. Published 2021 by WILEY-VCH GmbH.

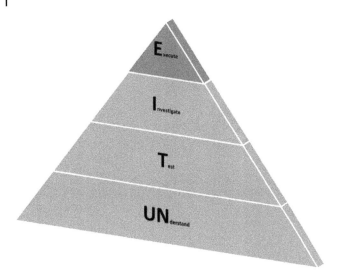

Figure 3.3.1 The UNTIE pyramid. Source: CUP Laboratorien.

The starting point was the "simple" adoption of a customer method from the Asian region for the determination of organic acids in a product with a high matrix load. This consisted of two IC methods that were not transferable 1 : 1 due to the equipment used. The customer's goal was to develop a method for determining the content and purity of all nine organic acids in the product, including suitable sample preparation.

Based on "Application Notes" of an instrument manufacturer for ion chromatography, an attempt was made to find a suitable separation method for the desired analytes. This was not successful for all analytes, even after lengthy optimization of the IC method. As a result, the system was changed and the same optimization process was started for HPLC. Again, after many runs and optimization steps with the typical HPLC settings (separation column, composition of the eluent, flow, temperature, etc.), the result was not satisfactory. On the contrary, the HPLC method proved to be much less sensitive than the IC method and thus unsuitable for determining purity. The next step was therefore to work on the IC issue again. The determination was now divided into two methods and thus, after another lengthy optimization, the appropriate separation for the desired analytes was achieved.

At the end of a very long, frustrating, and time-consuming and cost-intensive process, all analytes could then be sufficiently separated by means of suitable sample preparation and two different analytical methods, and the methods were finally validated. However, the goal originally formulated by the client, the development of only one method, could not be achieved and the overall project was considerably delayed.

Frequently blamed for such unsatisfactory project progress are the lack of diligence or scientific competence of the staff involved and/or inadequate equipment with analytical technology, lack of resources in the laboratory, and insufficient use of existing instruments for method optimization. According to the experience of the CUP laboratories, this is not applicable. The key is the project management.

3.3.3 How Do We Make It Better? - The UNTIE® Process of the CUP Laboratories

From the experience of the CUP laboratories, the success of an efficient method development and validation project can best be achieved through a process with up to four steps, which is derived from the Stage-Gate® model [1] and whose individual steps can be planned and calculated and are linked by decision gates. In individual cases, individual steps can be omitted or skipped.

The CUP laboratories understand project success here to mean that the customer also gets a difficult method developed and validated at predictable costs and within the agreed time frame. However, it is also considered a success if the customer receives clear scientific proof in good time that his goal cannot be achieved under the current conditions or under the assumptions made (analysis procedures, specification), and he can therefore make the right decisions in good time.

3.3.4 Understanding Customer Needs

The most important and at the same time the most expensive step (if not implemented) in the UNTIE® process is to understand the needs of the customer, which are sometimes not clear to the customer himself.

Unfortunately, specifications and analytical procedures are often set under time pressure and by people who are strong in regulatory affairs, but who sometimes lack the analytical background. In addition, when defining parameters and specifications, the maximum achievable is usually aimed for and the really necessary is lost sight of. Finally, very practical limitations are often not considered in this design phase. A current example is the determination of residual solvents in novel peptides that are very difficult to synthesize. Although these are administered to a person in therapy only very rarely and in small quantities, the methods must be "FDA-compliant" and thus, according to the usual understanding, correspond exactly to the relevant monograph <467> of the United States Pharmacopoeia. The fact that the methods described here work with limit values and sample quantities that are based on the PDE values of the respective solvents is often disregarded here in the risk analysis. One can easily imagine the enthusiasm of a synthetic chemist who is supposed to provide 10 g of such a peptide for a method development and validation.

The CUP laboratories therefore like to start such projects with a discussion about the real requirements of the customer and of course true to the motto "One day in the office saves one week in the laboratory!" also with an intensive study of the literature, in which application databases of common instrument and column suppliers are included as standard. An important aspect here is also the preparation of the samples with a view to eliminating unwanted matrix effects and the subsequent robust application of the method in everyday laboratory work.

In this first substep, in which a scientific expert deals intensively with the matter, very practical limitations such as sample quantity, processing time, availability, and costs of the materials used are also analyzed. The result of the "UN" substep is then a specification sheet for the method, which clearly describes and defines the parameters to be investigated, the formulation/composition of the sample, and the specifications.

In many cases, a method that is theoretically the most suitable can already be named at this point. Often, but by far not always, these are monograph methods or methods recommended by the R&D departments of customers. The specification sheet with the results of the literature research is made available to the customer and approved by the customer. It serves as an important basis for all further steps in the UNTIE® process.

3.3.5 The Test of an Existing Method

In the second substep of the UNTIE® process, either the method proposed by the customer or a method researched from the literature is then tested for validability. In either case, this is exactly one method and thus the customer receives clear feedback after a short time as to whether the problem can be solved quickly.

The activities within the scope of this feasibility are laid down in a plan and the scope is agreed in advance with the client. This scope depends very much on the "Requirement Specification" from the "UN" substep. In any case, however, it will already be determined according to the criteria of the ICH-Q2 guideline [2] is tested, i.e. for an HPLC content method, at least the specificity, linearity, working range, trueness, and precision of the method are determined at this stage. In addition, the stability of the samples, a point at which many validations later have failed, is already critically examined. All these measurements are already carried out under GMP conditions on qualified analytical equipment. This means that the results generated here can, in the best case, be used later in the validation report. In any case, all results of the feasibility are summarized in a report. This report is either the basis for validation if the method can already be validated in the best case, or it is the starting point for method development if the method does not yet meet the requirements.

In the past, it has been shown that it is necessary, especially in the case of unsuccessful feasibility studies, to prove the functionality of the complete system (autosampler, column, HPLC, detector) by performing suitable system suitability tests. This is not trivial at this point, due to the limited information about the method, and requires thorough planning and consideration.

3.3.6 Method Development and Optimization

The third and by far the most complex step in the UNTIE® process is the development or optimization of the method. The results of the first two steps are essential for efficient planning.

Many manufacturers of analysis systems currently offer automated solutions for "method scouting." The approach here is always similar. One combines up to six different columns with different organic and inorganic solvents and "runs" a program over a longer period in which the different possible combinations are tested and then ideally evaluated graphically. The approach of the company SHIMADZU [3] is shown as a schematic diagram in Figure 3.3.2.

With modern analysis systems, the schematic structure of the Nexera system from SHIMADZU IS SHOWN IN Figure 3.3.3, a large amount of data can be generated in a short time and with relatively little effort. Modern software systems also allow for efficient planning (see Figure 3.3.4) and evaluation (see Figure 3.3.5) of method screening runs.

It has proven to be a good idea to repeat this process several times, starting with the most diverse settings possible and then refining them iteratively. It is very important here that all tests, especially those that are not successful, are documented and evaluated in detail in the development report. In this way, the customer gains enormously helpful knowledge about his product, which can be of great importance during the entire life cycle or also in the case of modifications and further developments. In addition, this knowledge, as well as the information gained in the first two substeps, is later incorporated into the risk analysis for the analysis method, which is an important part of the validation plan.

A further, very important aspect of automated method development is that science becomes plannable in a certain way in terms of time and money, although of course not with a guarantee of success. However, after a few weeks, the clients will know whether or not a functioning method can be made available and they will also know the costs involved right from the start.

This approach also reduces the dependence of project success on the excellence, creativity, and knowledge of individual scientists. The procedure is thus scaled and, internally, vacation and downtime can be better compensated and the communicated timelines can still be met.

Figure 3.3.2 Schematic procedure of an automated "Method Scouting." Source: SHIMDAZU Corporation [3].

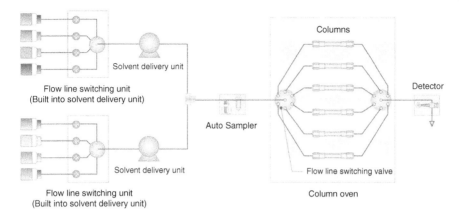

Figure 3.3.3 Schematic structure of the Shimadzu Nexera Method Scouting System. Source: SHIMDAZU Corporation [3].

With the presentation of the report on the development of the method to the customer and its approval by the customer, the creative "cloud phase" also ends and implementation in the "building block phase" begins.

3.3.7 Execution of the Validation

"Building block phase" means that from this point on, in contrast to the previous phases, the project managers are no longer open to all hints and suggestions from their own colleagues, from customers, from competitors, from suppliers and from science. Now the existing optimized method is implemented and even small details are not changed. For this reason, this last step of the UNTIE® process is called the "E" phase, or "Execute."

In the past, validation plans and validation reports at the CUP laboratories were created in the classical way with programs such as MS Word and spreadsheet programs such as MS Excel, and these were sent to the customer for discussion in advance. This worked sometimes better and sometimes worse and depended very much on the level of knowledge and experience of the customer's staff and of course the respective scientists of the CUP Laboratories. Sometimes, too much time was invested in coordination. The sad "highlight" in the history of the CUP Laboratories was the return of a 16-page validation report of a young scientist by the customer, an international pharmaceutical company, with **188!** comments, made by three different people, some of which were not directly related to the report, but reflected discussions within the customer.

It has therefore proved to be a good idea to use a system that is as rigid as possible in this phase, which deliberately gives the processor little flexibility. The wishes

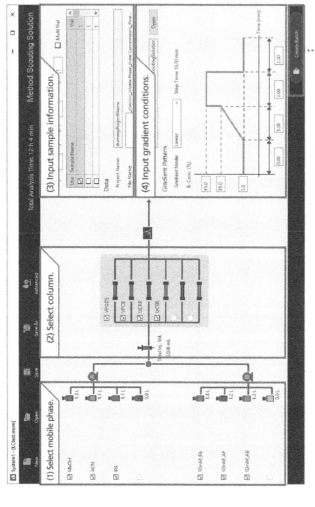

(1) Select a mobile phase and (2) select a column.

Select the column and mobile phase from those columns and mobile phases already added to the database. Methods will be generated automatically based on the number of selected conditions.

(3) Input sample information.

Injection volume and number of injections can be set for each sample.

(4) Input gradient conditions.

Gradient conditions can be configured simply by entering duration and concentration parameters. Investigations of gradient speed are also easy to configure.

(5) Create an analysis schedule.

An analysis schedule is generated automatically based on the conditions set above. This saves substantial labor over manual setting of conditions, and allows analysis to start immediately.

Figure 3.3.4 Easy creation of an analysis plan with the "Shimadzu Scouting System."

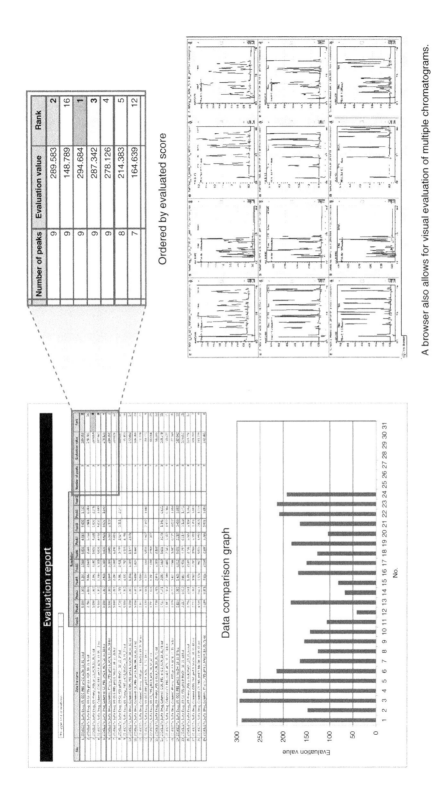

Figure 3.3.5 Automated evaluation of the "Scouting" runs and graphic display. Source: SHIMDAZU Corporation [3].

and requirements of the customer are, of course, very important and have therefore already been given due consideration in the three previous phases. At this stage, these have already been incorporated into the method to be validated and the accompanying documents.

The CUP laboratories use VALIDAT® for the documentation of validation plans and reports [4] as software of choice. This allows validation plans to be quickly and automatically created based on CFR-21-Part-11 templates. After approval by the customer, these can then be quickly implemented. Analysis data can be entered directly into the system by laboratory staff, and readymade text modules help project managers to concentrate their entire capacity on evaluating the data. Complex text checks for spelling and grammar and discussions about formulations are significantly reduced by the automated generation of validation reports.

3.3.8 Summary

The UNTIE® process is shown again schematically in Figure 3.3.6

The first step, the basis of the entire process, is the development of a common understanding with the customer for the requirements of the analysis method and a detailed theoretical consideration through literature research and, if necessary, simulation calculations. The result of this stage is a specification sheet with a proposed method. Afterwards, it is decided together with the customer whether it is worthwhile to start with the second stage, the testing of the method. At the end of the method test, in which only one method is tested, there is a feasibility report, which basically documents the suitability of the method proposed in stage 1 for the intended purpose. On this basis, a decision is then made as to whether it is already appropriate to start the validation, whether and which method developments are still necessary, or whether the selected method is generally not suitable for the intended use. In most cases, the development and optimization of the selected method now follows. This results in a detailed development report. Now it is

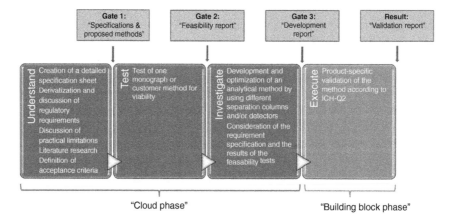

Figure 3.3.6 The UNTIE process of the CUP laboratories. Source: CUP Laboratorien.

decided, again together with the customer, whether the validation of the method can be started. Once this decision has been made, the implementation stage, the validation, begins. This is carried out according to a system that is as rigid as possible and is supported by VALIDAT® as a computer program for method validation.

All in all, consistent adherence to the UNTIE® process enables the customer to transparently show the various development stages of the method throughout the entire process and gives him the opportunity to exert influence at defined points in time. This significantly increases the predictability of method developments and validations and makes it easier to resolve the knot that the process is often perceived as in difficult projects.

Acknowledgments

The CUP Laboratories would like to thank Dr. Stavros Kromidas for his suggestions in the context of the development of the UNTIE® process and the SHIMADZU Corporation for providing the image material for "Automated Method Scouting."

References

1 Cooper, R. (2008). Perspective: the Stage-Gate® idea-to-launch process – update, what's new, and NexGen systems. *Journal of Product Innovation Management* 25(3):213–232.
2 EMEA (1995). *ICH Topic Q2 (R1) Validation of Analytical Procedures: Text and Methodology* ICH-Guideline. European Medicines Agency europa.eu.
3 SHIMDAZU Corporation. (2018). *High Performance Liquid Chromatograph – Method Scouting System*. Shimadzu Europa GmbH, Albert-Hahn-Straße 6-10 D-47269 Duisburg.
4 ICD. (2020). *VALIDATE*. Retrieved from https://icd.eu/produkte/validat/methodenvalidierung/ (accessed 23 February 2020).

3.4

Optimization Strategies in HPLC

Bernard Burn

INTERLABOR BELP AG R&D, Aemmenmattstrasse 16, 3123, Belp, Switzerland

The following explanations were written under the aspect that optimization is understood as the best possible solution to an analysis problem. Derived from this simple definition, the following two questions must be answered very carefully at the beginning:

(1) What is the analysis problem?
(2) What exactly should be optimized?

At the beginning of each method, optimization is the definition of what exactly is to be achieved or optimized. Even a simple high performance liquid chromatography (HPLC) method comprises a large number of parameters that can generally be optimized. Many of these sizes are dependent on each other, others are given from the outset and cannot be varied or only with great effort. In many cases, however, it is not even necessary to perfect all influencing variables down to the last detail. It is often sufficient to optimize only the most important parameters, while default values can be used for the rest. Ultimately, the aim is to develop a procedure to the extent that it provides precisely the information required for the issues in question with the necessary precision, within the time available and at the budgeted cost. Especially for service laboratories, the last two points often represent the greatest challenge. A typical task for such service providers is to find an adequate solution or answer to an acute problem or question of an authority as quickly and cost effectively as possible. This often requires HPLC analyses for which either new methods have to be developed or existing methods have to be adapted. A systematic approach and a targeted procurement of all necessary information are decisive for success. It is also very helpful to be able to draw on a wealth of experience. The following explanations therefore pursue three objectives:

(1) To show which information needs to be obtained at the beginning of a project and which preliminary considerations can be useful.
(2) To show simple ways how to obtain the HPLC-relevant data of a substance to be analyzed as quickly as possible.

Optimization in HPLC: Concepts and Strategies, First Edition. Edited by Stavros Kromidas.
© 2021 WILEY-VCH GmbH. Published 2021 by WILEY-VCH GmbH.

(3) To pass on extensive experience, which can be helpful in obtaining optimum analytical methods.

Many of the instructions described below are rules of thumb for which exceptions can be found too. Nor is it claimed that the concepts listed are all-encompassing and that the procedures described are the only ways to achieve the goal. Rather, it is intended to enable readers to optimize their personal workflow and to achieve success more quickly with the one or other useful information or rule of thumb.

3.4.1 Definition of the Task

Often, the task appears clear. For example, the content of a substance in a tablet is to be determined. A closer look reveals that even simple questions contain a great potential for misunderstandings. It is also helpful to clearly define the expectations or requirements of the customer. From a technical point of view, the main focus is on obtaining all necessary information on the analyte and the matrix, as well as on defining the requirements for the method and the exact intended use. These clarifications may include the following points (Table 3.4.1).

Based on this information, a checklist can be created that can be adapted to your own needs and processes, which can be very helpful for the exact definition of the task and goals. Only when the project framework is clearly defined can the actual work of method development and optimization begin.

3.4.2 Relevant Data for the HPLC Analysis of a Substance (see also Chapter 1.8)

To achieve optimal results, it is necessary to know the most important properties of the substance to be analyzed. The following points are important for HPLC analysis (Table 3.4.2).

These data can either be taken from the literature or estimated by means of the structural formula using simple rules of thumb.

3.4.2.1 Solubility

For HPLC analysis, it is optimal if a solvent is available, in which the substance to be tested dissolves at room temperature in a concentration of at least 1 mg/ml. If this is possible with water or a water-miscible solvent, the first major hurdle for a successful HPLC analysis is taken. Often, however, one finds only few data on this in the literature. In these cases, some preliminary tests are necessary. For this purpose, 10 mg of the substance are weighed into a test tube and 1 ml of the solvent concerned is added. If streaks form in the solvent, this is an unmistakable sign that the substance in question is sufficiently soluble. If not, immerse the test tube in an ultrasonic bath for one minute and swing it over from time to time. If the substance is still not sufficiently dissolved, add a further 9 ml of solvent and dissolve again in an ultrasonic bath. If the substance still cannot be dissolved in this way, the solvent

Table 3.4.1 Important points for the definition of the task.

Subject	Description
Substances to be tested	Four points are helpful with regard to the exact definition of the connections to be tested
	(1) CAS number (if available, otherwise sum formula) (2) Designation (how should the substances concerned be designated in the documentation and during discussions) (3) Structural formula (4) Special features - Polymorphism (if known) - Encapsulated active ingredients
Matrix	How are the substances to be tested available? Are these pure substances, active ingredients in a pharmaceutical formulation, or impurities in a product? For technical products, the matrix components should be disclosed as far as possible
Expected values	If no expected values are available, they must be estimated based on literature data. Expected values are decisive for the method design. If the analyte is only present in very low concentrations, enrichment steps may be necessary during sample preparation
	For toxic ingredients (for example, aflatoxins in plant drugs), any legal limits represent the expected values to which an analytical method must be oriented, also with regard to the required sensitivity (detection and determination limit)
	Another point that should be considered is the definition of the expected value or the nominal value for technical products such as pharmaceuticals. • Is the nominal value specification related to a volume or mass unit of the respective product? Is the reference quantity a dose unit such as a tablet or the contents of a dosage spoon? • Is the nominal point referring to the molecule in question, or to a corresponding salt (hydrochloride) or hydrate
Permissible measurement uncertainty	This point is often neglected. The permissible measurement uncertainty is often decisive for the design of an analytical method and the effort required for method development. In principle, the permissible measurement uncertainty is always a measure of the risk one is prepared to take to make a wrong decision based on the generated measurement data.
	If specifications exist for the substance to be tested (for example, a concentration range), the permissible uncertainty of measurement can be easily estimated from the specification width

(*continued*)

Table 3.4.1 (Continued)

Subject	Description
Measuring range	An important parameter for the design of a method is the measuring range to be covered. It makes a significant difference whether, for example, measurements are to be made only in the range of a limit or whether an exact measurement result is to be provided over several decades of concentration. Even with simple content methods for products with known concentrations, the definition of a measuring range makes sense. For example, should the same method be used to check both the content of the active ingredient and the concentrations of degradation products. Or the method should also be used in release tests of the active substance. These and numerous other factors define the measuring range of the method and should be determined in advance
Further information on the intended use or background of the analysis	An optimal method design depends very much on the intended use. A few keywords: Routine suitability: Is the method intended to be used for routine testing of a large number of samples or is it a one-time measurement? In the first case, much more attention will certainly be paid to the robustness of the method, the analysis time, the effort required for sample preparation, and the cost of the reagents. For one-off analyses, these points are irrelevant. Suitable for stability tests: If the method is to be used for stability studies, great care must be taken to ensure the selectivity of the method so that any degradation products can be separated from the analyte and, if desired, also detected Method transfer: If it is intended to transfer the method to the customer or to another laboratory of the company, any restrictions or requirements on the equipment must be clarified in advance

in question is not suitable. Based on the functional groups of a molecule and other information, the solubility of a substance can be estimated quite well. The following rules of thumb help here:

Substances with octanol–water partition coefficients (log P_{OW}) ≤ 1.5:

Compounds with log P_{OW} ≤ 1.5 are usually sufficiently water soluble (see Chapters 1.2 and 1.8).

Liquid substances:

Substances that are liquid at room temperature are soluble in almost all common solvents, at least in small quantities. For example, the solubility of benzene in water at room temperature is approximately 1.8 mg/ml. If the solubility limit is reached, two phases are formed. This can easily be detected visually (small drops that do not dissolve even after prolonged shaking).

Organic salts (hydrochlorides, alkali salts):

These compounds are usually sufficiently soluble in water or methanol.

3.4.2 Relevant Data for the HPLC Analysis of a Substance (see also Chapter 1.8)

Table 3.4.2 Data of a substance relevant for HPLC analysis.

Subject	Description
Solubility	For HPLC analysis, the analyte must be in dissolved form. A solvent must therefore be found in which the substance in question is sufficiently soluble
Acid–base constants	The majority of the compounds used as active pharmaceutical ingredients react alkaline or acidic. This property influences the separation by reversed-phase HPLC in several ways. On the one hand, the polarity of a compound is essentially dependent on its state of charge and, on the other hand, secondary interactions of the analyte with the stationary phase can lead to problems with the peak shape
Octanol–water partition coefficient	The octanol–water partition coefficient is a measure of the ratio between the lipophilicity (fat solubility) and hydrophilicity (water solubility) of a substance. A value greater than one means that the substance is more soluble in apolar solvents such as n-octanol. If this value is below one, solubility in water predominates. The octanol–water partition coefficient is usually given as a decadic logarithm. If the octanol–water coefficient is known, it is possible to estimate the amount of organic solvent in the mobile phase that is required to achieve the desired retention factor
UV absorption	If the molecule shows light absorption in the range of approximately 200–700 nm, detection is, in principle, possible using UV. If this is not the case, an alternative detector must be found. However, this is often associated with compromises in terms of sensitivity or reproducibility. This information is not relevant for trace analysis, since detection is almost always done by MS or MS–MS anyway. In this case, however, it is useful to research MS data. This concerns, in particular, the exact molecular mass and multiple reaction monitoring (MRM) transitions, as well as information on possible adducts
Stability in dissolved form	A solution must be prepared for HPLC analysis. The dissolved molecule should be stable for at least a few hours to allow HPLC analysis. Very reactive molecules that form stable degradation products via reproducible reactions can also be analyzed via these decomposition products

Carbohydrates (sugar, glycosides):

These polar compounds are usually water soluble. In the case of glycosides, depending on the compounds, the addition of methanol or ethanol is sometimes necessary.

Hydrocarbons and alkylglycerides (vegetable fats):

These apolar substances are usually sufficiently soluble in 2-propanol or tetrahydrofuran (THF). For chain lengths > C20, sometimes ethyl acetate or *tert*-butyl ether must be added.

Compounds with acidic functional groups (carboxylic acids, phenols):

These substances are usually soluble in dilute alkalis (0.01 M NaOH) or methanol. However, it should be noted that the stability of many compounds in alkaline solutions is limited. This applies in particular to phenols. Methanolic solutions of carboxylic acids tend to form esters. The addition of approximately 10% water has a stabilizing effect.

Compounds with alkaline functional groups (aliphatic and aromatic amines):

Substances with alkaline groups are almost always soluble in dilute acids (0.01 M formic acid) or in methanol.

Plastics:

Many plastics (except polyolefins) are at least partially soluble in THF. Plastics dissolved in THF can often be precipitated again by adding methanol. This is used when ingredients in plastic materials such as phenolic antioxidants need to be determined.

Polyolefins (polypropylene, polyethylene, etc.):

Polyolefins can be at least partially dissolved in toluene by heating.

Amino acids, peptides, and proteins:

Amino acids, peptides, and proteins are usually sufficiently soluble in water or an aqueous buffer. However, if the pH of the buffer is in the range of the isoelectric point of the compound concerned, the solubility may be limited. The solubility of this substance class is very low in organic solvents such as methanol or acetone if no lipophilic side chains are present.

Polynuclear nitrogenous aromatics:

Medium-polar compounds of this type are often difficult to dissolve. If sufficient solubility is not obtained with common solvents such as dilute aqueous acids or alkalis, methanol, and THF, the use of dimethyl sulfoxide (DMSO) or dimethylformamide often leads to the desired result. This is used to prepare a stock solution, which is then diluted with solvents more suitable for HPLC. Injection of solutions containing high concentrations of DMSO or dimethylformamide is not recommended, as peak deformations are often observed. An example of a substance that is very difficult to dissolve is riboflavin.

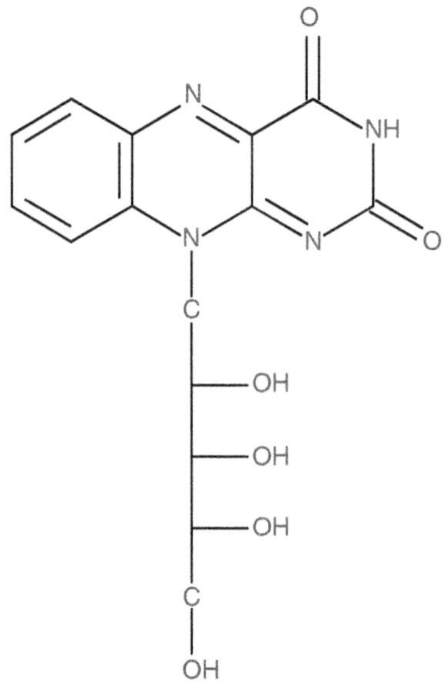

There are several crystal structures of riboflavin, which differ slightly in solubility. What they all have in common, however, is their overall very low solubility in water, ethanol, diethyl ether, and many other solvents. A maximum of only 0.1 mg/ml is also soluble in DMSO. Riboflavin dissolves quite well in 0.1 M sodium hydroxide solution and diluted hydrochloric acid. However, the substance decomposes very quickly in alkaline solution. The following method is used to obtain solutions, which, when protected from light, can be kept for several weeks. 20 mg riboflavin are weighed into a 200-ml volumetric flask and mixed with approximately 190 ml water. Add 1 ml acetic acid, heat to 50 °C with occasional stirring until the entire substance is dissolved, and after cooling add water to reach the mark.

3.4.2.2 Acidity Constants (pK_a)

The alkaline or acidic reaction of a substance must be considered in several ways when optimizing an HPLC method.

(1) The polarity of acids or bases is strongly pH-dependent.
(2) Protonation or deprotonation of molecules can have a great influence on UV absorption (maxima and intensities), which must be taken into account if UV detection is to be used.
(3) Acidic or alkaline functional groups can interact with the stationary phase in undesired ways, which can have an unfavorable effect on the peak shape (tailing).

3.4.2.2.1 Polarity of Acidic or Alkaline Substances (see also Chapter 1.8)

Numerous compounds show an acidic or alkaline reaction in aqueous solution by either giving off or taking up protons. This is always associated with a change in the charge of the molecule, which, in turn, influences the polarity. Charged molecules are present in aqueous solution as ions and are therefore much more polar than the corresponding uncharged particles. Since the retention of a compound in reversed-phase chromatography depends primarily on polarity, it is helpful for the development and optimization of HPLC methods to have as detailed information as possible on the charge state of the molecules to be tested as a function of pH. An important parameter in this context is the acid constant (pK_a). The stronger an acid, the more it has the tendency to release protons and the smaller the pK_a. Its counterpart is the base constant, which describes the tendency of a compound to take up protons. In aqueous solutions, the sum of the acid constant (pK_a) and the base constant (pK_b) of a substance always gives the value 14.

These relationships can be illustrated by means of a simple example. Aniline shows the following balance in water (Figure 3.4.1).

The acid constant pK_a of this reaction is 4.63. From this, the distribution of the proportions of acid (means a base with i hydrogens [HiB]) and conjugated base (B) as a function of the pH value can be calculated and graphically displayed (Figure 3.4.2).

This graphic shows the following important properties:

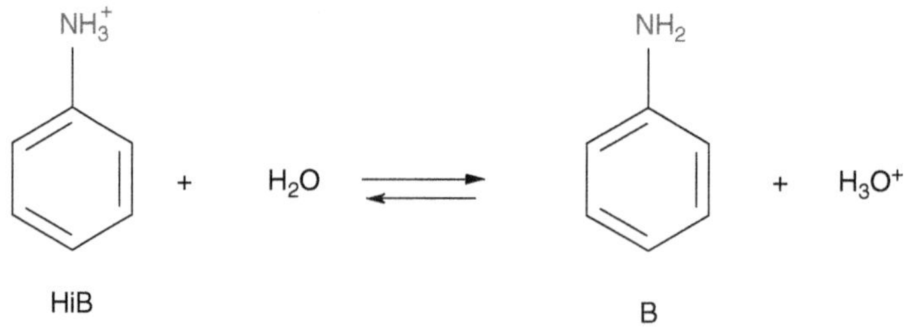

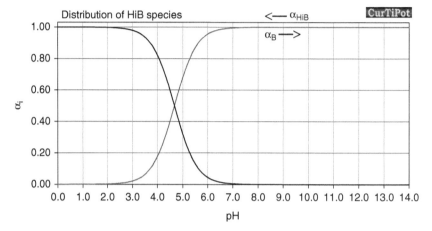

Figure 3.4.1 Protonation equilibrium of aniline in water.

Figure 3.4.2 Proportion of the two aniline species as a function of pH value (calculated with the CurTiPot program).

(1) In the pK_a range (pH 4.63), the aniline contained in a solution is present in equal parts in protonated form and without an additional proton. This means that about half of the aniline molecules carry a charge, while the rest are uncharged. In this range, even small changes in pH have a great influence on the ratio between the protonated and nonprotonated molecules.
(2) Approximately two pH units below the pK_a (pH 2.6), the molecule is almost completely present in protonated or charged form.
(3) Approximately two pH units above the pK_a (pH 6.6), practically no protonated molecules are contained. Most of them are not protonated and therefore do not carry a charge.

The following chromatograms show the effects of these laws in practice. The upper chromatogram shows the separation of aniline and phenol on a classical reversed phase at pH 7. Aniline is a base, while phenol reacts weakly acidic. Both molecules are uncharged at pH 7 and show similar retention times with a 20% acetonitrile content in the mobile phase. The situation is completely different with pH 2.3. Under otherwise-identical conditions, aniline already elutes very early, while the retention time of phenol hardly changes. Aniline has a pK_a of 4.6 and is therefore fully protonated at pH 2.3. As a result, the retention due to the non-polar stationary phase

Table 3.4.3 Plate number and USP tailing of two bases of different strengths.

Substance	pK_a	pK_b	Plate number	USP tailing
Aniline	4.6	9.4	1986	1.84
Benzylamine	9.3	4.7	1320	2.90

decreases markedly, which leads to the observed massive reduction of the retention time (Figure 3.4.3).

Column	Nucleosil 120-3 C18, 125 × 4 mm
Flow	1.0 ml/min
Temperature	40 °C
Mobile phase	0.01 M phosphate buffer/acetonitrile 80 : 20 (V/V)
Injection	1 µl (1 mg/ml)

3.4.2.2.2 UV Spectra

Previous chromatograms also show the influence of the charge state of a molecule on UV absorption. While the uncharged phenol has an identical peak size at both pH 7.0 and 2.3, a significant reduction in absorbance at 254 nm can be observed by protonation of the aniline. Accordingly, the UV spectra of aniline differ markedly at pH 7.0 and 2.3 (Figure 3.4.4).

While aniline at pH 7.0 shows two distinct maxima at 231 and 281 nm, only weak absorption bands are observed at pH 2.3 in the range above 215 nm.

3.4.2.2.3 Influence on the Peak Shape

The acid or base strength also has a significant influence on the peak shape. Classical silica-based reversed phases (silicon dioxide) contain more or less free silanol groups depending on the manufacturing process. These are slightly acidic and can interact with basic analytes, which can lead to peak tailing. This peak deformation considerably impairs the reproducibility of the method, so that symmetrical, narrow peaks are aimed for during method optimization. The following tables shows the plate number and United States Pharmacopeia (USP) tailing of two bases of different strengths on a classical C18 phase at pH 2.3 (Table 3.4.3 and Figure 3.4.5).

Column	Nucleosil 120-3 C18, 125 × 4 mm
Flow	1.0 ml/min
Temperature	40 °C
Mobile phase	0.01 M phosphoric acid/acetonitrile 95 : 5 (V/V)
Injection	1 µL (1 mg/ml)

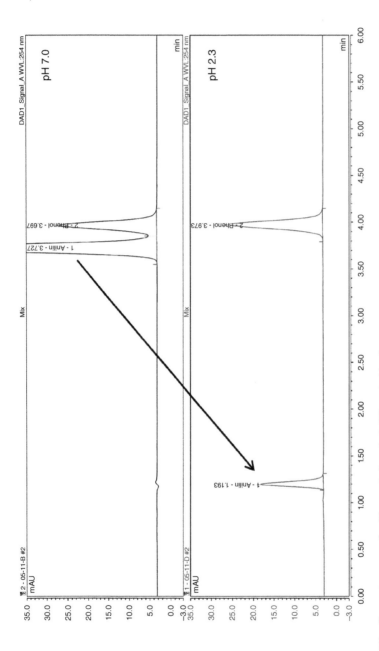

Figure 3.4.3 Chromatograms of aniline and phenol at different pH values.

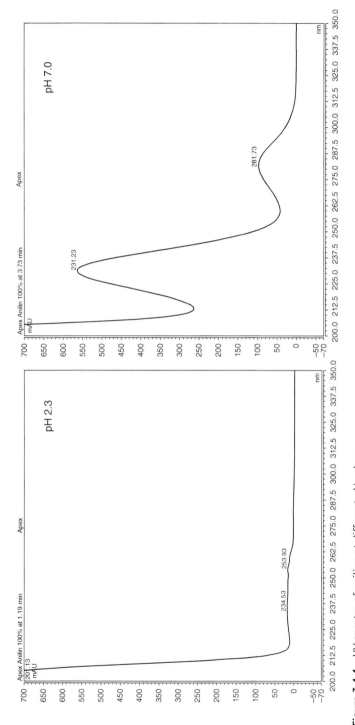

Figure 3.4.4 UV spectra of aniline at different pH values.

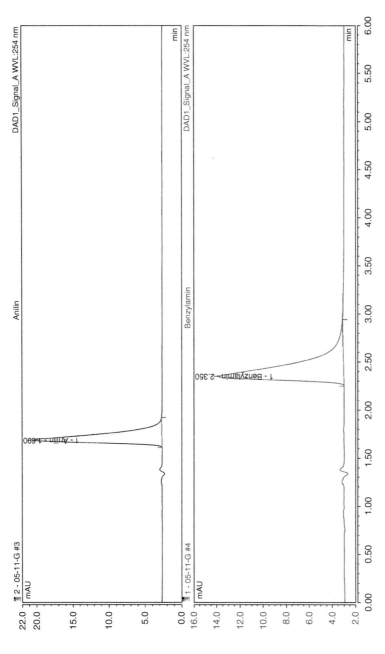

Figure 3.4.5 Chromatograms of aniline and benzylamine at pH 2.3.

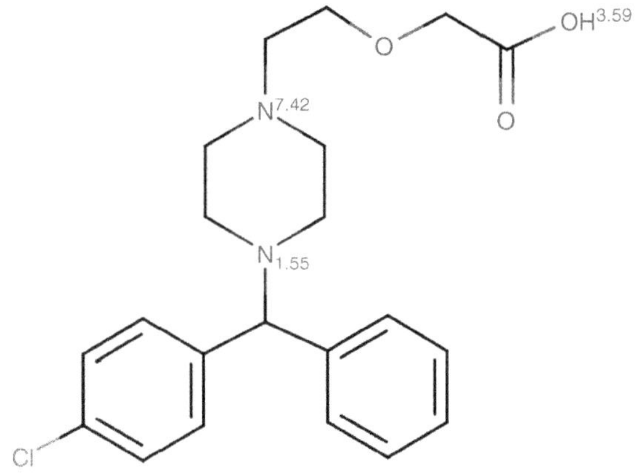

Figure 3.4.6 Acidity constants of cetirizine calculated with MarvinSketch.

The more alkaline benzylamine shows a much stronger peak tailing than aniline.

3.4.2.2.4 Acid Constant Estimation

If no information on acid constants is found in the literature for the compound in question, a quick estimate is possible based on the molecular structure. The following tables contain information on acid constants of typical organic acids and bases (Tables 3.4.4 and 3.4.5).

The values listed were taken from various compendia and may vary slightly depending on the source. Acid constants can also be calculated using suitable software based on the molecular structure. This is especially helpful if a molecule has several acidic and/or alkaline groups. The program MarvinSketch is very well suited for these calculations (Figure 3.4.6).

3.4.2.3 Octanol–Water Partition Coefficient

The octanol–water partition coefficient ($\log P_{OW}$) is an important parameter for medicine, toxicology, and the environment. For this reason, octanol–water partition coefficients have been determined for a very large number of substances, which are easy to find in literature. The octanol–water partition coefficient allows a simple estimation of the retention behavior of a substance in reversed-phase–high performance liquid chromatography (RP–HPLC). If some values of suitable calibration substances are available for the respective separation column, the proportion of organic solvent in the mobile phase can be estimated directly, which is necessary to achieve the desired retention factor for the substance to be tested. This is very helpful for people who often have to develop or adapt an HPLC method for different substances. It should be noted, however, that the octanol–water partition coefficient only applies to substances in the uncharged state. This can be achieved by adjusting the pH value of the mobile phase accordingly. The following table (Table 3.4.6) shows some examples and the amount of acetonitrile in the mobile phase required to obtain a retention factor (k') of about 2 for the substance in question when using Nucleosil 120-3 C18 as the stationary phase (Table 3.4.6).

Table 3.4.4 Acid constants of typical organic acids.

Class	Example	Acid	pK_a	Base	pK_b
Sulfonic acids (and sulfonic acid esters)	Benzene sulfonic acid	C₆H₅–SO₂–OH	−2.5	C₆H₅–SO₂–O⁻	16.5
Halogenated carboxylic acids	Trichloroacetic acid	CCl₃–COOH	0.66	CCl₃–COO⁻	13.4
Phosphones	Fosfomycin	(epoxide)–P(O)(OH)₂	2.0	(epoxide)–P(O)(OH)(O⁻)	12

Aromatic carboxylic acids	Benzoic acid	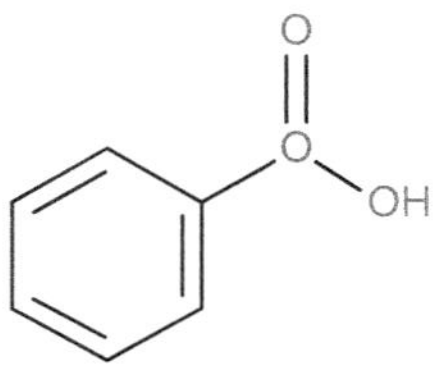	4.2	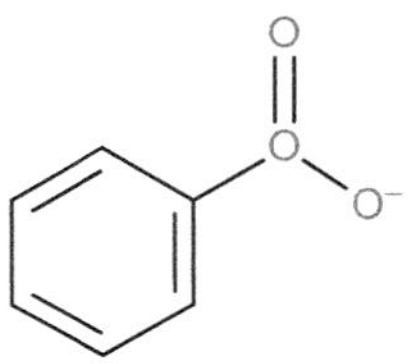	9.8
Enole	Ascorbic acid	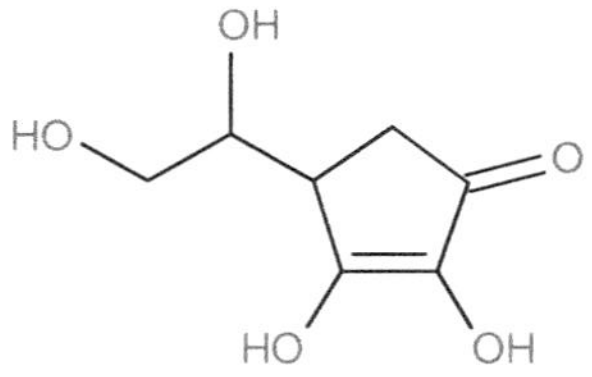	4.2	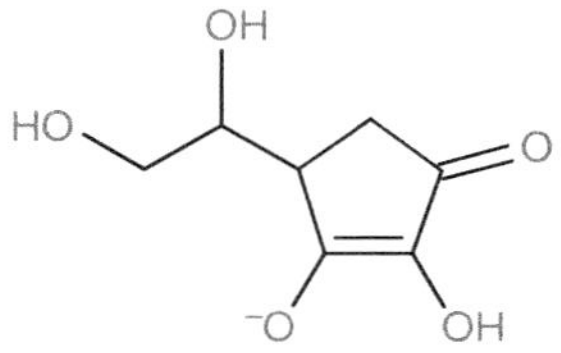	9.8
Carboxylic acids	Acetic acid	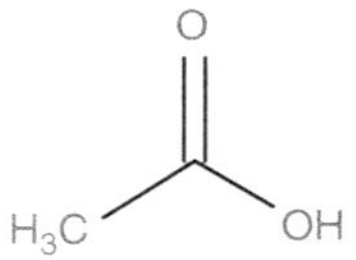	4.8	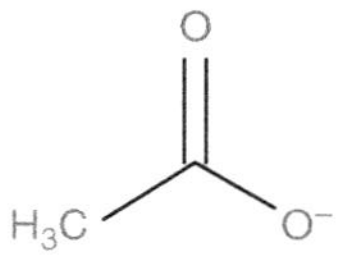	9.2
NH-acid compounds	Phthalimide	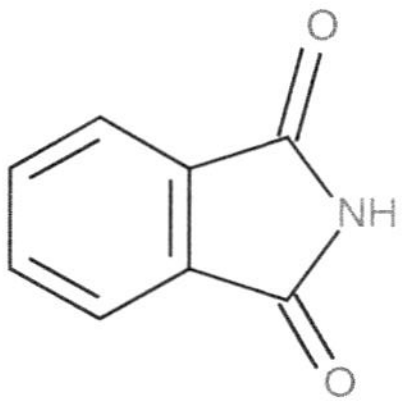	8.3	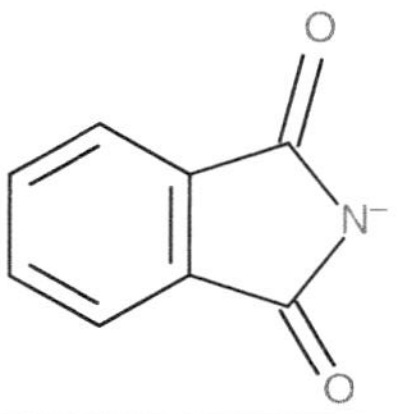	5.7

(*continued*)

Table 3.4.4 (Continued)

Class	Example	Acid	pK_a	Base	pK_b
Chlorophenols	o-Chlorophenol	(2-chlorophenol, OH)	8.5	(2-chlorophenoxide, O⁻)	5.5
Phenols	Phenol	(phenol, OH)	9.9	(phenoxide, O⁻)	4.1
Thiols	Ethyl sulfide	H₃C—C—SH	10.5	H₃C—C—S⁻	3.5

Table 3.4.5 Acidity constants of typical organic bases.

Class	Example	Acid	pK_a	Base	pK_b
Diphenyl-amines	Diphenylamine	(Ph)$_2$NH$_2^+$	0.8	(Ph)$_2$NH	13.2
Aromatic amines	Aniline	Ph-NH$_3^+$	4.6	Ph-NH$_2$	10.4
Aromatic-bound nitrogen	Pyridine (conjugated acid)	pyridinium	5.2	pyridine	8.8
Amines in benzyl position	Benzylamine	Ph-CH$_2$-NH$_3^+$	9.4	Ph-CH$_2$-NH$_2$	4.6

(continued)

Table 3.4.5 (Continued)

Class	Example	Acid	pK_a	Base	pK_b
Tertiary amines	N,N,N-Triethylamine	H₃C–N⁺(H)(CH₂CH₃)–CH₂CH₃	10.8	H₃C–N(CH₂CH₃)–CH₂CH₃	3.2
Primary amine	Ethylamine	H₃C–CH₂–NH₃⁺	10.8	H₃C–CH₂–NH₂	3.2
Secondary amines	N,N-Diethylamine	H₃C–CH₂–N⁺H₂–CH₂CH₃	11.1	H₃C–CH₂–NH–CH₂CH₃	3.9
Cyclic amines	Piperidine	piperidinium (N⁺H₂ ring)	11.2	piperidine (NH ring)	3.8

Table 3.4.6 Octanol–water partition coefficient of some compounds and acetonitrile content for k' of 2.

Substance	Log P_{OW}	Acetonitrile content for k' of 2 (%)
Propionic acid	0.33	1
Benzyl alcohol	1.1	19
Benzene	2.1	47
Ethylbenzene	3.2	60
Biphenyl	4.0	64
Dibenzyl	4.8	70

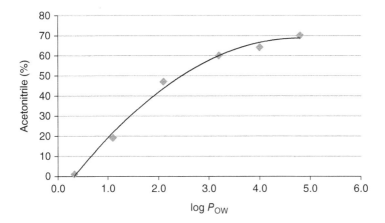

Figure 3.4.7 Proportion of acetonitrile in the mobile phase to obtain k' of 2 (Nucleosil 120-3, C18).

If a graph is plotted with the acetonitrile content (% V/V) against the octanol–water distribution coefficient (log P_{OW}), a characteristic curve is obtained for the corresponding combination of stationary- and mobile phase (Figure 3.4.7).

If the log P_{OW} of a substance is available, this graph can be used to estimate the acetonitrile content in the mobile phase, which is necessary to quickly obtain a first useful chromatogram. Depending on the complexity of the sample, further experiments may be necessary to further optimize the separation.

Similar curves can be generated with little effort for almost any retention coefficient and additional stationary phases and solvent mixtures.

The same concept can also be used to estimate the retention time of a compound with a gradient. For this purpose, a mixture of substances is simply chromatographed with the relevant method and the retention time is applied against log P_{OW}. The following graph shows the retention times of the substances listed above, which were obtained with a gradient method (Figure 3.4.8).

If a substance with a log P_{OW} of 2.5 is chromatographed with the method in question, a retention time of approximately 11 minutes can be expected.

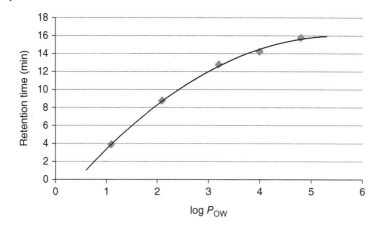

Figure 3.4.8 Retention times as a function of the log P_{OW} (column: Nucleosil 120-3 C18, 125 mm × 4 mm; flow 1.0 ml/min; gradient in 20 minutes from 10% to 90% acetonitrile).

Octanol–water partition coefficients can also be calculated based on the molecular structure. MarvinSketch is very well suited for this purpose. In connection with the octanol–water partition coefficient, the following limitations and rules can be formulated.

(1) The estimates described above work satisfactorily in the range of log P_{OW} 0.3 to approximately log P_{OW} 6.
(2) If a compound has a log P_{OW} of less than 0.3, the compound is too polar for reversed-phase chromatography. At best, an attempt can be made to achieve sufficient retention by adding ion-pair reagents. However, it may be more advantageous to use a different separation mechanism such as HILIC (hydrophilic interaction chromatography).
(3) Using methanol as solvent, compounds with log P_{OW} up to approximately 5 can be eluted from C18 phases. With acetonitrile, the limit is at log P_{OW} of about 6.
(4) If compounds have octanol–water partition coefficients of more than 6, it is advantageous to work with C8 materials. Alternatively, chromatography can be performed with "stronger" solvents such as THF, 2-propanol, and *tert*-butyl methyl ether (not water-miscible).
(5) After calibration of an HPLC method with substances with known log P_{OW}, it is also possible to determine the octanol–water partition coefficient of further substances via the respective retention factor.

3.4.2.4 UV Absorption

UV spectrometry is by far the most important detection technique for HPLC. For technical reasons, however, the usable wavelength range is limited from 200 to 700 nm.

Based on the structure of a molecule, it can be estimated whether a substance can be detected by UV and, if so, at about which wavelength. The following rules of thumb can be applied:

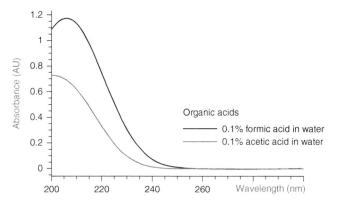

Figure 3.4.9 UV spectra of formic and acetic acid.

(1) Compounds without double bonds cannot be detected by UV. This also applies to substances with only one or more isolated C—C double bonds (for example alkenes). These absorb only in the range of 170–175 nm. There are, however, exceptions where certain substituents shift absorption maxima to above 200 nm and thus make the compound in question accessible for UV detection.
(2) Functional groups such as alcohols, ethers, or amines generally absorb UV light only at wavelengths close to 200 or below 200 nm and are therefore not easily detectable with UV.
(3) Carbonyl groups have a UV maximum at about 285 nm. However, the absorption there is so low that detection is usually not possible in this area. A further, stronger maximum is observed at about 188 nm. The spectra of carbonyl compounds are strongly influenced by substituents in α or β position. These so-called auxochromes can increase the wavelengths of the absorption maximum as well as amplify the actual absorption (bathochromic shifts and hypertomic effects). Typical substituents with these effects are hydroxyl and alkoxyl groups. As a result, for example simple carboxylic acids and their esters in aqueous solution have an absorption maximum at about 208 nm and can thus be detected by UV detection, albeit with limited sensitivity (Figure 3.4.9).
(4) Molecules that have an aromatic ring show a UV maximum in the range of 250–300 nm depending on the substituents. In addition, a strong absorption in the range of 220 nm is almost always observed. For bicyclic compounds (molecules based on naphthalene), the maxima are about 30 nm higher.
(5) For compounds without aromatic groups, it roughly applies that if the molecule has at least three conjugated double bonds (C=C or C=O), UV detection with wavelengths greater than 230 nm is possible.
Example:
Prednisolone has three conjugated double bonds and has a UV maximum at 243 nm (Figure 3.4.10).
(6) Compounds with two conjugated double bonds can usually be successfully detected at UV wavelengths below 230 nm. However, the sensitivity is sometimes somewhat limited.

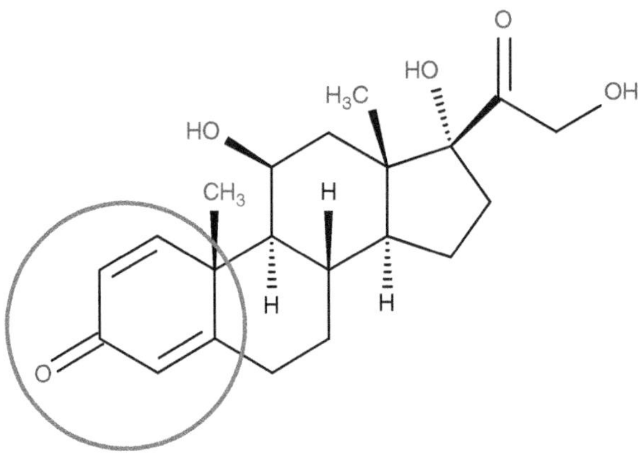

Figure 3.4.10 Structural formula of prednisolone.

(7) If wavelengths below 210 nm are required, it is worth considering alternative detection techniques such as evaporative light scattering detector (ELSD) or refractive index (RI), or at best mass spectrometry (MS).

Of course, the position of the UV maxima is also influenced by numerous other influences such as substituents, charge state (protonated, nonprotonated), and also the solvent. However, these simple rules provide completely sufficient results in most cases for first tests. If a device with a diode-array detector is available for method development, further optimization of the detection wavelength is possible without problems. If you want to make a more precise estimate, you can use the very extensive Woodward rules [1].

3.4.2.5 Stability of the Dissolved Analyte

Sufficient stability of the dissolved analytes is essential for HPLC investigations. This is particularly important if the purity of a substance has to be tested. If the substance decomposes too quickly in solution, there is a high probability that in this case the test will be falsified by the degradation products formed in solution. For routine tests, it is optimal if the solutions are stable for a few days. This way, even large sequences can be processed and it is not a disaster if the HPLC analysis is not performed immediately after sample processing. When developing methods, it is therefore helpful if the stability of the solutions is checked at an early stage. In many cases, it is possible to sufficiently stabilize the solutions by taking appropriate measures. Below some common decomposition reactions and measures to slow them down are listed (Tables 3.4.7 and 3.4.8).

Further notes:

Often, it is not possible to produce sufficiently stable solutions for longer HPLC sequences. In these cases, a coolable autosampler is helpful, so that the sample vials can be stored at 4–10 °C.

Numerous constituents of the sample matrix can shift the pH value of the sample solution into ranges unfavorable for the stability of the dissolved analyte. It is therefore advisable to also check the pH value of the sample solution. This is the only way to detect a possible lack of buffer capacity of the sample solvent.

Table 3.4.7 Typical decomposition reactions of dissolved substances.

Reaction	Example	Prevention
Hydrolysis	Esters such as 4-hydroxymethylbenzoate (methyl paraben) are cleaved in solutions with pH values above 6.5	Slightly acidify sample solvent or buffer to pH < 6.5
Water-sensitive compounds	3-Glycidoxypropyltrimethoxysilane (glymo) is completely hydrolyzed in a weakly acidic solution within a short time	Glymo and similar compounds are not stable in aqueous solution. These molecules can only be stored in apolar solvents like pentane for a certain period of time. In weakly acidic or alkaline aqueous solutions, the molecules, including the epoxy ring, react completely and form very stable hydrolysis products. For extremely water-sensitive substances, it is often advantageous to hydrolyze them completely and then perform the HPLC analysis using the hydrolysis product

(*continued*)

Table 3.4.7 (Continued)

Reaction	Example	Prevention
Oxidation	Captopril is easily oxidized, forming dimers via a disulfide bridge	Numerous compounds are sensitive to oxidation. However, there are effective measures to reduce oxidative degradation: • Removal of oxygen in the sample solvent (blowing through helium) • Addition of antioxidants Ascorbic acid has proved to be effective for aqueous media and ascorbyl palmitate for organic solvents. There are also numerous other antioxidants that are also well suited • Addition of EDTA to remove any heavy metals, which may catalytically promote oxidation If methanol is to be used as sample solvent, an addition of 5–10% water prevents the formation of methyl esters
Esterification	Although carboxylic acids generally dissolve well in methanol, they easily form the corresponding methyl esters in the absence of water. A typical example of this is tetrahydrocannabinoleic acid	

Table 3.4.8 Typical decomposition reactions of dissolved substances.

Reaction	Example	Prevention
Epimerization	Tetracyclines in dissolved form tend to form epimers (Tetracycline to epimerization at C4)	Epimerization can be prevented by acidifying the solvent to approximately pH 2. It should be noted, however, that this favors the formation of anhydrous products. Tetracyclines usually have a very limited shelf life in dissolved form due to the numerous reactive groups
Photoisomerization	Chlordiazepoxide shows an intramolecular rearrangement by photochemical excitation. The reaction can be reversed by heating the solution	Photoisomerization can be prevented by light protection. This can be achieved by working with amber glass vessels. Amber glass vessels, however, contain iron oxide (the brown glass is obtained by adding iron oxide). Traces of this compound can get into the sample solution and thus catalytically promote oxidation. To prevent photoisomerization, however, it is often sufficient to protect the solutions from direct sunlight

(continued)

Table 3.4.8 (Continued)

Reaction	Example	Prevention
Thermal isomerization	Cholecalciferol isomerizes noticeably to pre-cholecalciferol already at room temperature. This reaction can be reversed with sunlight	Thermal isomerization can practically only be slowed down by cold storage
Adsorption	Quaternary ammonium compounds like benzalkonium tend to adsorb on glass or metal surfaces	Absorption on surfaces can be prevented by acidifying the sample solvent. Often, an addition of organic solvents such as acetonitrile or water (at least 10%) helps. Depending on the connection, it is often helpful to use plastic vessels (incl. HPLC vials). The addition of approximately 0.005 M tetramethylammonium hydrogen sulfate has also proven to be very effective. The higher the dilution of the analyte, the more losses due to adsorption are significant

Rearrangement of ester groups Corticosteroid 17-monoesters easily rearrange themselves to 21-monoesters. A typical example is betamethasone 17-valerate Corticosteroid 17-monosteroids can be stabilized by slight acidification (approximately pH 4)

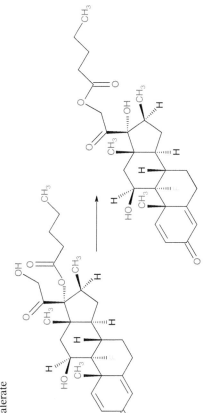

If buffers are mixed with acetonitrile or methanol, the pH can change considerably depending on the mixing ratios. As a rule of thumb for acidic solutions, the pH increases between 0.15 and 0.2 per 10% solvent.

The complete removal of dissolved oxygen in the sample solvent by adding antioxidants may take some time. It is therefore advantageous to prepare the solution the evening before and let it react overnight in a tightly closed container.

As a tendency, many substances (not all) are more stable in slightly acidic solutions than in the neutral or alkaline range. If no information on the stability of a substance is available, a weakly concentrated acetate buffer with pH 4.7 is usually a good starting point.

Within the framework of these considerations, possible negative effects of the contents of the sample matrix on the stability of the dissolved analytes must always be taken into account. A typical example is multivitamin tablets, which also contain metallic trace elements. When dissolved, these metals can greatly accelerate the decomposition of vitamins that are sensitive to oxidation.

3.4.3 Generic Methods

HPLC methods from literature or public compendia such as pharmacopoeias are very poorly standardized. A large number of stationary phases and countless buffer systems are used. This makes, in particular, the analysis of individual samples sometimes very complex. Unless there are regulatory reasons against it, it is advisable for each laboratory to standardize HPLC analyses as far as possible. This is especially helpful if numerous methods are in use or if developments or adaptations have to be made constantly for smaller projects. Clever standardization of HPLC methods can usually significantly improve instrument utilization and staff productivity. Standardization in this case means that all analyses can be performed with the same combinations of stationary phase (column) and mobile phases (solvent, buffer). This eliminates the time-consuming conversion of the HPLC system before each new analysis, which usually consists of rinsing the previously used column, installing the new column, changing the solvent, rinsing the solvent channels, and reconditioning the entire system. Meanwhile, many devices have column switching valves that allow the installation of several columns, as well as multichannel pumps that allow the selection of up to six mobile phases. Thus, it is conceivable to perform the majority of the HPLC analyses of a laboratory without the need for any conversion. Only for real special cases are separate processes and column types required.

3.4.3.1 General Method for the Analysis of Active Pharmaceutical Ingredients

The general method described here is well suited for the analysis of a very large number of different active pharmaceutical ingredients in all possible products and dosage forms. Even strongly alkaline and relatively polar compounds as well as peptides and proteins can be chromatographed in this way. In addition, this method is a good basis for targeted modifications that can be used to significantly expand the range of applications. This method has been designed for classical HPLC systems as well as for ultrahigh performance liquid chromatography (UHPLC) (Table 3.4.9).

3.4.3.2 Extensions of the Range of Application

The method described here can be easily adapted for further applications (Table 3.4.10).

3.4.3.3 Limits of this General Method

Strongly polar molecules such as carbohydrates or inorganic ions cannot be chromatographed with the method described here because their retention on reversed phases is too weak. Likewise, the method is only conditionally suitable for the separation of very large molecules with molar masses greater than approximately 100 000 Da, even with suitable separation columns. For both cases, there are better suited separation techniques, such as HILIC or ion exchange. It is also possible that certain isomers can only be separated with the aid of special columns (for example, C30).

3.4.3.4 Example, Determination of Butamirate Dihydrogen Citrate in a Cough Syrup

3.4.3.4.1 Basic Data
Table 3.4.11.

3.4.3.4.2 Expected Difficulties
Based on the available information, the following difficulties are expected for HPLC analysis (Table 3.4.12).

3.4.3.4.3 HPLC Method
Table 3.4.13.

3.4.3.4.4 Example Chromatogram
Figure 3.4.11.

3.4.4 General Tips for Optimizing HPLC Methods

Often, it is simple things that lead to a significant gain in effectiveness and efficiency in the laboratory. The following are some general tips for optimizing steps typically performed in HPLC analyses.

Table 3.4.9 General method parameters and rationales.

Parameter	Selection	Rationale
Stationary phase	C18 column based on hybrid material with a polar group on the alkyl residue (polar embedded) and a pore size of about 130 Å. Suitable are for example Waters XBridge Shield RP18 or Waters Cortecs Shield RP18	The selected material is based on a hybrid phase, which can be used up to pH 12 in contrast to the classic silica gel columns. These hybrid materials are also very robust and therefore well suited for routine operation
		The polar group on the alkyl radical (polar embedded) suppresses the tailing of alkaline substances. In many cases, good selectivity is also observed when separating substances with polar functional groups. This is due to the fact that, in addition to the typical reversed-phase separation mechanisms, polar interactions also contribute to separation
		A pore size of approximately 130 Å allows the analysis of compounds up to a molecular weight of approximately 1000 Da
Particle size	HPLC: 3.5 μm UHPLC: 1.7 or 2.6 μm	Particles with 3.5 μm or smaller offer numerous advantages over particles with larger diameters
Column diameter	HPLC: 2.1, 3.0, and 4.6 mm UHPLC: 1.0 or 2.1 mm	The column diameter depends on the required sensitivity or loading capacity
Column length	HPLC: 150 mm UHPLC: 100 mm	In many cases, a column length of 100 or 150 mm represents a good compromise between resolution and analysis time. The back pressure is also kept within limits even when using 1.7-μm particles. Increasing the column length from, for example, 100 to 150 mm only marginally improves the resolution (factor 1.2). In many cases, a 50-mm column would also be completely sufficient. However, this places significantly higher demands on the equipment in terms of dead volumes and gradient delays

Table 3.4.9 (Continued)

Parameter	Selection	Rationale
Mobile phase A (aqueous)	0.1% trifluoroacetic acid in water	Basic components are protonated. The trifluoroacetic acid anion acts as an ion pair, which slightly increases retention times and also suppresses peak tailing of basic substances
		Carboxylic acids and other weakly to moderately acidic components are present uncharged and can therefore be easily chromatographed in most cases. Trifluoroacetic acid can also be used with detectors that require volatile mobile phases (ELSD, CAD, or MS)
Mobile phase B (organic)	0.1% trifluoroacetic acid in acetonitrile	Compared to methanol, acetonitrile gives the higher plate numbers. Mixtures with water also have lower viscosities, which massively reduces the back pressure. These advantages more than outweigh the higher price in most cases. Trifluoroacetic acid is also added to the organic phase to avoid possible baseline drift at lower wavelengths
Blend mode	Isocratic or using gradients	Often an isocratic way of working is completely sufficient. For more complex analyses, acetonitrile gradients can be useful
		By using multichannel pumps, the proportion of organic solvent can be varied without any problems and optimally adapted to the respective analyte
Flow rate	0.1–1.5 ml	The flow rate depends on the column diameter. The following flow rates have proven to be effective:
		0.1 mm: 0.05–0.1 ml/min
		2.1 mm: 0.2–0.4 ml/min
		3.0 mm: 0.4–0.8 ml/min
		4.6 mm: 1.5 ml/min
Column temperature	40 °C	For reproducible results, it is recommended to keep the temperature of the separation column constant by means of a column oven. Forty degree Celsius can hold even column ovens without active cooling at normal ambient temperatures. Any separation problems can often be solved very easily by increasing or decreasing the column temperature by 5 or 10 °C

Table 3.4.10 Enhancements to the generic method.

Extension for	Adjustment
Proteins	The standard pore sizes of HPLC phases are 60–120 Å. This allows optimal separation of molecules with molar masses up to approximately 1000 Da. For larger molecules, stationary phases with larger pores should be used. Typically, many reversed phases are also available with 300 Å. This allows molecules with molecular masses in the range of 500–100 000 Da to be chromatographed without interference. It should be noted that the upper molar mass limit also depends on the geometric shape of the molecule in question. If substances with even higher molar masses are to be analyzed, the use of materials with 1000 Å is indicated
Highly polar compounds	If no sufficient retention of the analyte on the stationary phase is obtained, it may be useful to work with ion-pair reagents. Together with the analytes, which are preferably present as ions, these form ion pairs that are much less polar than the original molecule. MS-compatible ion-pairing agents are heptafluorobutyric acid or long-chain amines. If UV detection is used, classical reagents such as alkyl sulfonic acids or quaternary amines can also be used. By clever selection of the chain length of these counter ions, retention times of the ion pairs can be adjusted as desired for a given proportion of organic solvent in the mobile phase
Strongly nonpolar connections	Very nonpolar compounds such as triglycerides have a very high retention and may not be able to be eluted from the column with acetonitrile. The elution power of acetonitrile can be increased by adding methyl *tert*-butyl ether. However, it should be noted that methyl *tert*-butyl ether is not water-miscible and the solution mixture always contains at least 50% acetonitrile as solubilizer.
UV detection in the deep wavelength range	With trifluoroacetic acid, detection at wavelengths below 215 nm is difficult. If the method is not also used for MS detection, trifluoroacetic acid can be replaced by methane sulfonic acid. With this, wavelengths well below 210 nm are possible. If it is possible to do without the ion-pair formation by the trifluoroacetic acid, it can also be replaced by phosphoric acid (0.1%).
Improvement of MS detection	With mobile phases containing trifluoroacetic acid, ionization by means of electro spray in positive mode usually works without problems, especially with alkaline substances. For other molecules and in negative mode, however, formic acid may be more favorable. In this case, the trifluoroacetic acid is replaced by equal concentrations of formic acid. However, it should be noted that the retention may be reduced and the peak shape may become worse. An addition of 0.0025 to 0.005 M medronic acid (methylene diphosphonic acid, CAS 1984-15-2) may possibly prevent the decline of the peak shape
	Depending on the substance, the sensitivity can be much higher in the neutral or alkaline range. If this is the case, replace the trifluoroacetic acid with ammonium formate or ammonium carbonate (pH 8.0–10.5). However, both ammonium formate and ammonium carbonate are only slightly soluble in acetonitrile, so that the acetonitrile content should not exceed 80–90% depending on the concentration

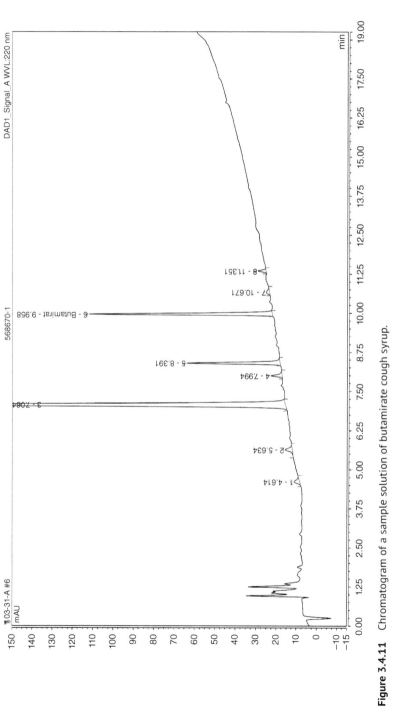

Figure 3.4.11 Chromatogram of a sample solution of butamirate cough syrup.

Table 3.4.11 Basic data for the analysis of butamirate in cough syrup.

Objective	Determination of the concentration of butamirate dihydrogen citrate in a cough syrup
Scope	One-off testing of some samples as part of a product control
Specification	95.0–105.0%
Expected value	1.5 mg/ml calculated as butamirate dihydrogen citrate (0.15% m/V)
Ingredients of the product	The product contains mainly extracts of ribwort, sundew, and hibiscus, as well as sucrose and, as a preservative, parabens. The concentration of sucrose is in the range of 50%
Structure	(chemical structure of butamirate)
pK_a value	9.4 (calculated)
Solubility	Due to the alkaline character of the compound, at pH values lower than 7.4, it is predominantly present in protonated form. A solubility of more than 1 mg/ml is therefore to be expected in neutral or acidic media
$\log P_{OW}$	3.4 (unloaded)
UV maxima	Due to the molecular structure, UV absorption in the range of 250 nm and in the deep wavelength range <230 nm can be expected
Stability	Due to its molecular structure, the substance in dissolved form should be sufficiently stable at pH values below 7. However, the oxidation of the tertiary amine to the corresponding N-oxide is conceivable

3.4.4.1 Production of Mobile Phases

3.4.4.1.1 Reagents

All reagents used should be as pure as possible. Particular attention must be paid to insoluble components and organic impurities. The former can lead to clogging of filters and columns, whereas organic impurities are often the cause of ghost peaks and baseline disturbances in gradient methods. The water used for the HPLC is expediently freshly prepared on site with a suitable purification system. The most important thing is the rigorous removal of all organic contaminants, which is done for example by reverse osmosis and subsequent purification with suitable absorption cartridges or UV light. It is helpful to check the total organic carbon (TOC). This should typically be less than 5 ppb measured online.

Table 3.4.12 Possible difficulties with the HPLC analysis of butamirate in cough syrup.

Difficulty	Solution
Since the formulation contains extracts from several plants, the presence of numerous secondary plant constituents must be expected. This could lead to problems regarding sufficient separation of the analyte from matrix compounds	To ensure sufficient separation efficiency, a column 150 mm long and packed with 3.5-μm particles is used. In addition, a gradient is applied in order to obtain the best possible resolution over a wide retention range.
Butamirate is relatively highly alkaline because of the ternary amine group, which could lead to peak tailing	A combination of a well deactivated stationary phase (for examples XBridge Shield RP18 and a mobile phase containing trifluoroacetic acid as an ion pairing agent effectively reduces any peak tailing
The expected concentration of butamirate dihydrogen citrate is 1.5 mg/ml. Due to the high content of sucrose, the sample must be diluted by at least factor 25. Based on the molecular structure, no particularly high absorption coefficient is to be expected even at low wavelengths. Therefore, lack of detector sensitivity could be a problem	At a detection wavelength of 220 nm and a flow rate of 0.4 ml with a column diameter of 2.1 mm, sufficient sensitivity is achieved (signal level approximately 100 mAU [milli Absorption Unit])

3.4.4.1.2 Vessels and Bottles

All used vessels and bottles, including caps, are prerinsed with a small amount of HPLC water or the appropriate solvent immediately before use. In this way, any dust, residues from cleaning, or other impurities can be largely removed.

3.4.4.1.3 Measurement of Reagents and Solvent

It is recommended to prepare all solutions and solvent mixtures as far as possible by weighing directly in the solvent bottle. This eliminates the inconvenient handling of volumetric flasks and measuring cylinders and the pouring into the solvent bottle. Also, solvent mixtures are ideally gravimetrically produced, taking into account the density of the individual components. This procedure is much more reproducible and faster than measuring with measuring cylinders. In addition, there is no need to clean these glass devices.

In this way, for example, a 0.01-M phosphoric acid solution can be prepared by pouring 1000 g of water directly into the solvent bottle and adding 1.15 g of phosphoric acid 85%.

Table 3.4.13 HPLC parameters for the analysis of butamirate in cough syrup.

Parameter	Selection	
Column	Waters XBridge Shield C18, 150 mm × 2.1 mm, 3.5 µm	
Mobile phase A	0.1% trifluoroacetic acid in water	
Mobile phase B	0.1% trifluoroacetic acid in acetonitrile	
Flow	0.4 ml/min	
Detection	220 nm	
Temperature	40 °C	
Gradient		
Time (min)	Mobile phase A (%)	Mobile phase B (%)
0.0	90	10
1.0	90	10
16.0	30	70
16.5	95	10
20.0	95	10

3.4.4.1.4 Preparation of Buffer Solutions

In general, the use of buffer solutions whose pH value must be adjusted should be avoided. The preparation of this type of solution is usually quite time consuming. In addition, impurities can easily be introduced by the pH electrode, which can lead to disturbances of the baseline (ghost peaks, humps), etc., especially in analyses with solvent gradients. In many cases, an exact adjustment of the pH value is also unnecessary. If, for example, the use of a phosphate buffer with pH 3.0 is prescribed, the separation often works just as well if 0.01 M phosphoric acid is used. If it is essential to work at precisely defined pH values, it is advantageous to prepare the corresponding buffers in such a way that no adjustment of the pH by adding acids or alkalis is necessary. This is done, for example, by weighing salts and their conjugated acids in the appropriate molar ratios so that the resulting solution automatically has the desired pH value. The corresponding molar proportions can be calculated using the pK_a values. For example, 1 l of a 0.02-M acetate buffer with pH 4.6 can be prepared by adding 0.60 g acetic acid and 1.36 g sodium acetate trihydrate to 1000 g water.

3.4.4.1.5 Filtration of Solvents and Buffer

A controversial topic is the filtration of mobile phases before use. It is undisputed that sooner or later, particulate impurities in mobile phases lead to problems with filters installed in the instrument or with the separation column. However, filtration steps always involve the risk of introducing contaminants from the filter membrane or equipment. Moreover, the filtration step may be quite time consuming. It is therefore often better to skip filtration and focus more on the quality of the reagents.

While solvents and acids are largely unproblematic with regard to particles, inorganic salts such as phosphates may contain small amounts of insoluble components. However, this can easily be checked by filtering 1000 ml of the solution in question through a membrane filter and then microscopically checking the filter for particles. If necessary, the reagent in question must be obtained in better quality or purified by recrystallization.

3.4.4.1.6 Degassing of Mobile Phases

There are also many opposing opinions on the degassing of mobile phases. The fact is that mixtures of aqueous solutions with methanol and acetonitrile tend to form bubbles. This is because the solubility of nitrogen and oxygen in water is much lower than in many organic solvents. When mixtures are made, the solubility of the gases decreases and the excess forms bubbles. Bubbles can lead to problems with HPLC pumps (low-pressure systems) or to spikes in detectors with flow cells. There are various methods for degassing, whereby modern online degassing with vacuum devices is preferable to all other techniques. Less recommended is the frequently practiced degassing in an ultrasonic bath. Besides low efficiency, the formation of small glass splinters was also observed. The degassing with helium, which was also frequently used in the past, is very effective. When changing the solvent bottles, however, there is always the risk that the fresh solvent will be contaminated by residues from the frits. Extensive degassing of mixtures can also change the mixing ratio and Helium is quite expensive. In summary, efficient degassing with suitable well-maintained vacuum equipment is definitely recommended.

3.4.4.2 Blank Samples

It is recommended to prepare a blank sample with each measuring sequence. With a blank sample, potential interference (additional peaks) caused by contamination from the reagents or contamination of the instrument or column can be detected. This can be very important in some circumstances to identify problems and avoid false results for example in the purity testing of drugs or active ingredients. It is not enough to simply fill the sample solvent into a vial and inject it. Rather, the blank sample should represent the entire analysis procedure, including any filtration steps, etc. Only in this way is it possible to detect interferences from insufficiently cleaned glassware or from extractable ingredients from filters or other plastic material. Such a blank sample is generated by simply carrying out the entire analysis procedure without weighing the sample. If a placebo is available (sample without active ingredient), it can also be processed exactly like the sample.

The following chromatograms show the pure solvent (methanol) and a blank sample in which a 100-ml volumetric flask was filled with methanol and filtered with the aid of a plastic syringe (10 ml) through a 0.45-μm syringe filter into an HPLC vial after discarding the first milliliter of the filtrate. In contrast to the pure solvent, the chromatogram (total ion current (TIC), electrospray ionization (ESI) positive) of the blank sample shows numerous peaks representing impurities from the glassware, syringe or filter.

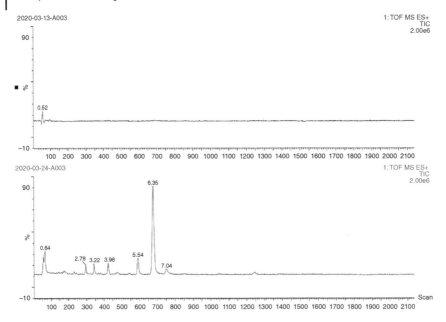

3.4.4.3 Defining Measurement Wavelengths for UV Detection

The following rules apply when determining the detection wavelength:

(1) Detection wavelength should preferably be set to a UV maximum. This increases the robustness of the method, since deviations in the wavelength accuracy of the detector have less effect on the magnitude of the detector signal than if measured in the edges. But sometimes it can be helpful to set the wavelength to a minimum (between two maxima). For example, with highly concentrated solutions, it is possible to avoid measuring outside the linear range.

(2) If a substance has several maxima, it is often advantageous to set the measuring wavelength to the maximum with the highest wavelength, since this is the best way to achieve the selectivity of the method. This can be shown using the example of pendimethalin, a pesticide from the dimethylaniline group. Pendimethalin has two UV maxima at 238 nm and about 440 nm. At 238 nm, the method is relatively unspecific, since numerous other compounds are detected at this wavelength. In contrast, only pendimethalin is detected at 430 nm (Table 3.4.14 and Figure 3.4.12).

3.4.4.4 UV Detection at Low Wavelengths

A prerequisite for optimal UV detection is that the corresponding mobile phase is largely transparent at the selected wavelength. However, at wavelengths below approximately 235 nm, this is no longer the case for many reagents used for mobile phases. In these cases, the following problems may occur:

- Restricted linear range due to high absorption of the mobile phase itself

Table 3.4.14 Structural formula and UV spectrum of pendimethalin.

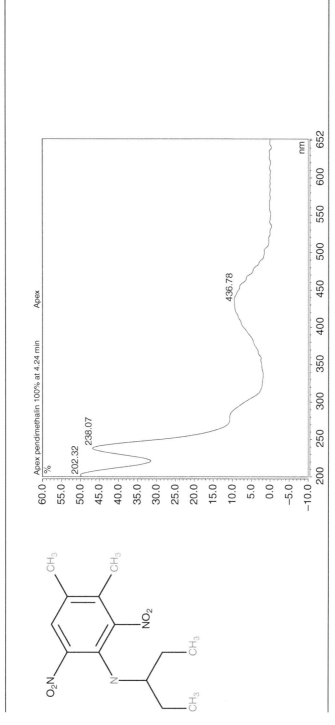

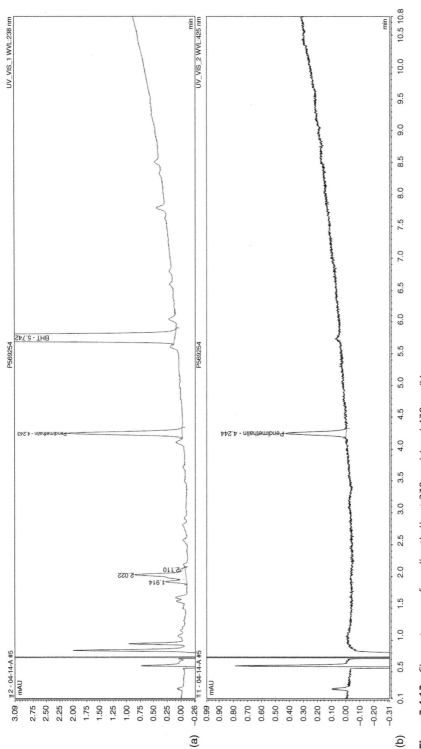

Figure 3.4.12 Chromatogram of pendimethalin at 238 nm (a) and 430 nm (b).

Table 3.4.15 UV cut-off of some solvents.

Solvents (best quality)	UV cut-off (nm)
Water	190
Acetonitrile	190
Hexane	195
Heptane	200
Methanol	205
2-Propanol	205
Methyl *t*-butyl ether	210
THF	212
Dichloromethane	233
Ethyl acetate	256
Acetone	330

Source: Burdick and Jackson.

- Limited sensitivity due to increased baseline noise caused by pump pulsation or small variations in mixing ratios with multichannel pumps
- Pronounced baseline drifts at solvent gradients

There are numerous detection techniques based on principles other than the absorption of UV light, which in principle would be a good alternative to UV detection. Nevertheless, the range of application of these techniques is often very limited. Be it because of the insufficient sensitivity as with RI detectors or the relatively high measured value dispersion as well as the small linear range of ELSD or charged aerosol detection (CAD) devices. Therefore, it is often unavoidable to work with UV detection even at wavelengths below 235 nm. To get the optimum results, there are a few points to keep in mind.

3.4.4.4.1 Solvents

Decisive for the suitability of a solvent for use at low wavelengths is the so-called UV cut-off. The UV cut-off is the wavelength below which the absorption of the relevant solvent in a 1-cm cuvette exceeds 1 AU (Table 3.4.15).

The table shows that the most frequently used solvents for both the reversed phase (water, acetonitrile, methanol) and the normal phase (heptane, 2-propanol) have a UV cut-off at less than 210 nm and are therefore in principle well suited for UV detection at low wavelengths.

The following graph shows the absorption curves of solvents typically used for reversed-phase chromatography (Figure 3.4.13).

While acetonitrile can also be used at 200 nm without hesitation, methanol and THF already show considerable self-absorption below 215 and 230 nm, respectively. Thus, the practical applicability of most solvents is 5–15 nm above the UV cut-off.

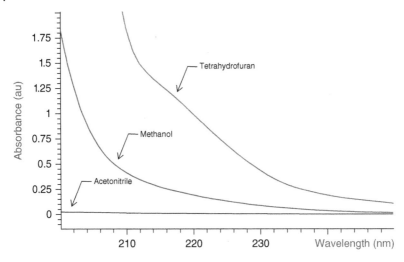

Figure 3.4.13 UV spectra of some solvents (measured against water).

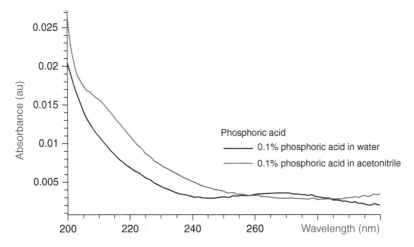

Figure 3.4.14 UV spectra of phosphoric acid.

The UV transparency of a solvent also depends on its purity. Acetonitrile is particularly sensitive in this respect and therefore also quite expensive.

3.4.4.4.2 Acids and Buffer Additives

In many cases, it is necessary to adjust the pH of the mobile phase. This is done by adding acids, bases, or suitable buffer salts. However, many of the substances commonly used for this purpose show considerable absorption of UV light at wavelengths below 235 nm. This lack of UV transparency can considerably limit the range of application of the substances concerned. The following spectra show the UV spectra in the wavelength range 200–300 nm (measured against water) of some frequently used acids and buffer additives (Figures 3.4.14–3.4.17).

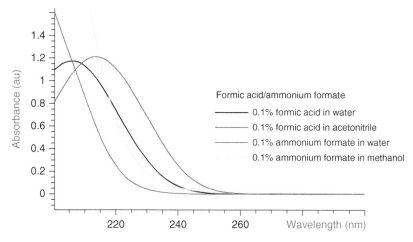

Figure 3.4.15 UV spectra of formic acid and ammonium formate.

Phosphoric acid and its salts show a very good UV transparency even in the low-wavelength range. For this reason, phosphoric acid or phosphate buffers are generally very well suited for detection using UV. However, certain pH ranges cannot be covered by phosphate buffer. Moreover, neither phosphoric acid nor its salts are sufficiently volatile to operate detectors in which the mobile phase has to be nebulized and/or vaporized. This can become problematic if, for example, UV and MS are to be detected one after the other.

In contrast to phosphoric acid, formic acid has distinct absorption maxima in the range of 210–220 nm, depending on the solvent (Figure 3.4.15). As a result, mobile phases containing formic acid are only conditionally suitable or not suitable for detection wavelengths below 240 nm. Interestingly, however, the maxima of ammonium formate solution are at lower wavelengths, so that at least with the aqueous solution it is still possible to work well at wavelengths around 220 nm. Ammonium formate is only slightly soluble in acetonitrile; therefore, the spectrum of a methanolic solution is shown here. In this respect, it should be noted that formic acid dissolved in pure methanol is also frequently used. However, experience shows that such solutions are not stable. Dimethyl ether is formed after a short time, which can lead to different retention times. However, the addition of 10% water can prevent this decomposition.

Acetic acid and ammonium acetate show similar spectra to formic acid and ammonium formate, respectively (Figure 3.4.16). However, the absorption maxima are about 5 nm lower. Thus, acetic acid and ammonium acetate are somewhat better suited for detection in the low-wavelength range than formic acid and ammonium formate, respectively.

In contrast to formic acid and acetic acid, trifluoroacetic acid, which is also frequently used, has relatively favorable properties with regard to UV transparency (Figure 3.4.17).

A solution in acetonitrile shows a UV maximum at 220 nm. However, the absorption is significantly weaker than with formic or acetic acid with comparable

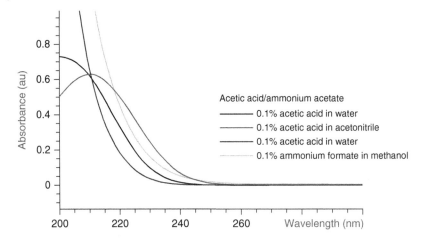

Figure 3.4.16 UV-spectra of acetic acid and ammonium acetate.

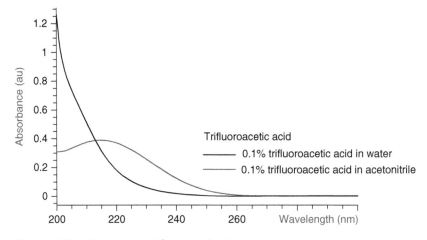

Figure 3.4.17 UV spectrum trifluoroacetic acid.

concentrations. Therefore, trifluoroacetic acid up to wavelengths around 215 nm can generally be used without any problems.

3.4.4.4.3 Drift at Solvent Gradients

While a lack of UV transparency of the mobile phases in isocratic separation methods primarily leads to a limited linear range and increased baseline noise, a further problem can arise with solvent gradients. If the UV transmittance of the two mobile phases differs considerably, this leads to strongly rising or strongly falling baselines. Under certain circumstances, this may result in poorer reproducibility for quantitative measurements, because it is more difficult for the integration algorithm of the evaluation software to calculate the start and end of the peak. This problem can be solved, or at least reduced, by skillful selection of the measuring wavelength (or at most variation of the concentrations of the additives to the

3.4.4 General Tips for Optimizing HPLC Methods | 295

Table 3.4.16 HPLC parameters for gradient method with trifluoroacetic acid.

Parameter	Setting	
Column	Nucleosil 120-3 C18, 125 mm × 4 mm	
Mobile phase A	0.1% trifluoroacetic acid in water	
Mobile phase B	0.1% trifluoroacetic acid in acetonitrile	
Flow	1 ml/min	
Gradient		
Time (min)	Mobile phase A (%)	Mobile phase B (%)
0.0	90	10
0.5	90	10
20.5	10	90
21.5	90	10
25.0	90	10

individual mobile phases). The following figure shows the chromatograms at different wavelengths of a typical gradient with mobile phases containing 0.1% trifluoroacetic acid (Figure 3.4.18).

An aqueous solution of 0.1% trifluoroacetic acid has a higher absorption below about 216 nm than a solution of the same concentration in acetonitrile. For wavelengths greater than 216 nm, the opposite is true. This results in a decreasing baseline at wavelengths, while wavelengths above 216 nm result in a baseline increase. At 216 nm, however, a practically horizontal baseline is achieved. When working with trifluoroacetic acid, a wavelength of approximately 216 nm is therefore well suited. The self-absorption of the mobile phase is still within limits, numerous molecules show very sensitive detection at this wavelength, and no significant baseline drift is to be expected with gradients. If one wishes to measure at slightly different wavelengths, the concentrations of trifluoroacetic acid in the two mobile phases can be varied so that the same compensation effect is achieved (Table 3.4.16).

3.4.4.5 Avoidance of Peak Tailing

Although steady improvements in stationary phases have greatly reduced undesired secondary interactions between analyte and column material that can lead to peak tailing, techniques for avoiding tailing or other peak deformations have by no means become obsolete. Alkaline compounds in particular can still be a problem. In practice, it has been shown that even using modern stationary phases, alkaline substances with pK_a above 7 tend to peak tailing to a disturbing extent. To avoid this problem, there are numerous measures that have different effects. The following step-by-step procedure has proven to be effective for reversed-phase chromatography (Table 3.4.17).

An example will be used to show the effects of the individual steps. For this purpose, benzylamine (pK_a 9.4) was chromatographed on a modern reversed phase

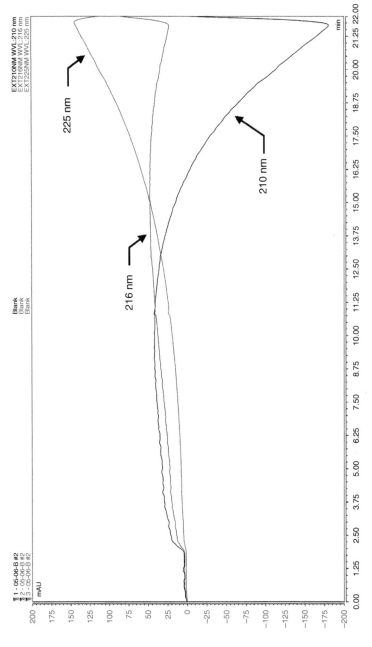

Figure 3.4.18 Chromatogram of a blank injection at different wavelengths.

Table 3.4.17 Step-by-step procedure to reduce peak tailing.

Step	Technology	Explanation
1	Acidification of the mobile phase (0.1% phosphoric or formic acid)	The mobile phase is acidified by adding phosphoric or formic acid (approximately 0.1% each). This suppresses interactions of the now completely protonated base with free (acidic) silanol groups of the stationary phase
2	Addition of trifluoroacetic acid (approximately 0.05–0.1%) to the mobile phase	With the addition of trifluoroacetic acid, secondary interactions with the stationary phase are suppressed due to protonation of most alkaline substances on the one hand, and on the other hand the trifluoroacetation forms an ion pair with the protonated bases, which significantly reduces the polarity of the compound and improves the peak shape
3	Addition of an ion pairing agent to the mobile phase	If the analyte is available in ionic form, ion pairs can be formed by adding suitable reagents, which are much less polar than the original analyte. In addition, ion pairing reduces the tendency of the acidic or alkaline groups of the target molecule to interact with the stationary phase. There are numerous suitable substances that can be used as ion-pair formers for HPLC. Sulfonic acid salts are often used for bases and quaternary amines for acids. However, the pH of the mobile phase must be adjusted so that the analyte is present as an ion (protonated or deprotonated)
4	Use of alkaline mobile phases (pH 8–12)	Whereas older stationary phases based on silica gel were only sufficiently stable in the range of pH 2–7, modern materials can be used up to pH 12. This allows bases with a pK_a of up to about 10 to be completely deprotonated and thus be uncharged, which effectively suppresses secondary interactions with active groups of the stationary phase

separation material with different mobile phases and the plate number and tailing factor (calculated according to the method of the American Pharmacopoeia USP) were determined.

Chromatography conditions	
Column	XBridge Shield RP18 150 × 3 mm, 5 µm
Flow	0.6 ml/min
Temperature	40 °C
Injection	1 µl (1 mg/ml benzylamine dissolved in acetonitrile/water 1 : 1)

Results (Table 3.4.18).
Chromatograms (Figures 3.4.19 and 3.4.20)

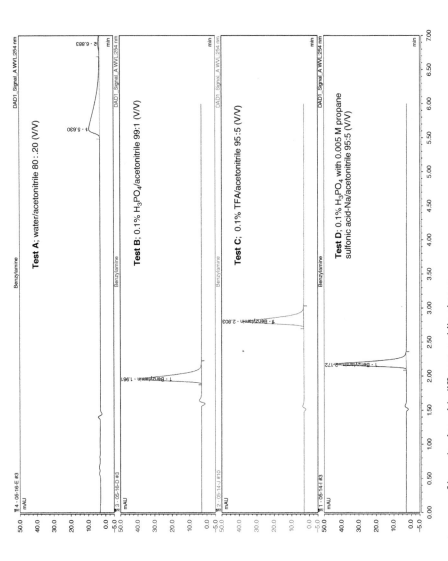

Figure 3.4.19 Chromatograms of benzylamine with different mobile phases.

Table 3.4.18 Results plate number and USP tailing with different mobile phases.

Trial	Mobile phase	Plate number	USP tailing
A	Water/acetonitrile 80 : 20 (% V/V)	1068	5.06
B	0.1% phosphoric acid/acetonitrile 99 : 1 (V/V)	2720	1.68
C	0.1% trifluoroacetic acid/acetonitrile 95 : 5 (V/V)	5358	1.55
D	0.1% phosphoric acid with 0.005 M propane sulfonic acid-Na/acetonitrile 95 : 5 (V/V)	4767	1.54
E	0.01 M ammonium hydrogen carbonate pH 10/acetonitrile 80 : 20 (V/V)	517	0.91
F	0.1% ammonia (approximately pH 11)/acetonitrile 80 : 20 (V/V)	7995	1.51

Test A; water/acetonitrile 80 : 20 (V/V)

Even with a very modern stationary phase, which is based on organic–inorganic hybrid particles and has embedded carbamate groups to shield any residual silanols, benzylamine shows very strong tailing using only a water/acetonitrile mixture as mobile phase.

Test B; 0.1% phosphoric acid/acetonitrile 99 : 1 (V/V)

The addition of phosphoric acid results in a marked improvement in the peak shape and the plate number. However, protonation also increases the polarity of the molecule, so that the acetonitrile content must be greatly reduced.

Test C; 0.1% trifluoroacetic acid/acetonitrile 95 : 5 (V/V)

A further improvement is achieved with trifluoroacetic acid. Both the plate number and the tailing are positively influenced. The formation of ion pairs between the trifluoroacetate anion and the protonated benzylamine reduces the polarity of the analyte, which is reflected in increased retention.

Test D; 0.1% H_3PO_4 with 0.005 M propane sulfonic acid-Na/acetonitrile 95 : 5 (V/V)

Due to the pentane sulfonic acid, a marked improvement in the plate number and tailing is observed compared to test B. Like trifluoroacetic acid, the pentane sulfonic acid anion forms an ion pair with the analyte. Retention could be further increased with longer-chain ion pairs such as heptane sulfonic acid.

Test E; 0.01 M ammonium carbonate pH 10/acetonitrile 80 : 20 (V/V)

Increasing the pH value to 10 produces a much better peak shape. However, the peak is very wide and therefore the bottom number is low. This phenomenon is sometimes observed when the pH of the mobile phase is in the range of the pK_a value of the analyte.

Test F; 0.1% ammonia pH 11/acetonitrile 80 : 20 (V/V)

The best values regarding plate number and peak tailing are obtained with 0.1% ammonia solution at pH 11. With the appropriate stable column material, even strongly alkaline mobile phases can be used without problems in routine analysis. The disadvantage is, however, that depending on the make, the windows

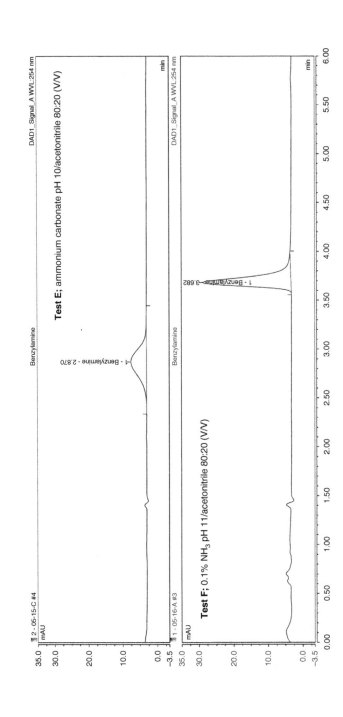

Figure 3.4.20 Chromatograms of benzylamine with alkaline mobile phases.

of measuring cells for UV detectors can only be used up to approximately pH 10. If routine work at higher pH values is to be carried out, it is recommended that detector cells specified for this range be obtained. Furthermore, many compounds are not very stable in the alkaline solvents, which can lead to decomposition even during chromatographic separation.

Further notes on the subject of peak tailing:

Less critical with regard to tailing are acidic substances. With carboxylic acids, it is often sufficient to acidify the mobile phase. As a result, both the acidic silanol groups and the analyte itself are protonated and are therefore uncharged, which effectively prevents undesired interactions. Well suited as mobile phases are 1% acetic acid or 0.1% phosphoric acid. It was also observed that chromatography of carboxylic acids with methanol often yields better peak shapes than with acetonitrile. In contrast to bases, ion-pair reagents are mainly used for the analysis of acids to increase retention. For example, it is possible to improve the retention of very polar ascorbic acid ($\log P_{OW}$ −1.88) with long-chain ion-pair reagents such as hexadecane trimethylammonium bromide in such a way that reversed-phase chromatography is possible without any problems.

Trifluoroacetic acid is almost a universal aid for improving the peak form of bases and is also suitable for screening methods where detection is performed both in deep UV and by MS due to its volatility and good UV transparency. Trifluoroacetic acid forms strong ion pairs, which could theoretically lead to difficulties in ionizing these compounds for MS detection. However, this problem has never been observed by the author when working with electro spray in positive mode. If only UV detection is used, methanesulfonic acid is a very good alternative to trifluoroacetic acid.

For the alkaline range (pH 8 to approximately pH 10.5), ammonium hydrogen carbonate is very suitable. It is compatible for MS detection and has a UV cut-off of 190 nm.

In addition to molecules with amine groups, substances with highly complexing properties often show peak deformations. This problem is caused by traces of heavy metals in the stationary phase or insufficiently inert metal surfaces (capillaries, columns, etc.). The following measures will remedy the situation:

(1) Acidification of the mobile phase (the complexation of metals by organic molecules is often weaker at low pH values).
(2) Addition of phosphates to the mobile phase
    Phosphates also have complexing properties and can thus suppress or attenuate the interactions of the analyte.
(3) Acid rinsing
    If acid-stable stationary phases are used (e.g. Zorbax Stable Bond), the complete HPLC device, including column, can be rinsed with a mixture of 1% phosphoric acid/acetonitrile 90 : 10 for 24 hours. This removes a large part of any heavy metals contained.
(4) Addition of strong complex reagents to the mobile phase.
    Ethylenediaminetetraacetic acid (EDTA), acetylacetone, etc. are able to complex interfering metals and thus suppress interactions with the analyte. An addition

of 0.0025 to 0.005 M medronic acid (methylene diphosphonic acid, (CAS no. 1984-15-2) can also help.
(5) Inert materials
Capillaries made of polyetheretherketone (PEEK) are often a good choice. Problems are sometimes caused by corroded surfaces of metal capillaries or injection needles. High-purity stationary phases (possibly also polymer-based) and reagents, as well as metal-free capillary compounds are very helpful
(6) Analysis of the target compound as a metal complex
Strong complexing agents can be analyzed as heavy metal complexes. For example, it is possible to chromatograph EDTA as a copper or iron complex after the addition of Cu^{2+} or Fe^{3+}.

3.4.4.6 Measurement Uncertainty and Method Design

In many cases, a sample has to be tested against specifications that have already been defined in advance. Typically, this is a content specification for an ingredient or a limit for an impurity or contaminant (substance that has unintentionally entered the sample). From this information, the requirements for the method with regard to the measurement uncertainty can be derived directly. If no specification information is available, this estimation can be made based on the intended use of the method.

For example, if a method is intended for the determination of the content of an active substance in a pharmaceutical preparation, it can be assumed that the tolerance range for the concentration of this substance is within ±5% of the target value. To allow a statistically sound evaluation of the result, the relative measurement uncertainty should therefore be less than 2%. This measurement uncertainty includes all systematic and random errors that occur during the preparation and HPLC measurement of the standard and sample solution. It is therefore easy to see that the design of a method to be used for the described purpose must be done very carefully to keep the measurement uncertainty as small as possible. For this purpose, the entire analysis from sample handling to the calculation of the results must be taken into account and all analysis steps must be coordinated as well as possible.

3.4.4.6.1 Weighing in or Measuring

The weights or volumes of both the reference substance and the sample should be as large as possible to keep the weighing or pipetting error small and to compensate for small inhomogeneities within the sample.

For volumetric dosing, volumes below 1 ml should be avoided. If samples are measured volumetrically, pipettes made of glass are preferable to the frequently used piston–stroke pipettes. Piston–stroke pipettes have a very good reproducibility for aqueous solutions, but their tolerances for the absolute deviations (deviations from the nominal value) are comparatively high. Piston–stroke pipettes are definitely not suitable for organic solvents and viscous samples such as vegetable oils. However, pipettes based on the displacement principle and using syringe-like tips are becoming more and more available. Such devices represent a significant improvement over piston–stroke pipettes in terms of correct dosing.

Weighing in of substances is not always unproblematic either. Again and again, one observes drifting results due to volatile compounds or absorption of water from the ambient air. If you select the weights large enough and set the weighing parameters optimally, the errors associated with these problems are usually small and negligible. If fine powdery substances are weighed, static electricity can also cause considerable deviations. This can be remedied by appropriate discharging devices or by increasing the relative humidity to over 40%. When weighing into volumetric flasks with volumes of more than 50 ml, it is advantageous to use weighing boats. Although modern balances can cope well with high tare values (heavy vessels), handling large volumetric flasks on the balance always involves the risk of weighing errors. This may be due to small temperature differences of the volumetric flask and balance or due to the large surface area for air flows. In addition, weighing in weighing boats is much more convenient than circulating substances through narrow bottlenecks. Very practical are small weighing boats made of solid aluminum foil or thin glass, which can be placed directly into the volumetric flasks after weighing. This avoids any losses due to the transfer (insufficient rinsing) and the whole thing is very time saving. Since these weighing scoops have only a very small volume, the influence on the nominal volume of the volumetric flask is usually negligible. If necessary, this error can also be compensated for by calculation. It should be noted, however, that weighing boats made of aluminum are truly frayed by ultrasonic baths, which makes a filtration step necessary. In addition, they are partially dissolved in strongly acidic or alkaline media, which can be undesirable depending on the analyte.

3.4.4.6.2 Dilutions

For HPLC analyses, the reference substance and the sample are dissolved in a suitable solvent and the solution is adjusted to a defined volume. However, dilution steps (measuring a defined volume of the stock solution and diluting it with a suitable solvent) should be avoided as far as possible, as these operations are often associated with relatively large errors. A typical dilution step almost always causes a relative error in the range of at least 0.5% and is also quite time consuming.

Weighing and dilutions must always be considered as a whole, since the total error in preparing the solutions is made up of the weighing error and the errors in dissolution to a defined volume and any dilutions. In many cases, it has therefore proved to be best if reference and sample solutions are prepared in one step (if possible, without dilutions). Of course, this procedure may involve a rather large consumption of solvents. On the other hand, however, there are time savings and a generally much lower measurement uncertainty. It is often helpful to estimate the errors of the individual steps (for example, based on the error tolerances of the devices used) and to calculate different variants. It is also possible to take into account the costs of solvents and the amount of work involved to find an optimal solution for the respective analysis.

3.4.4.6.3 HPLC Analysis

Another part of the overall measurement uncertainty of a method is the HPLC determination. Here too, measurement errors can be minimized by skillful selection of the measurement parameters. Important sizes are:

Peak Shape/Separation The basis for a good reproducibility of the peak areas is a symmetrical signal that is as sharp as possible and completely separated from all interference peaks (from the matrix or the reagents). If at all possible, a resolution of at least 2.0 should be aimed for, to avoid problems regarding robustness of the method. Experience shows that critical separations must always be tested on two columns with separation materials from different production batches. It is also very helpful to check the influence of slight variations in the composition of the mobile phase or the temperatures on the resolution.

Detection The detection principle has a great influence on the reproducibility of a measurement. Experience has shown that detection techniques in which the liquid eluent flows continuously through a measuring cell (UV, RI, fluorescence detection (FLD)) provide significantly more reproducible measured values than methods such as ELSD, CAD, or MS, which require spray formation. It is important in all procedures to optimize the settings so that the best possible reproducibility is achieved. This means, among other things, that the concentration or quantity of the analyte should be adjusted to the optimal range of the respective detector. Working in the upper- or lower-measuring range of the detector is always associated with limitations in terms of reproducibility and possibly further problems. It is also important to know and use the technical possibilities of the detector and to choose an optimal acquisition rate or response time. Although smoothing of chromatographic signals is frowned upon by many people, these mathematical functions are, if used with care, quite helpful and useful.

Calibration–Calibration Model The influence of the applied calibration model or the calibration technique on the measurement uncertainty should not be underestimated. Detectors that show a linear dependence of the signal magnitude on the concentration are unproblematic. These include UV, refractive index detector (RID), and FLD. If one knows approximately the expected value of the sample and does not work in the extremes of the detector range, a one-point calibration is completely sufficient. It is more difficult with detectors that do not work linearly. The following rules of thumb apply here (which are of course also helpful for linear functions):

- At least five calibration points.
- Do not select a too wide calibration range. As a rule, 50–150% of the expected value is completely sufficient.
- Do not extrapolate.
- If a very large calibration range is required, increase the number of calibration points and work with suitable weighting functions.

- A suitable calibration model can be selected with the modern software packages. Mathematical operations to obtain a linear function (logarithms) are not necessary and do not give better results.

3.4.4.6.4 Internal Standards

While internal standards are indispensable in trace analysis to compensate for all errors caused by the often-complex sample preparation and the measurement of very small amounts of substances, they are rarely used in classical HPLC analysis. Since the reproducibility of HPLC measurements is usually very good and sample preparation involves only a few steps, there is no need to work with internal standards. Moreover, the addition of a precisely defined quantity of an internal standard is always associated with a mistake. However, there are cases where the use of an appropriate internal standard can be useful. This is the case, for example, when extensive dilution steps cannot be avoided or when very precise measurements are required. In the latter case, however, the internal standard must then be added gravimetrically (by weighing). Internal standards are also useful if sample preparation is to be automated and only small volumes can be processed or if precise dosing of the solvent and exact dilutions are avoided to save time.

3.4.4.7 Column Dimension and Particle Sizes

An important parameter in the development of an HPLC method is the column diameter and the particle size of the stationary phase. These two parameters offer a lot of room for optimization. In recent years, great progress has been made in the development of stationary phases. Almost all manufacturers now offer numerous materials with particle sizes below 2 µm. These small particles offer numerous advantages. Thus, the separation performance is much better with comparable column lengths and, in addition, higher linear flow velocities are possible without deteriorating the resolution. In addition, the trend is toward smaller column diameters. This allows to work with much lower flow rates, which reduces the consumption of solvents massively and increases the sensitivity of concentration-sensitive detectors such as UV significantly. While until a few years ago columns with a diameter of 4.6 mm and particle sizes of 5 or 10 µm were standard, columns with a diameter of 2 mm with particles smaller than 2 µm are increasingly being used. This configuration undeniably offers many advantages where sensitivity combined with good resolution and high throughput are required. However, there are also some disadvantages. In addition to the greater demands on equipment (dead volumes, compressive strength), thinner columns have a considerably lower loading capacity, because smaller quantities of separation material can be filled in for the same lengths. This means that either smaller injection volumes are used or that the reference and sample solutions must be diluted more strongly. Both increase the measurement uncertainty of the method. Due to the high demands on the instrument technology, a lower robustness of the methods cannot be excluded. Therefore, depending on the task at hand, it is often still useful to work with column diameters of 4.0 or even 4.6 mm. It is ideal when identical separation material is available in different particle sizes

Table 3.4.19 Conditions cannabinoid analysis.

Parameter	HPLC-DAD	UHPLC–HRMS
Column	Cortecs Shield RP18, 100 mm × 4.6 mm, 2.7 µm	Cortecs Shield RP18, 100 mm × 2.1 mm, 1.6 µm
Column temperature (°C)	40	40
Flow (ml/min)	1.5	0.35
Mobile phase	Acetonitrile/0.01 M phosphoric acid 60 : 40 (V/V)	Acetonitrile/0.1% formic acid 60 : 40 (V/V)
Detection	UV at 210 nm	HRMS, ESI positive (base peak intensity [BPI])

and different column formats. This scalability can be very helpful to find optimal solutions for different tasks with the same analyte.

This can be shown by the example of a stability study of a cannabinoid. Routine testing is carried out using classical HPLC. The HPLC column has a diameter of 4.6 mm. After several months of storage of the active substance, several unknown degradation products were detected. To obtain more detailed information on the possible identity of these degradation products, further investigations were carried out using UHPLC coupled with a high-resolution mass spectrometer. Because the separation material used is also available in a typical UHPLC format, method transfer was very easy. Identical chromatographic profiles were obtained with both methods, which make the task much easier (Table 3.4.19; Figures 3.4.21 and 3.4.22).

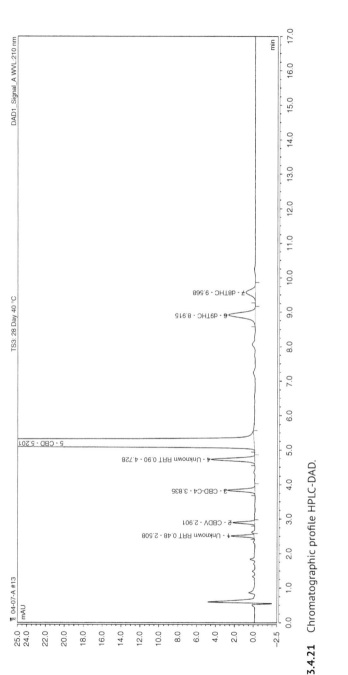

Figure 3.4.21 Chromatographic profile HPLC-DAD.

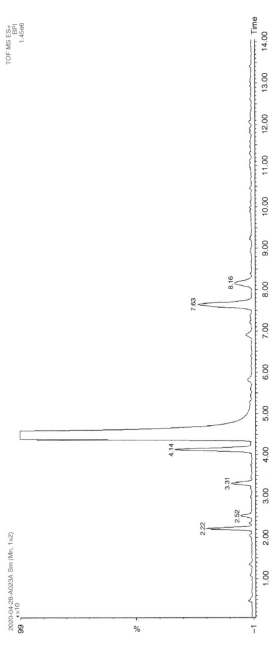

Figure 3.4.22 Chromatographic profile UHPLC–HRMS (BPI, ESI positive).

Reference

1 Woodward, R.B. (1942). Structure and absorption spectra. III. normal conjugated dienes. *American Chemical Society* 64 (1): 72–75.

Part IV

Current Challenges for HPLC Equipment Suppliers

4.1

Optimization Strategies with your HPLC – Agilent Technologies

Jens Trafkowski

Product Specialist HPLC, Agilent Technologies Switzerland GmbH, 4052 Basel, Switzerland

The last two decades have seen many developments in liquid chromatography, and the readers of this books are aware that these developments cannot be reduced only to the important factor of increasing the system pressure to more than 1000 bar. Agilent has contributed to these developments in many ways; in this chapter, we will discuss the dedicated possibilities to get the best out of an Agilent high performance liquid chromatography (HPLC) and ultrahigh performance liquid chromatography (UHPLC) system.

Increasing performance means something different to different people. Asking experienced system users, you often hear the wish to increase sensitivity or selectivity. But these two generic expressions can be split into various subtopics, and often also other aspects play a role, when looking at the performance of a single HPLC system or a complete HPLC lab. Direct instrument users and responsible managers are mainly interested in:

– Absolute sensitivity
– Separation performance and selectivity
– Linearity
– Simplicity
– Speed of analysis, throughput, and the time to results
– Width of analytical possibilities
– Controllable operational costs

These different points are addressed by different technical tools, and besides the increase at the high end of the performance scale, the overall performance of an HPLC laboratory can be heavily determined by increasing the performance of the standard systems. Many of the next possibilities will also or especially address the optimization of the "basic" (U)HPLC systems.

Optimization in HPLC: Concepts and Strategies, First Edition. Edited by Stavros Kromidas.
© 2021 WILEY-VCH GmbH. Published 2021 by WILEY-VCH GmbH.

4.1.1 Increase the Absolute Separation Performance: Zero Dead-Volume Fittings

Sometimes small changes have huge effects. When looking into a complete LC system, the dead volume can become an important factor. Agilent InfinityLab Quick Connect UHPLC Column Fittings deliver truly dead-volume-free fluidic connections to any HPLC column, independent from the vendor. As the well-known different depths of receiving port connections in different HPLC columns can lead to insufficient fitting [1], these fittings are designed to avoid exactly these potential causes of errors. The spring-loaded design is truly delay-volume free and enables to connect different column types 200 times and more. Like this, problems with broken parts (e.g. overscrewed column fittings), false results and instrument downtime are minimized.

These fittings ensure the reliable dead-volume free fit, whenever changing the columns, and ensure that no leaks will occur after unconnecting and reconnecting. *Source:* Modified from Dwight R. Stoll (2018), Fittings and Connections for Liquid Chromatography—So Many Choices!, (May 1, 2018): 304–311.

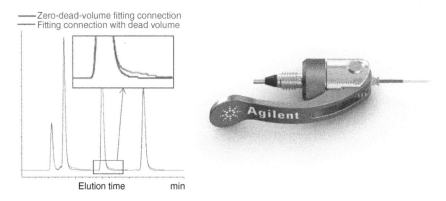

4.1.2 Separation Performance: Minimizing the Dispersion

As the 1290 Infinity II LC is designed for a large power range of 1300 bar and up to 5 ml/min flow rate, many applications are running with low-flow rates and narrow-bore columns, where the dispersion can become a more and more important factor, especially in isocratic analysis. To address also these isocratic applications at low flow rates, Agilent has developed a dedicated Ultra-Low Dispersion Kit to minimize the capillary volumes before and after the column. With the overall optimized system dispersion achieving narrower and higher peaks, resolution and sensitivity will be enhanced.

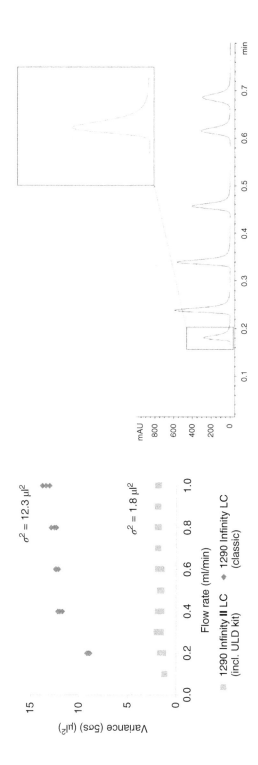

4.1.3 Increasing the Throughput – Different Ways to Lower the Turnaround Time

For many laboratories, sample throughput is of highest interest. The pure analysis time is only one factor, which is determined by the separation time on the column, but the throughput can be further enhanced with different techniques:

- *Alternating column regeneration* enables the reconditioning of a second column, while the other column is in use. For this, an easy setup with an additional valve and an isocratic pump is required to condition the column for the next injection. A simple 2/10 valve can be used, but the best choice is the 2D-LC valve with two absolute symmetrical flow paths.

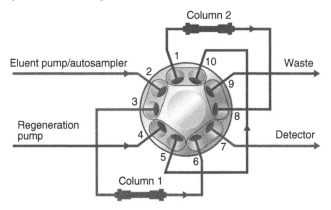

- *Dual-needle multisampler for ultrafast injection cycles:* If the pure injection time has already reached its minimum, the use of a second injection needle enables overlapping steps with the alternating needle. Like this the first needle can be cleaned, while the second one is running the injection of the next sample. So, while the actual run is not ready, the next run can already be prepared resulting in injection times down to five seconds.

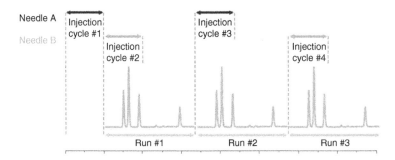

- *StreamSelect – the ultimate use of a mass spectrometer with different LC runs in parallel:* If a high-end mass spectrometer shall be occupied as much as possible, StreamSelect allows overlapping runs of up to four LC systems and detecting them in a single mass spectrometer. With this technique "dead times" in the mass spectrometer, when the peaks have not eluted, when the LC separation is already over and/or the column equilibration starts, can be reduced to a minimum, and the use of the mass spectrometer is maximized.

- *High capacity for thousands of samples with the Multisampler:* For high-speed analysis with analysis times below one minute, the sample capacity of the instrument can become the limiting factor. The flexibility of the Multisampler allows the usage of up to 8 drawers hosting 16-well plates in a single stack. With this setup and 1532 samples in 96-well format and 6144 samples in 384-well format can be subsequently analyzed, e.g. for a long sample set over a weekend.

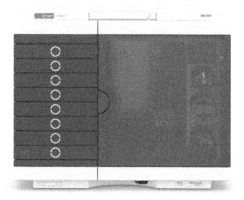

4.1.4 Minimum Carryover for Trace Analysis: Multiwash

A limiting factor for the overall system sensitivity can be the carryover. Multiwash enables the use of three different external solvents and a washing procedure, including the inner and outer needle wash, as well as the active back flushing of the needle seat, reducing carryover to 9 ppm (specified for Chlorhexidine, typically it is much lower). This can be a deciding factor for accurate and precise analysis of small traces or residues.

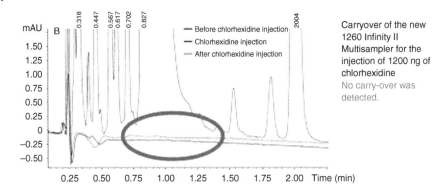

Carryover of the new 1260 Infinity II Multisampler for the injection of 1200 ng of chlorhexidine
No carry-over was detected.

4.1.5 Increase the Performance of What you have got – Modular or Stepwise Upgrade of Existing Systems

It is always possible to add special features or enhance the performance of an existing system with the exchange or the addition of newer modules. As Agilent offers a variety of different modules, a stepwise upgrade can enhance the performance in smaller or larger steps:

- Exchange the pump and autosampler to ramp up the pressure range from 400 to 600 bar, even 800 or 1300 bar. This allows the next step toward high-end UHPLC.
- Gain sensitivity with a new DAD with MaxLight technology for maximum light absorption and minimum peak dispersion. Just exchange the detector and profit already from this gain.
- Upgrade your system with different valves to one of the various valve-based solutions, such as:

    o Method development system
    o Multimethod system
    o Online SPE system
    o 2DLC system
    o SFC–UHPLC hybrid system

- Automize method transfer by emulating existing HPLC systems
- Increase linearity using HDR

Base of the easiest upgradeability is always the modular design of Agilent's LC, enabling to mix systems out of different modules, also from different system generations. So, even an existing Agilent 1100 from the 1990s can be upgraded to a full working 2D-LC system. The system upgrades to different solutions will be described in the next section.

4.1.6 Increase Automation, Ease of Use, and Reproducibility with the Features of a High-End Quaternary UHPLC Pump

Automation can increase work performance, also in the lab. Besides delivering more reliable and repeatable results, it can also facilitate a laboratory scientist's work by taking over manual tasks.

One of the (often daily) tasks of an HPLC scientist is the preparation of eluents. Depending on the method, various eluents must be prepared, in RP–HPLC mainly based on water, methanol, and acetonitrile, often adjusted with a limited choice of additive like buffer salts and/or acids. In many labs, often the same additives are used, but in different concentrations, leading to a choice of different prepared eluents, which are only used for dedicated applications. If one of the applications is not often used, but requires a dedicated buffer or acid concentration, the eluent is just prepared for this application and disposed, when the analyses are over.

Quaternary pumps in general offer the possibility to blend an eluent out of the pure solvent and a stock solution, to realize the required additive concentration. Here, the two tasks are

(1) the calculation of the quaternary gradient from two stock and two solvent solutions to achieve the binary gradient,
(2) the precision and accuracy for the solvent mixing to be highly accurate, so that the solvents are mixed together always correctly.

The design goal of the Agilent 1290 Infinity II Flexible Pump was to deliver the utmost performance in terms of accuracy and precision, enabling the pump to deliver always the right amount of solvent, also of the critical stock solution, which is pumped in much smaller amount [2].

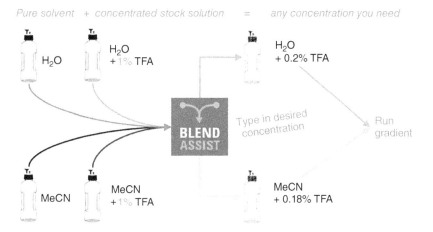

Furthermore, the easy software tool BlendAssist has been implemented to facilitate the operation by a simple interface, where the user just has to type in the solvents and the concentrations of the stock solution and can then just program a simple binary gradient as it is given in that method.

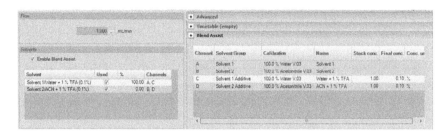

The combination of the pump's high accuracy and precision together with a simple programming interface, which is available in any software controlling this pump, makes it also a very valuable task for method development, when additive concentrations can be a possibly deciding method parameter.

Even without the use of a software tool, facilitating the interface for the solvent mixing, it can be of high interest to use the good mixing performance to achieve more reliable gradients and even isocratic runs. Agilent has shown this with two technical notes, in which the repeatability of analysis with manually premixed solvents have been compared to the mixing performed by the quaternary pump [3]. The picture below illustrates the differences in terms of retention time deviation between the pump-mixed (dynamically) and manually (premixed) mobile phases.

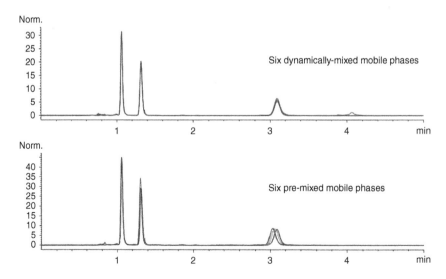

As described at the beginning of this section, also in this case, automatization can not only make life of a chromatogapher easier, but also lead to better results.

Table 4.1.1

Autosampler	Calibration means	Calibration function	Correlation R^2	Deviation control sample (%)
Agilent 1260 infinity II multisampler with 100 μl analytical head and 100 μl loop	Manual Autosampler	$y = 27.85^*x + 0.53$ $y = 27.64^*x - 0.18$	0.99999 1.00000	1.08
Agilent 1260 infinity II Vialsampler with 100 μl analytical head and 100 μl loop	Manual Autosampler	$y = 27.76^*x + 0.43$ $y = 27.65^*x - 0.73$	0.99999 0.99999	0.79
Agilent 1260 infinity II multisampler with 900 μl analytical head and 900 μl loop	Manual Autosampler	$y = 27.69^*x + 0.54$ $y = 27.53^*x + 0.02$	0.99999 1.00000	0.72

4.1.7 Increase Automation: Let your Autosampler do the Job

Another manual step in quantitative analytical workflows is the pipetting of samples. Just a simple calibration curve requires time for preparation, manual work, and is a potential source of errors. Since launch of the Agilent 1100 HPLC system, autosamplers can already do pipetting steps in the module, but with the requests for more automatization also this possibility is getting more interest. It has been described in a Technical Overview [4], leading to comparable, maybe even slightly better results than manual pipetting, as shown in the Table 4.1.1.

4.1.8 Use Your System for Multiple Purposes: Multimethod and Method Development Systems

Every system can be used for different HPLC methods with different parameters: the gradient itself can be programmed in the method, but the column and/or the solvents require hardware changes. This manual process does not cost a big amount of time, but the need for the manual interaction prevents the seamless use of different analysis. An easy way to overcome this problem is the use of valves to increase the automation.

Agilent Technologies offers different valves for column selection, starting with 2-position valves for 2 different columns and going up to 8-column selection valves. Furthermore, every solvent channel can be equipped with a solvent selection valve, offering the choice of up to 12 different solvents. With this setup, different methods with different physical conditions (columns and solvents) can be run unattended, e.g. overnight or over a weekend.

The setup is simple – just upgrade your system with a valve and the corresponding valve drive, if you don't have it already (e.g. built in into the column oven), and you can use the system with a higher flexibility.

With the described hardware setup for different columns and different solvents, the system can be upgraded to a true method development system, optimizing the LC method with various combinations of gradients, temperatures, columns, and eluents. To support this operation and deliver a good data analysis, which method suits best for your separation task, Agilent has developed the Method Scouting Wizard. In this software you just determine how the different parameters shall be changed, the software just runs all the different combinations and sums up the data. Different reporting tools enable you a fast data analysis, based on the most important criteria for your dedicated separation, whether it is the resolution or the number of peaks. A helicopter view on the different runs illustrates rapidly the good combinations for the best separation.

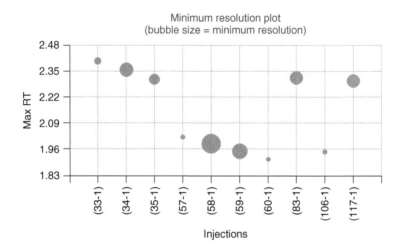

4.1.9 Combine Sample Preparation with LC Analysis: Online SPE

A time-determining and time-consuming step in the whole analysis of a sample is the sample preparation. For various tasks, robotic systems have already taken over this important step, which has been done manually for years. When it is not about a big number of samples with shortest analysis time, online SPE in an HPLC system can be an attractive alternative.

For online SPE, you can add existing parts of the portfolio in and program your method accordingly. A system needs to be upgraded with an additional pump to deliver the solvents for the solid-phase extraction and at least one valve for the extraction functionalities itself.

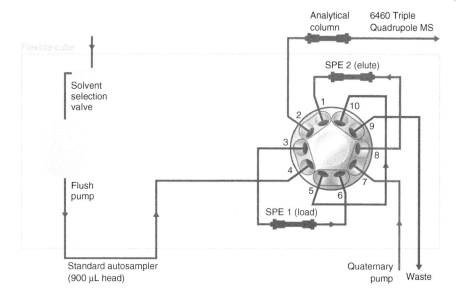

For larger sample amounts, e.g. typically for water analysis, the autosampler should be equipped with a larger analytical head of 900 µl for a faster injection. Depending on the needs, the system can be further equipped with:

- A valve to switch between the SPE path and a direct LC part (for the analysis of standards and analyses without SPE)
- A valve to switch between different SPE columns
- A valve with additional sample loop to enable even larger injection volumes

With these parts, you have various possibilities to enhance your standard HPLC system to a highly automated online SPE system.

4.1.10 Boost Performance with a Second Chromatographic Dimension: 2D-LC (see also Chapter 1.1)

More complex sample matrix or more complex analyte spectra are still increasing the need for better chromatographic performance. With the development of UHPLC, the van-Deemter maximum is already in reach and now high engineering hurdles must be taken to increase the performance, by lowering particle sizes in chromatographic columns and therefore dealing with higher back pressures. But the outcome is counted in single- or low double-digit increase of peak capacity.

2D-LC goes a totally different way by offering the possibility of an additional separation after the first separation step. This independent separation usually has a different selectivity, which is the key to success to achieve a separation, which was not accomplished in the first step. There are different techniques leading to better chromatographic separation: Besides comprehensive 2D-LC, which is often used to increase absolute peak capacity, mainly Multiple Heart-Cutting 2D-LC and

High-Resolution Sampling 2D-LC are applied to have the deepest look into the samples, with the highest chromatographic resolution.

Agilent Technologies offers by far the most comprehensive portfolio for 2D-LC in terms of variability, technical performance, and ease of use. Of course, it is always most straightforward to have a completely new 2D-LC system, but it is important that also existing systems can be upgraded. Any Agilent HPLC system starting with the 1100 series (which has been launched in 1995) can be used as part of a 2D-LC system. In most of the cases, the complete HPLC system can be used as it is, just the main parts for the second chromatographic dimension have to be added. Besides the 2D pump, the 2D-LC valve as interface and sampling unit for the second dimension, a 2D detector and the 2D-LC software are required, while the column oven can be used for both dimensions. 2D-LC is for sure the most straightforward way to address complicated separation tasks.

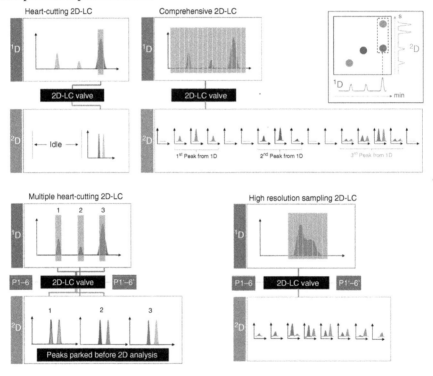

4.1.11 Think Different, Work with Supercritical CO_2 as Eluent: SFC – Supercritical Fluid Chromatography (see also Chapter 1.6)

Changing mobile phase(s) is, beside changing the stationary phase, one of the most common ways to achieve different selectivity for a separation problem. An unusual mobile phase is carbon dioxide, which becomes under certain physical circumstances first liquid, then supercritical, and can be used as highly apolar solvent to achieve a completely different selectivity and separation. Depending on the setup, an existing HPLC can be upgraded with an additional pump, autosampler and an additional control module to an SFC–HPLC hybrid system. With this system, normal-phase separations can be run in a much faster way, highly reducing analysis time and especially solvent costs, sometimes to less than 10% of your actual costs. Additionally, SFC is a truly green technique, highly increasing your ecologic footprint.

As alternative to normal-phase HPLC, SFC is often used for high-speed chiral analysis (see below the separation of propanolol enantiomers in less than four minutes!) and fast separations of nonpolar compounds, e.g. fat-soluble vitamins or in the petrochemical analysis. SFC is a versatile technique with the rule of thumb: everything, which you can dissolve in methanol, can be analyzed in SFC.

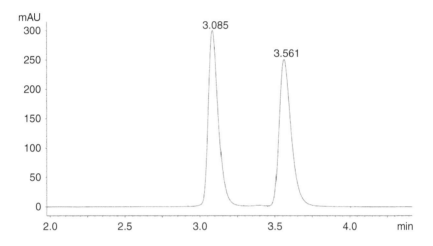

4.1.12 Determine Different Concentration Ranges in One System: High-Definition Range (HDR) HPLC

In various formulations, concentration ranges differ by more than a factor of 10, often up to factor of 100. This leads to differences in UV signal intensity and causes problems, if HPLC users must decide for the best sensitivity or the linearity to higher concentrations. Often, users have to do one analysis each for the high and one for the low concentrated analytes.

By combining the signal of 2 DADs with different optical path lengths, Agilent's 1290 Infinity II High Dynamic Range Diode Array Detection (HDR-DAD) Impurity Analyzer Solution enables to detect and quantify analytes in high and low concentration ranges in a single run. In this setup, one DAD uses a 60-mm UV cell for the highest sensitivity, while the other DAD is equipped with the 3.7-mm UV cell for the detection of high concentrated substances without saturation.

Smart firm- and software combine the signals and enable the simultaneous analysis of high-dose, low-dose, and trace compounds in a single run. The combination of these tools leads to a massive increase of sensitivity on the low-concentration side and to a better upper limit on the higher-concentration side.

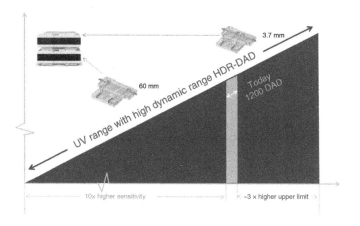

4.1.13 Automize Even Your Method Transfer from other LC Systems: Intelligent System Emulation Technology (ISET)

Methods for conventional HPLC analysis have often been already developed years or even decades ago, while nowadays more and more methods are developed for high-end UHPLC analysis. The idea of using a system for conventional analysis and high-end UHPLC analysis has also led to the development of ISET, Agilent's Intelligent System Emulation Technology. Knowing the characteristics of different HPLC and UHPLC systems, the high-end UHPLC systems can mimic the real behavior at the most important point of the whole separation, the column head. This gradient behavior is not only determined by the delay volume, but also by the mixing behavior based on different flow architectures.

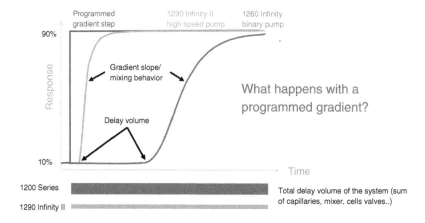

With this knowledge of the different systems and the superior performance, the other LC systems can be emulated. Emulating other systems enables the users to work in two directions:

(1) Take over existing methods from other systems to the new system and enhance these methods, if necessary.
(2) Develop methods for other systems with only one single system.

ISET incorporates the characteristics of various LC systems, not only from Agilent Technologies, but also from third-party vendors. With this technology, you enable your systems to act as any of those systems, making it a the most versatile system for method transfer or method development. Furthermore, the system allows you to run high-performance UHPLC methods and start increasing your method performance by a stepwise transfer of existing HPLC methods to real UHPLC methods.

4.1.14 Conclusion

The term "optimization" is multidimensional and there are many directions, in which you can optimize your Agilent (U)HPLC system. Depending on your needs, smaller or bigger changes are required to optimize your system. Some of them have not been addressed in this chapter, as they do not directly contribute to your system performance, but still can make the HPLC scientist's or the lab manager's life easier:

- *Implementation of long life and/or easy maintenance pump heads*: These pump heads for the 1290 Infinity I and 1290 Infinity II pumps deliver a guaranteed liter performance to ensure your lab will stay in working mode before the next planned maintenance. Easy maintenance pump heads enable even normal users to do some maintenance steps.
- *InfinityLab LC companion*: The HPLC system can be controlled by a browser application on every device, which is connected to the same network. Usually tablet PCs are used (and also offered directly from Agilent), but also smartphones or PCs can be used to have direct access to your system. On the one hand, it is possible to control or monitor one system from several devices, on the other hand one device can control several systems. Like this, you and everybody can have the full overview of all your HPLC systems in the lab.
- *Column identification*: With the possibility to use up to 8 columns in a single column oven, it can be interesting to have all the columns tagged and track their history. This can be easily done with the column identification kits, which are using open and programmable tags, to be used with any column, independent from the manufacturer.

References

1 DWIGHT STOLL. www.chromatographyonline.com/fittings-and-connections-liquid-chromatography-so-many-choices?pageID=1.
2 Retention time precision of Agilent 1290 Infinity Quaternary Pump. Agilent Technical Note 5991-0525EN.
3 Agilent 1290 Infinity Quaternary Pump. Comparing premixed isocratic conditions with dynamically-mixed conditions. Agilent Technical Note 5991-0098EN.
4 Let Your Autosampler Do Your Pipetting. Agilent Technical Note 5994-1704EN. https://www.agilent.com/cs/library/technicaloverviews/public/technicaloverview-injector-program-1260-infinity-ii-vialsampler-multisampler-5994-1704en-agilent.pdf

4.2

To Empower the Customer – Optimization Through Individualization

Kristin Folmert and Kathryn Monks

KNAUER Wissenschaftliche Geräte GmbH, Applications & Academy, Hegauer Weg 38, D-14163, Berlin, Germany

4.2.1 Introduction

The decisive bottleneck between all the possible optimization options and the resulting outcomes and improvements is always the user (Figure 4.2.1). The philosophy of the user determines whether an investment or adaptation is worthwhile. Because only if the mindset of the personnel is open to changes can an optimization in high performance liquid chromatography (HPLC) unfold its full potential. The user should confidently subject the HPLC system to their own requirements and expectations and make use of the flexibility gained by small changes. This section gives useful advice and application examples on how everyone can implement this maxim for themselves.

4.2.2 Define Your Own Requirements

4.2.2.1 Specification Sheet, Timetable, or Catalogue of Measures

No matter whether a new HPLC system is planned or an existing system is to be optimized, a specific consideration of the needs and requirements should always be the beginning of such a process. As versatile as HPLC can be, there is no system that can do everything perfectly at the same time. A major optimization initiative should therefore always be planned and carried out at the right parts of the system. It is usually advisable to draw up a specification sheet in which goals, their priority, and desired time frame are recorded, as well as by which measures these goals are going to be achieved (Table 4.2.1). The user should regularly ask themselves which parameter has the higher priority. For example, if the purity of a production process is routinely checked on an HPLC system and occasionally smaller research projects are carried out, the speed, robustness, and precision of the routine method probably has a higher priority than the desired sensitivity or flexibility for the infrequent research projects. Often, the user has to decide between parameters like maximum speed and

Optimization in HPLC: Concepts and Strategies, First Edition. Edited by Stavros Kromidas.
© 2021 WILEY-VCH GmbH. Published 2021 by WILEY-VCH GmbH.

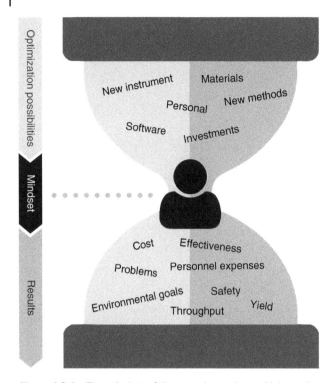

Figure 4.2.1 The mindset of the user determines which results are obtained from an optimization.

Table 4.2.1 Exemplary specification sheet for optimization measures on the HPLC system.

Priority	Goal	Measure	Deadline
1	Reduce LOD	Smaller tubings, optimize method, adjust injection volume, reduce matrix/noise	1. February
1	Increase efficiency	Replace manual injection by autosampler, adapt method, train affected employees	1. February
2	Prevent manipulation	Customize software and create individual authorizations for each employee	1. March
3	Accelerate troubleshooting	Create logbook, have two employees trained	3. Quarter of the year

best possible resolution before optimization. Perhaps a cost-effective method is also more important than maximum recovery rates.

Optimization can be approached at various levels. A differentiation is usually made between the personnel, equipment, and method levels (Figure 4.2.2). For smaller optimizations or measures at the personnel level, it is usually sufficient to solely provide a time schedule for implementation. It is much more important to involve all affected employees or colleagues in the planned optimization to achieve

Figure 4.2.2 Optimization levels that are taken into account.

the best possible acceptance. A simple catalogue of measures can be created by each user to keep an overview of which measures can be bundled or which ideas could have contrary effects. In this way, results can often be achieved more quickly.

4.2.2.2 Personnel Optimization Helps to make Better Use of HPLC

The personnel level is often the deciding factor in any optimization. Ultimately, all changes must be accepted and implemented by the user. An expensive investment is not worthwhile if the new equipment is subsequently not or reluctantly used. An open working philosophy of the user with a lot of curiosity for new things is the cornerstone for every optimization. Managers can promote this by encouraging employees to suggest improvements themselves or by planning them together with them, so that the resulting benefits are in the foreground and not the perhaps exhausting changeover. Every change means overcoming a threshold of inhibition and there are various ways to reduce this inhibition. Apart from good communication, the user friendliness of products is usually decisive. Decision makers should therefore prefer devices, software, or methods that are as practical and intuitive as possible. The acceptance of employees can also be increased through training. Training can also increase the efficiency of users. The knowledge gained in method optimization training can be passed on to colleagues and, of course, used to evaluate and optimize existing methods. Troubleshooting trainings usually pay off by increasing the sensitivity of the employees for the needs of HPLC and reducing maintenance costs and reaction times in case of problems. Some HPLC manufacturers offer training as well as various tools to increase user friendliness. In addition to good, online available user manuals and service hotlines, explanatory videos, and application examples are also useful offers, which show the user friendliness of a supplier. Furthermore, the design of the HPLC should be chosen so that the user can carry out simple repair and replacement work themselves.

4.2.2.3 Mastering Time-Consuming Method Optimizations in a Planned Manner

A time-consuming method optimization is usually only worthwhile for routine applications and requires the most experience of all adjustments from the user. Finding the optimal column is usually still quite easy by screening with an overview gradient. If, however, a particularly high robustness is required or the method is very complex, it is advisable to commission method optimization from experienced experts. The basic requirement here is to work in an especially planned manner. There is a lot of literature and training courses available that explain the different strategies of HPLC method optimization in detail. Common to all of them is that a great deal of time must be invested by the user in the optimization tasks. Therefore, it is strongly recommended to prepare a specification sheet with priorities in advance to avoid unnecessary use of valuable resources such as chemicals, time, solvents, and personnel. If method optimizations are carried out regularly, the process itself can be made considerably more effective by supporting software such as *DryLab [MOLNÁR-INSTITUTE for applied chromatography]* or *ChromSwordAuto®Developer [ChromSword]*. Part of a method optimization should always include the subsequent robustness test, a test for reproducibility, and the determination of the recovery rates. A method validation is again dependent on the requirements of the method as defined in the specifications.

4.2.2.4 Optimizations at Device Level do not Always have to Mean an Investment

At device level, for example, an optimization in the form of a maintenance contract can increase the reliability and longevity of a system. In addition, for almost every component that is in the flow of the system, there is an optimal solution for each specific application. Flexibility here means maximum adaptability to the application or the needs of the user. Optimizations can range from very obvious measures, such as adjusting the sample loop volume, to new acquisitions, such as additional valves or a more sensitive detector. Also, very small step things like other septa materials for sealing the sample vials or the search for the best washing liquid in the autosampler are possible optimizations. An often-forgotten possibility is the adaptation of the wetted materials in HPLC, which is described in more detail in Section 4.2.4. The software, too, offers room for numerous improvements, whereby processes and habits that have been established over a long period of time are sometimes in the way and can only be overcome with an open mindset of the user. Some examples of software optimization are described in more detail in Section 4.2.5. A perfect optimization is more successful, the more freedom a manufacturer gives the end user to make changes to the system or the software. A good example of this are LC docking stations for small pumps, detectors, and valves, so-called assistants, which can be individually configured by the user. LC docking stations are explained further in Section 4.2.3.

4.2.3 An Assistant Opens Up Many New Possibilities

4.2.3.1 If the HPLC System must Simply be able to do more in the Future

Optimization can reach its limits if an HPLC system has to meet more and more requirements over time. In addition to the originally planned analyses, more and more requests for more automation or more flexibility often arise, for example, when the user improves their own knowledge, or the company expands. These wishes include, for example, more speed, a wider range of applications through more automation or more flexibility. Unfortunately, a limited budget or space does not always allow the purchase of an additional HPLC system. An inexpensive additional assistant in the system can then provide the desired additional flexibility or even enable completely new HPLC applications. An HPLC assistant is an LC docking station for various HPLC modules and can be integrated into the HPLC tower as an extension by the user or used as a standalone simple HPLC instrument. The *AZURA* assistant *ASM 2.2L [KNAUER wissenschaftliche Geräte GmbH]* consists of a maximum of three modules that can be combined from a range of pumps, valves, and detectors according to the user's requirements (Figure 4.2.3). The user gains additional flexibility through the free choice of valve heads. A large number of different analytical and preparative valve heads can be attached to the assistant module with the valve drive by the users themselves. The user can install an analytical two-way valve today and a preparative multiposition valve tomorrow. The valve drive, in turn, automatically recognizes the valve head and passes this information on to the connected software. To configure a simple standalone HPLC instrument, for example, an isocratic pump, a UV detector, and a valve drive can be combined, either for fractionation or injection. These small assistant-based HPLC instruments are well suited for educational purposes and isocratic applications and can thus relieve existing systems. If the assistant is integrated into a system as an extension, it can perform many different tasks depending on the modules installed, such as eluent delivery, detection, sample and solvent selection, sample injection, column selection, or fractionation.

4.2.3.2 Individual Optimizations with an Assistant

The examples shown in the following are intended to provide suggestions on how new areas of application can be opened up by an assistant or how an existing system can be upgraded. Experienced users should free themselves from preformed expectations. With a little creativity and a precise knowledge of one's own HPLC needs, highly effective and innovative individual solutions can be achieved.

4.2.3.3 Automatic Method Optimization and Column Screening

Especially in research or application departments, new methods are regularly developed, also to make them available for the routine in other departments. As described in Section 4.2.2.3, method optimizations and column screenings are often very time consuming and require high personnel expenditure. If an HPLC system consisting of a gradient pump, a detector, and an autosampler is extended by an assistant with two-column switching valves and a column holder module, many manual steps can be automated and thus run overnight to save time (Figure 4.2.4). The system is then available for other applications during the day. If the columns are to be suspended in a column thermostat, an eluent selection valve can be used instead of the column holder module for even more flexibility. The same design is also suitable for column screenings in which the right column for an analyte is sought or the quality can be checked in a column production. The rapid change between methods with different columns is also extremely effective. Normally, a quite long equilibration phase is necessary to rinse the system. After a method, a column is then rinsed back to storage conditions so that it can be exchanged for another column by hand. For both steps, 20 minutes and personnel costs must be taken into account. With the additional assistant, however, five columns could be connected and a bypass installed, for example. With the bypass, the system can be flushed to a new start eluent in a few seconds at high flow rates. Since the column is not removed, it can remain on the starting conditions and does not have to remain equilibrated for long. In addition, this prevents installation errors, material fatigue, and air inclusions during constant

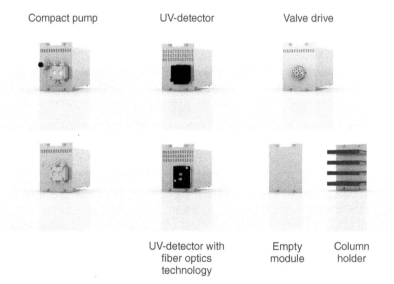

Figure 4.2.3 The assistant AZURA ASM 2.2L from KNAUER Wissenschaftliche Geräte GmbH can be individually assembled from three selectable modules according to the user's needs. The intelligent instrument automatically recognizes which module has been installed, so that users can easily change the modules themselves. This enables maximum flexible optimization for the respective application.

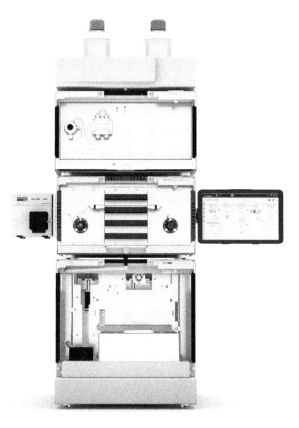

Figure 4.2.4 AZURA HPLC system with assistant with two-column switching valves and one-column holder. In addition, the Mobile Control tablet software [KNAUER Wissenschaftliche Geräte GmbH].

column changes. With this arrangement, long measuring sequences with hundreds of individual runs and constant method changes are possible without the need for an employee to be present.

4.2.3.4 A New Perspective at Fractionation, Sample Preparation, and Peak Recycling

If one wants to run a (semi-)preparative HPLC, usually a fraction collector and often also an autosampler is bought. But not always such expensive purchases are possible and necessary. An assistant with a fractionating valve, an injection valve, and a small feed pump can partially take over both functions and thus save a lot of money. With a feed pump, the loop in the injection valve can be automatically refilled from a storage vessel after each run. This approach is particularly worthwhile for samples with a large solution volume, where the same method has to be run several times in sequence. The sample storage vessel can also be placed in a cooling/heating bath. A fractionating valve allows up to 15 fractions and the waste to be automatically fractionated during separation.

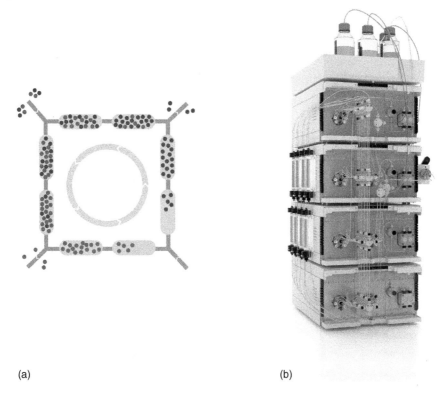

(a) (b)

Figure 4.2.5 Schematics of the fractionation valve (a) and continuous chromatogram of a peak recycling purification (b) performed with a compact preparative HPLC system consisting of a UV detector, a 50-ml pump, an injection, and a fractionation valve.

A special case in sample preparation is the so-called peak recycling (Figure 4.2.5). If the column does not have adequate separation capacity to sufficiently purify a sample in one run, for example because strong matrix effects prevent this, an intelligent arrangement allows a fraction of the sample to be recycled. Recycling maximizes the theoretical column length and thus improves the resolution of the peaks. For this purpose, a connection is made from the fractionating valve to the pump and the fraction is fed into the eluent. The pump acts as a mixing chamber with the eluent. This procedure can be repeated until the desired resolution is achieved. Peak recycling allows complex separations even with very simple means and a cost-effective compact preparative system in the form of a standalone assistant. Of course, a large preparative pilot or an analytical system can also be equipped with an additional valve to realize peak recycling.

4.2.3.5 Continuous Chromatography, a New Level of Purification

Sometimes, classical preparative HPLC reaches its limits. Especially when it comes to the effective purification of large sample volumes. If only two fractions are to be separated from each other in such a separation, continuous chromatography can be a significant improvement over classical HPLC. In simulated moving-bed (SMB)

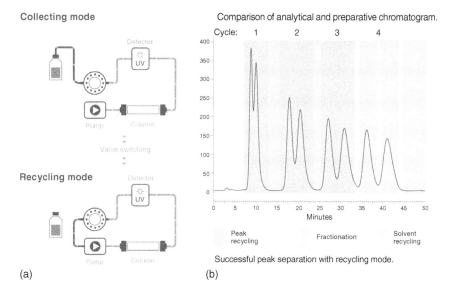

Figure 4.2.6 A continuous, simulated moving-bed (SMB) system from KNAUER Wissenschaftliche Geräte GmbH, consisting of four assistants and eight identical columns (a). The separation principle of the SMB is shown as a model (b). Source: KNAUER Wissenschaftliche Geräte GmbH.

chromatography, 4, 8, or 16 columns are circularly operated via four assistants, each with a multiposition valve and an isocratic pump via an innovative capillary system (Figure 4.2.6). Whereas in classical HPLC two fractions with different retention times are eluted and collected one after the other, rhythmic switching of the valves against the flow direction allows a two-dimensional and thus permanent, simultaneous separation. The faster eluting product A is transported together with the eluent, while the slower, more strongly interacting product B is transported with the column bed against the flow by switching against the flow. The result is not only a significantly improved resolution and a very high product yield in a short time, but also a saving in solvent consumption and very good recovery rates of up to 100%. Another advantage is that the products are not diluted compared to the initial concentration and can therefore often be processed directly. However, the many advantages of SMB chromatography can only be realized if some effort is invested in method development by SMB experts beforehand. It is also recommended that the manufacturer provides detailed training for the personnel.

4.2.4 The Used Materials in the Focus of the Optimization

4.2.4.1 Wetted vs. Dry Components of the HPLC

When optimizing HPLC materials, most users think of the solvents and column bed, but rarely of the materials used in an HPLC system. With the installed materials, a distinction is made between wetted and dry materials. The dry materials are usually hidden behind the housing wall and are not accessible to the user. Also, the optimization of the accessible dry parts is often limited to aesthetic aspects. For example, the housing can be protected from sunlight by foils to prevent fading. With wetted materials, i.e. all materials that come into contact with liquids, the possibilities for influencing the process are usually much greater.

4.2.4.2 Chemical Resistance of Wetted Components

In solution, numerous chemical reactions on the various surfaces of an HPLC system can lead to problems, poor performance, or faster wear and tear. Especially if different HPLC modes such as RP and NP or GPC are to be used in a system, the wetted materials must also be adapted to these requirements. Solvents typical for NP and GPC such as tetrahydrofuran (THF), *n*-heptane, or dimethyl sulfoxide (DMSO) can cause great damage in a system designed for RP. THF can soften PEEK materials. PEEK is not only very often used as a capillary material, but it is also frequently used in degassers, valves, or as screws. DMSO attacks not only PEEK components but also especially seals in pumps and valves and can lead to leaks and carryover. *n*-Heptane and other organic solvents can react in the check valves of pumps and destroy them.

In ion chromatography, aqueous acidic or basic eluents are often used. Strong acids such as sulfuric acid and hydrochloric acid can corrode metal components. The pump is particularly affected by this, as pressure and friction promote corrosion. Ceramic pistons in the pump should also be avoided if strong acids are to be used. In this case, pistons made of sapphire glass are well suited. Aqueous bases such as sodium hydroxide solution can attack glass surfaces such as sapphire glass or optical surfaces in measuring cells, and the rotor seals of valves are often particularly unstable due to bases. For all the above-mentioned parts, there are therefore spare parts made of other resistant materials. Depending on the experience of the user, the components can be replaced themselves. However, it should be noted that other materials are often less pressure or temperature stable. It can therefore often be useful to contact the device manufacturer to find out which wetted parts should be considered in combination with the chemicals used and which replacement material should be chosen. Excellent users with many changing eluents and analytes can create a catalogue which component is critical for which chemical. This can be shown as an example using the available materials for rotor seals and stator in the valve (Figure 4.2.7). Depending on the pH values, temperatures, or chemicals used, there are suitable materials available. Vespel®, for example, is a polyimide compound and is most often used for rotor seals. It is inert to almost all chemicals at temperatures up to 200 °C, but it is not suitable for pH values above 9.5 and below 2.3. It is also incompatible with concentrated mineral acids such as sulfuric acid, glacial acetic acid, DMSO, and THF. It is also degraded by nucleophilic substances like ammonia or acetates. As an alternative, Tefzel® is the perfect material for extreme pH values above 9 and below 3. It consists of ethylene tetrafluoroethylene (ETFE) and should not be used if chlorinated eluents or temperatures above 80 °C are required. In addition, Tefzel® rotor seals need to be replaced more often as the material is softer and wears out faster. On the other hand, the fluorinated polymer has a very high resistance to solvents in the neutral and basic range. A relatively rare material is the rather expensive PPS (polyphenylene sulfide). It has a good chemical resistance against almost all solvents as well as organic acids and all kinds of alkalis up to 200 °C, but at temperatures above 50 °C it can be oxidized by atmospheric oxygen. Oxidizing acids such as diluted hydrochloric, nitric, or sulfuric acid also cause instability of the material. Furthermore, PPS is comparatively brittle and can break more easily than the other materials described. PEEK consists of polyetheretherketones and is one of the most commonly used materials for HPLC components. Similar to Tefzel®, it is stable against almost all eluent materials at room temperature. Furthermore, it can be used in the entire pH range from 1 to 12.5 and temperatures up to 100 °C. However, it swells with some aliphatic hydrocarbons such as THF and DMSO, especially at higher temperatures. Also, very high concentrations of ACN have this effect. Many halogenated hydrocarbons such as DCM, HF, methylene chloride, or concentrated inorganic acids, especially nitric or sulfuric acid, lead to decomposition already at room temperature. Hydrochloric acid, however, can usually be used without problems. Furthermore, PEEK is not UV resistant and should therefore be protected from direct sunlight. The stator of a valve is usually made of PEEK or SST (stainless steel). SST is an iron alloy with 10–13% chromium content and is characterized by a very high mechanical, chemical, and thermal stability and almost no gas permeability due to the chromium oxide layer formed on the surface. To achieve a basic stability against chlorine ions, 2% molybdenum is often added to the steel alloy. The material is nevertheless corroded by higher percentage hydrochloric acid or other halogen acids. Due to its high stability and favorable production costs, SST is used proportionally in almost every HPLC component. However, SST is not chemically inert and can both release ions in the eluent and interact with various, especially charged analytes. If necessary, the much more expensive titanium is therefore used as an alternative for some HPLC components such as the pump. Titanium also forms an oxide layer on the surface, which makes the material largely chemically inert to almost all chemicals, including alkalis and acids. Diluted hydrochloric (7.5%), sulfuric (5%), and nitric (98%) acids can also be used in combination with titanium. In the case of organic acids, it should be noted that titanium is very sensitive to oxalic acid as part of the eluent and is also attacked by formic acid from a concentration of 50%. The mechanical and thermal stability of titanium is comparable to SST.

All resistances mentioned here refer to information provided by the material manufacturers and no guarantee can be given for general validity or completeness. In case of doubt, the user should always contact the manufacturer and make sure that they are prepared for each specific use.

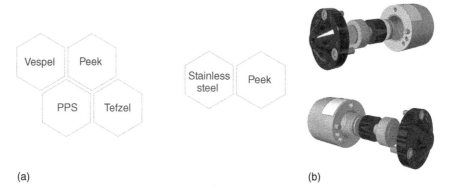

Figure 4.2.7 Shown are the possible materials for the rotor seal (a) and the stator (b). The rotor seal and the stator are the two most sensitive wetted components in a valve.

4.2.4.3 Bioinert Components

Besides the chemical resistance of the components, users should also consider possible interactions with analytes. Especially with biological samples such as antibodies and enzymes, where the protein conformation and thus the function should be maintained, ionic or metal surfaces like SST can be problematic. Smaller, particularly reactive organic molecules can also react with the metal surface and thus change their chemical state during chromatography. This often results in inaccurate results or low recovery rates in the preparative area. For this case, some manufacturers offer biocompatible HPLC solutions. Here, all wetted parts are exchanged for PEEK, titanium, or ceramic. HPLC columns are also available in PEEK or glass shells to exclude the described interactions.

4.2.4.3.1 Material Certification
In the regulated environments, wetted parts must be ordered together with a material certification including information concerning the quality, compliance and composition of materials; this should also be requested from the manufacturer.

4.2.5 Software Optimization Requires Open-Mindedness

With HPLC software, optimizations are often difficult to perform by the user and many people are reluctant to change the software even when buying a new HPLC, no matter how many problems there have been so far. If you are open-minded and compare the different software solutions, you can save money and find the perfect solution for your needs. Just as with the compilation of the specification sheet when buying hardware, the most important requirements should be collected first. The most important aspect is of course that the purchased or existing HPLC system can be integrated into the software. Most instrument manufacturers offer drivers for several different software packages. Sometimes, it is even possible to combine components from different manufacturers to one HPLC system and control them together. This service is offered by *ClarityChrom [Data Apex], Chromeleon [Thermo Fischer Scientific], OpenLab [Agilent Technologies], Empower [Waters], LabSolution [SHIMADZU]*, or *Mobile Control [KNAUER wissenschaftliche Geräte GmbH]*. HPLC and software manufacturers are in demand here to continue working on industry standards to simplify the networking of instruments from different manufacturers.

An intuitive user interface usually depends on the individual taste of the user. Therefore, it is advisable to have different software solutions demonstrated and tested. A few software vendors offer the possibility to personalize the user interface and to create customized solutions. With *Mobile Control*, for example, it is possible to personalize not only the visual appearance but also the access rights of the user or to adapt operating elements individually. This can be of great advantage in a regulated environment if a user can only start and stop a present method and the results should be output afterwards in an unmanipulable way. It also allows a great simplification of the software for less demanding or experienced users. The software is operated by finger touch on a tablet and the user can see visually on the screen how, for example, valve positions change. The tablet software also offers the possibility to monitor and control the HPLC system from the office, which can be a great relief for users with little time. Even for routine measurements, the processes can be easily adapted and automated.

4.2.6 Outlook

When looking to the future, the goal must be that highly qualified personnel should waste less time setting up the technology but can focus their attention on achieving and interpreting good measurement results and developing innovative ideas. Workarounds should be replaced by individualized HPLC solutions. Whether on the software/device level, in personnel deployment, or method optimization, users should be able to adapt HPLC to their needs. An important step in this direction is the Internet of Things (IoT), on which several working groups worldwide are currently researching. Many experts expect IoT to bring a decisive paradigm shift in the way laboratory equipment is organized and used. It is intended to enable the networking of laboratory instruments from different manufacturers, even across building and server boundaries. Up to now, the communication interfaces of the devices and the data formats, which are measured and output, for example, with HPLC systems from different manufacturers, are still different. In the future, the networkability of the laboratory instruments among each other should improve and data formats should be standardized. These innovations can be expected to appear on the market in the coming years. However, there is still a long way to go before networking at the process industry level comes true. For this purpose, the devices are first connected to the Internet and the collected data are assembled in manufacturer-independent clouds. It is conceivable, for example, that a higher-level laboratory information management system (LIMS) could monitor the system data and send a message to the user's smartphone if the values are close to the limit or if an unwanted trend occurs in the data. It would be conceivable, for example, to send warning messages if a slight increase in pressure is registered in routine analysis over time, or less directly, if a certain amount of solvent has been used up and this should be reordered. IoT could therefore not only facilitate quality assurance but also resource planning. If not only HPLC instruments, but also different laboratory instruments are connected via the IoT, quality assurance can reach a whole new level by checking the environment, such as laboratory and server rooms or reagent storage areas, for over/under temperature conditions. In this way, the compliance of safety and operating regulations can be monitored by employees via remote communication. The resulting data, whether environmental/sample data or chromatogram results, could be viewed and edited centrally and in real time. This development can pay off especially in large industrial or regulated environments, as communication paths could be shortened and information could be evaluated in a bundled manner. The developments for IoT are in full progress and several companies have taken first steps in this direction. With the *Nexera* systems *[SHIMADZU]*, for example, basic HPLC system data can be automatically examined for anomalies and simple processes can be automatically triggered as a reaction. The anomaly is automatically documented by the system and reported to the user. *WinLaisy [In-QM-Team-Software GmbH]* offers an example of an independent LIMS for central master data administration, data evaluation, and sample management. To use cross-system and cross-manufacturer solutions, laboratory managers will also have to focus more on problems such as data security against hacker attacks and confidentiality of results in the future. The big step toward digitalization of analytics will nevertheless be a central topic within the next few years.

4.3

(U)HPLC Basics and Beyond

Gesa Schad[1], Brigitte Bollig[1], and Kyoko Watanabe[2]

[1] *Shimadzu Europa GmbH, Duisburg, Albert-Hahn-Strasse 6-10, 47269, Germany*
[2] *Shimadzu Corporation, 1, Nishinokyo-Kuwabaracho, Nakagyo-ku, Kyoto 604-8511, Japan*

4.3.1 An Evaluation of (U)HPLC-operating Parameters and their Effect on Chromatographic Performance

The fundamental knowledge of high performance liquid chromatography (HPLC) equipment and the instrument's functionality does play a key role in producing quality data. Next to the standard settings, such as flow rate and oven temperature, there are several less well-known parameters that are also crucial in improving (U)HPLC performance. The compressibility value, for example, is essential for the pump to run an accurate flow rate according to the density of the mobile phase. Opposed to common assumptions, compressibility is not a linear function of the ratio of components mixed together. Peak shape, intensity, and carryover can be considerably improved by using an appropriate combination of dilution and rinsing options of the autosampler, and the slit width value of the cell in a photodiode array (PDA) detector affects both the chromatogram and resolution of the UV spectrum [1].

4.3.1.1 Compressibility Settings

The compressibility settings, according to the mobile phase used, have a significant impact on pump performance. Pressure and pressure fluctuations increase when an incorrect value is used (Figure 4.3.1), also pump mixing and flow rate accuracy can be impaired.

The parameter "compressibility" corrects the mobile phase flow produced by the pump depending on the solvent's density. However, compressibility doesn't have a linear relationship with the composition ratio of organic and aqueous solvent, as the relationship between the mixing ratio of water–organic solvent and its density is not linear up to 50% (Figure 4.3.2). Compressibility values of common solvents are listed in Table 4.3.1. To make life easier for users, correct compressibility settings are usually supported by modern control software. After specifying the mobile-phase solvent, the system will automatically use the corresponding default settings. If manual adjustment is required, the following values are recommended:

Optimization in HPLC: Concepts and Strategies, First Edition. Edited by Stavros Kromidas.
© 2021 WILEY-VCH GmbH. Published 2021 by WILEY-VCH GmbH.

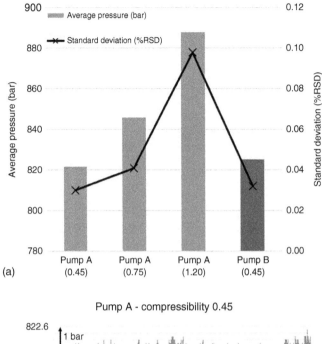
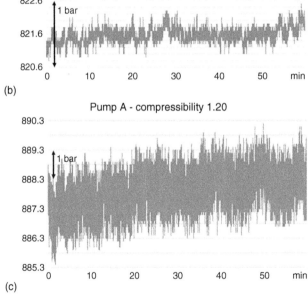

Figure 4.3.1 Effect of compressibility setting on pump pressure pulsation. (a) Changes in average pressure and standard deviation depending on compressibility setting, (b) pump A pressure trace showing pressure fluctuation with suitable, and (c) inappropriate compressibility setting.

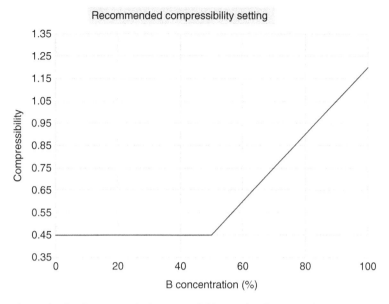

Figure 4.3.2 Recommended compressibility setting for a premixed mobile phase. *Source:* Shimadzu [2]. © 2019, Shimadzu Corporation.

Table 4.3.1 Compressibility of common (U)HPLC solvents.

Mobile phase	Compressibility $(GPa)^{-1}$
Water	0.45
Acetonitrile	1.20
Methanol	1.25
Ethanol	1.20
2-Propanol	1.20
Hexane	1.60
Heptane	1.25
Cyclohexane	1.25
Ethyl acetate	1.10
Chloroform	1.10
Benzene	1.00

- Single solvent

  Set the compression rate according to the values in Table 4.3.1.

- Mixture of water and organic solvent (Figure 4.3.2)

If the amount of organic solvent is 50% or less, it is recommended to set the compressibility value to "0.45" for pure water. If more than 50% organic solvent is used, adjust the value closer to the compressibility of the organic solvent according to the mixture ratio. For example, when the ratio between water and acetonitrile is 50 : 50 v/v, set "0.45." When the ratio between water and acetonitrile is 30 : 70 v/v, use "0.75" [2].

4.3.1.2 Solvent Composition and Injection Volume

UHPLC systems exhibit lower system dispersion volume than that of conventional HPLC configurations. They commonly include tubing with an inner diameter of 0.1 mm or less. Also, the internal diameter (I.D.) of columns used for UHPLC analyses is narrow, with 2.0–3.0 mm being the most common dimensions. Therefore, the effect of extra-column volume on peak dispersion is more severe compared to HPLC, where considerably higher flow rates are being used.

Smaller internal diameters result in less effective mixing of the sample and eluent within the tubing. If the sample solvent is organic and has a higher concentration and higher elutropic strength than that of the mobile phase, it can interfere with the adsorption of the sample on top of the column after injection and result in broad, misshapen peaks. A small column I.D. can emphasize this phenomenon [1], as depicted in Figure 4.3.3. The analytical conditions used in the example can be found in Table 4.3.2.

The strong eluting power of acetonitrile in the sample solution results in strong dispersion of the ASS signal in this example. The effect is further enhanced by an increase in injection volume. A simple solution to this problem is dilution of the sample, to create a solvent composition more similar to the mobile phase, with regards to concentration of organic solvent. Modern autosamplers feature an "auto-dilution" function that allows this step to be incorporated into the method. Figure 4.3.3 shows that even a 50 µl injection of the dilute sample (10×) gives a significantly higher signal than the 5 µl injection with no dilution. Alternatively, the so-called co-injection

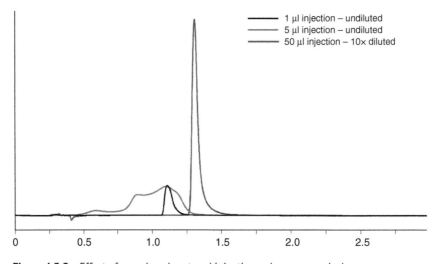

Figure 4.3.3 Effect of sample solvent and injection volume on peak shape.

Table 4.3.2 Analytical conditions used for analysis shown in Figure 4.3.3.

Mobile phase	0.085% Phosphoric acid in H_2O/MeOH (65 : 35 (v/v))
Flow rate	0.3 ml/min
Column	C18 (50 × 2.1 mm, 1.3 µm core–shell)
Column temperature	40 °C
Detection	UV 230 nm
Sample	Acetylsalicylic acid (ASS) in acetonitrile

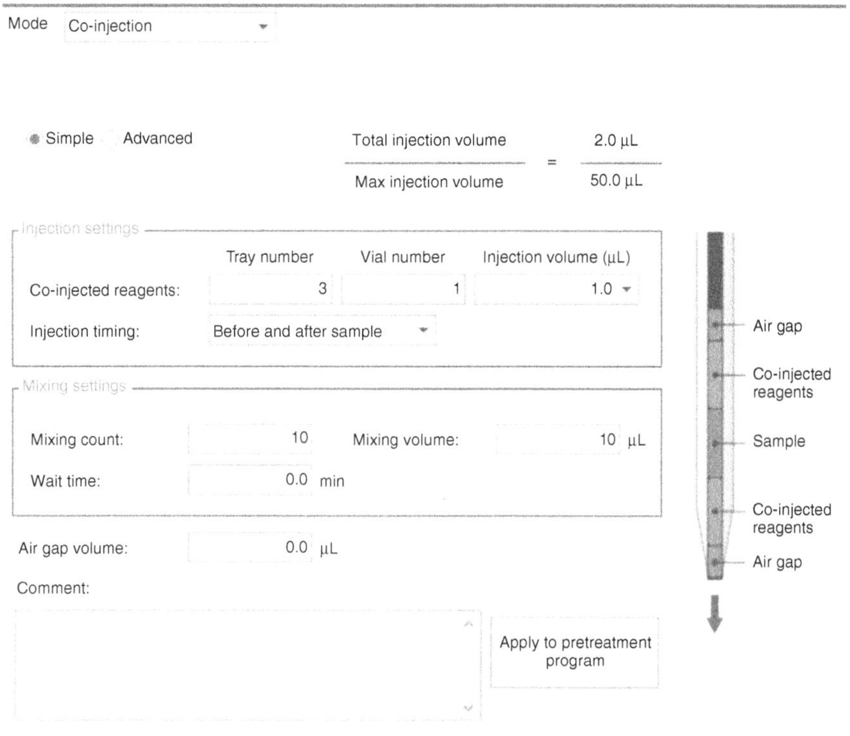

Figure 4.3.4 "Co-injection" settings in the LabSolutions software.

function can also be used. Next to the sample, it enables aspiration of an additional, weaker solvent, in this case water, (Figure 4.3.4) to dilute the detrimental effect of the strong solvent on peak shape. The analyte experiences a better focus in the stationary phase, which leads to a much sharper, higher signal, even with a larger injection volume [3] (Figure 4.3.5).

In the Shimadzu LabSolutions Chromatography Software (CDS), a "co-injection" can be easily set in the graphical user interface (Figure 4.3.4). An image of the injection needle illustrates the ratio of sample to added reagent.

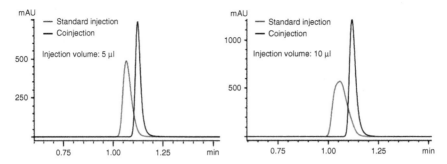

Figure 4.3.5 Effect of co-injection of water on peak shape of caffeine in methanol/water (60 : 40 v/v) using different injection volumes. *Source:* Osaka [3]. © 2018, Shimadzu Corporation.

4.3.1.3 Photodiode Array Detector: Slit Width

"Slit width" of the PDA detector is one of those parameters that are often fixed in "standard settings." The slit width controls how much light can pass through the flow cell, to reach the photodiode array, therefore affecting signal intensity and detector sensitivity. It can be useful to optimize this setting when trying to obtain lower detection limits, or better qualitative data by means of UV spectral data.

A narrow slit width provides improved spectral resolution for analytes to provide UV spectra with enough fine detail to be used for qualitative analysis. High spectral resolution will increase the confidence of library-matching search results when attempting to identify unknown peaks within a chromatogram. A wide slit width allows more of the light passing through the flow cell to increase signal intensity and detector sensitivity. At the same time baseline noise is reduced, resulting in an increase in the signal-to-noise (S/N) ratio. However, with a wider slit width the optical resolution of the spectrophotometer diminishes, as the wavelength of light falling on each diode becomes less specific with the light becoming more diffuse.

To emphasize the influence of the slit width setting, Table 4.3.3 states the S/N values obtained with one compared to 8-nm slit width, while Figure 4.3.6 shows the UV spectrum of pyrene, acquired with different slit widths [1].

As can be seen in Figure 4.3.6, a narrower slit width offers improved UV spectral data. The peaks show better resolution and λ_{max} values obtained with 1-nm

Table 4.3.3 Signal, noise, and S/N ratio obtained from analysis of pyrene.

	Slit width (nm)	Peak height	Noise	Signal/noise
$\lambda_{max}1$ (240 nm)	8	485 642	103.47	4 694
	1	542 232	504.63	1 075
$\lambda_{max}2$ (334 nm)	8	259 512	61.19	4 241
	1	299 460	280.27	1 068

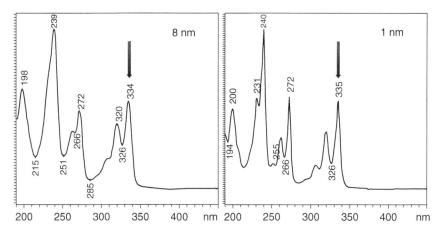

Figure 4.3.6 Visualization of the effect of slit width on the UV spectrum of pyrene. *Source:* Watanabe et al. [1]. © 2017, hplc2017-prague.org.

slit width were proven to be in accordance with spectral data recorded with a UV spectrophotometer (UV-2600, Shimadzu). Some of λ_{max} values obtained with 8-nm slit width showed a small deviation from the accurate result (e.g. λ_{max} at 334 instead of 335 nm). Results in Table 4.3.3 clarify the relationship between the amount of incident light and detector sensitivity – a larger slit width results in a higher signal. It is important to note – if qualitative data for identification of substances is recorded, it is better to choose a narrow slit width, whereas a wide slit width will increase detector sensitivity for quantitative analyses.

4.3.2 "Analytical Intelligence" – AI, M2M, IoT – How Modern Technology can Simplify the Lab Routine

4.3.2.1 Auto-Diagnostics and Auto-Recovery to Maximize Reliability and Uptime

The latest generation of Shimadzu (U)HPLC systems offer intelligent auto-diagnostics, as well as auto-recovery functionality to support users in their day-to-day workflow. As an example, air bubbles can, in rare cases, form by mixing of solvents and cause problems when wandering through the flow line. New sensor technology in the Shimadzu Nexera systems recognizes the resulting anomalies in the pressure trace as an error and triggers the appropriate counter measures (Figure 4.3.7). After automatically pausing the batch and rinsing of any remaining sample off the column, the system will apply a corrective purge to remove the air. This is followed by re-equilibration of the column and, if the pressure and detector base line remain stable, re-injection of the sample, so that the batch can be finished without the risk of producing inaccurate data, due to air bubbles and pressure fluctuations.

Thanks to latest technology, also the common issue of a system running dry, as the solvent reservoir wasn't adequately filled before starting the batch, can be reliably

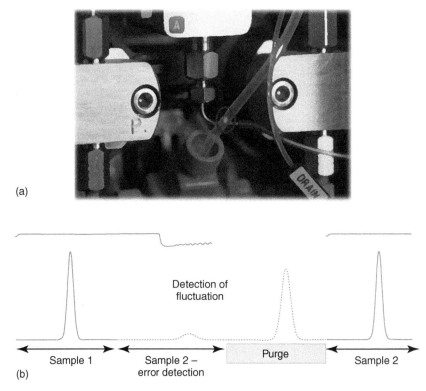

Figure 4.3.7 (a) Air bubble trapped in flow path. (b) Illustration of auto-diagnosis and corrective measures.

solved. Reservoir tray weight sensors enable measurement of mobile-phase levels in real time (Figure 4.3.8). The system uses flow rate, method runtime, and number of samples to calculate the amount of eluent required for the analysis. If the remaining amount of mobile phase is not sufficient to finish this sequence, the user will be warned before starting the first run (www.shimadzu.eu/nexera-series).

4.3.2.2 Advanced Peak Processing to Improve Resolution

Despite advances in separation technologies, chromatographic method development, and optimization tools, coelution of compound peaks is still a common issue. Especially in complex samples with a large number of analytes, or with isomers that can't be distinguished by different mass and exhibit similar elution behavior, achieving baseline separation can be a challenge. There are several integration options for overlapping peaks, such as the valley-to-valley integration, perpendicular drop or peak skim method, as visualized in Figure 4.3.9, but none of these will offer truly accurate quantification of the single compounds. To cite John Dolan's LC trouble shooting advice: "Integration of poorly resolved peaks is only an estimate of the more accurate results you would get when the peaks are baseline resolved" [4].

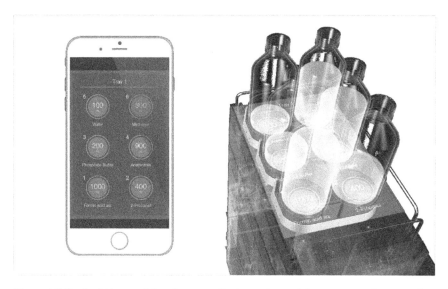

Figure 4.3.8 Real-time mobile-phase monitoring, using weight sensors and a smartphone app.

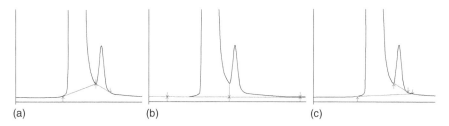

Figure 4.3.9 Integration example of overlapping peaks using (a) a valley-to-valley integration, (b) perpendicular drop, or (c) the peak skim method.

Despite best efforts, sometimes baseline separation is more easily said than done, and achieving that last extra bit of resolution will considerably increase the time and effort, hence cost, of method development.

An alternative is the use of intelligent peak deconvolution analysis (i-PDeA), where multivariate curve resolution-alternating least squares (MCR-ALS) is applied to PDA detector data, so that single analytes can be extracted from coeluting peaks [5]. This novel approach enables visualization and detection, e.g. of a minor single impurity even when the impurity is coeluting with a main component. It therefore facilitates the individual, accurate quantification of hard-to-separate peaks through computer processing, and deconvolution of spectral information, thereby reducing the effort required to further optimize separation parameters. Figure 4.3.10 shows a schematic image of the peak deconvolution process.

While digital peak processing to improve data quality is widely accepted in the field of spectroscopy, separation scientists have so far been hesitant to adapt

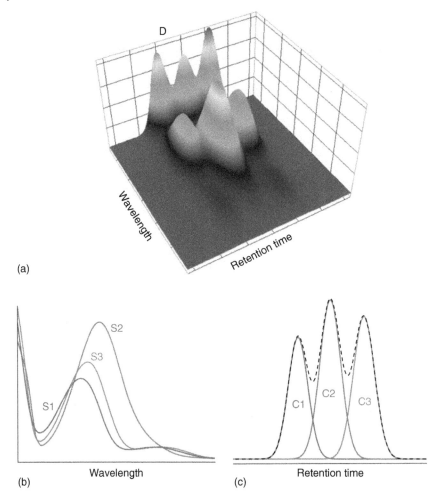

Figure 4.3.10 Data matrix (D) showing spectral data (S) and peak profiles (C) from a three-component mixture. *Source:* Yanagisawa [5]. © 2017, Shimadzu Corporation.

mathematical manipulations to enhance peak resolution and evaluation [6]. The Shimadzu i-PDeA intelligent peak deconvolution algorithm, embedded in the chromatographic data system, is an easy-to-use powerful tool to aid in the interpretation of data from overlapping peaks and allow for accurate quantification of single components without the need of time-consuming method optimization to achieve baseline resolution. Figure 4.3.11 illustrates the benefits of this technology for the accurate integration, hence quantification of a main component and a partially overlapping impurity.

Deconvolution and visualization of single peaks can easily be applied to coeluting peaks using the LabSolutions software. It offers a cost-effective alternative to mass spectrometry, especially for analysis of isomers with similar retention behavior and identical m/z values [7].

4.3.2 "Analytical Intelligence" – AI, M2M, IoT – How Modern Technology can Simplify the Lab Routine | 353

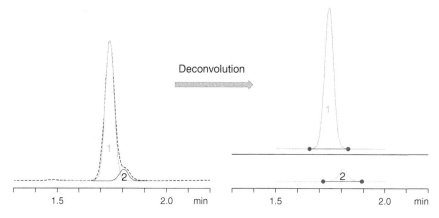

Figure 4.3.11 Illustration of the i-PDeA function for the individual quantification of a main component (1) and a partially overlapping impurity (2).

4.3.2.3 Predictive Maintenance to Minimize System Downtime

Another important part of laboratory management is the maintenance of running analytical instruments. Also here, modern technology can simplify planning to save costs and avoid unnecessary system failure and downtimes. In a well-run routine laboratory, regular, preventive maintenance is part of the workflow. Consumable parts are exchanged as standard and system modules get checked for performance. With use of the latest network technology, also known as IoT (Internet of Things), maintenance intervals and service resources can be organized much more efficiently (Figure 4.3.12).

By connection of the analytical systems with modern software, such as the Shimadzu Smart Service Net, the instrument status can be constantly tracked. Information on device usage is collected, as well as the amount of mobile phase that was pumped, the number of injections on a specific column or the switching intervals of a valve. The data are recorded in a database and consumable part replacement

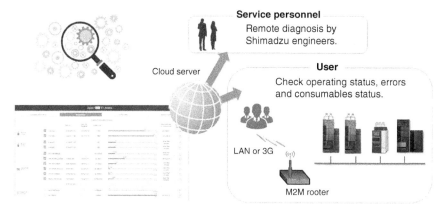

Figure 4.3.12 Use of IoT for resource optimization in an analytical laboratory.

can be scheduled according to the amount of use, instead of replacing pistons, seals, filters, etc. once a year by default, or worse, after a system failure. Based on the available data, warning intervals can be defined at which the instrument alerts the user about required maintenance, including article numbers of replacement parts. This not only offers the advantage of resource optimization in the laboratory, but also for the service provider who can plan appointments in advance to minimize waiting times. The larger the amount of data collected in the database, e.g. the column stability for a specific method, the more precisely it is possible to plan the reordering of spare parts and avoid unnecessary downtime – preventive maintenance becomes predictive maintenance.

References

1 Watanabe, K., Imamura, S., Fujisaki, S., Tanaka, K., Watanabe, M., Schad, G., Euerby, M., Hedgepeth W., Yamaguchi, T., Tomita, M. (2017). *(U)HPLC basics and beyond: An evaluation of (U)HPLC operating parameters and their effect on chromatographic performance*. Poster presented at the 45th International Symposium on High Performance Liquid Phase Separations and Related Techniques, Prague (21 June 2017), Poster presentation at HPLC 2017 Prague, 18–22 June 2017, Czech Republic, hpl2017-programme.pdf (hplc2017-prague.org).

2 Shimadzu (2019). *Instruction Manual for LC-40B XR Solvent Delivery Module (PN 228-92323)*, 41. Kyoto, Japan: Shimadzu Corporation.

3 Osaka, Y. *Shimadzu Application News No. L522 – Peak Shape Improvement Using the Auto-Pretreatment Function*, vol. 2018. Kyoto, Japan: Shimadzu Corporation.

4 Dolan, J. (2009). Integration problems. *LCGC North America* 27 (10): 892–899.

5 Yanagisawa, T. (2017). *Technical Report C191-E042: New Data Processing Method for Photodiode Array Detectors*. Japan: Shimadzu Corporation https://www.shimadzu.com/an/literature/hplc/jpl217011.html.

6 Wahab, M.F., Hellinghausen, G., and Armstrong, D.W. (2019). Progress in peak processing, recent developments in HPLC. *Supplement to LCGC North America* 37 (6): 24–32.

7 Schad, G. (2019). Advanced peak processing to reduce efforts in method optimization. *Chromatography Online* 15 (12): 22.

4.4

Addressing Analytical Challenges in a Modern HPLC Laboratory

Frank Steiner and Soo Hyun Park

Thermo Fisher Scientific, Dornierstr. 4, 82110, Germering, Germany

In today's efficiency-driven communities, optimization of facilities and processes is a common requirement. Typical improvement areas for analytical laboratories are workflows to shorten time from sample to analytical information, minimizing failure and unplanned downtime of analytical equipment, and rugged transferability of analytical processes for global scalability. While these are common themes also for high performance liquid chromatography (HPLC)-based analyses, not all criteria apply for every analytical task. It is up to the user to identify specific requirements to the HPLC equipment and methodology. One type of criteria is related to performance aspects of methods and devices, e.g. analysis speed and/or sample throughput, theoretical peak capacity or capability to address elevated sample complexity, and finally the detectability of target analytes as well as the accessible concentration range. Small-molecule characterization differs substantially from biomolecule characterization, both in respect of methodology and equipment requirements. While some extra performance on all these criteria can be advantageous, methods and equipment should not be substantially better than required, as this is always jeopardizing cost effectiveness. Moreover, it is hardly possible to optimize performance, robustness, and ease of use of methods and tools simultaneously, but it is more likely a decision to be made depending on the effective requirements as well as the education and experience level of the lab staff. Another kind of criteria depends on the type of laboratory. A research lab has different challenges than a routine lab, e.g. in quality assurance (QA)/quality control (QC) (see Chapter 1.9), and a regulated environment is also different from a nonregulated one. Sometimes HPLC methods are needed for a single use on a smaller sample set without specific further demands, sometimes they need full validation and documentation for regulated routine use, including the transferability to other laboratories that do not necessarily use identical equipment. A general industry trend to outsourcing research activities to contract organizations is another important driver for utmost flexibility requirements of today's typical labs, paired with demanding compliance and documentation demands. The Thermo Scientific Vanquish (U)HPLC platform offers solutions to meet all these needs, but it is helpful to understand the specific capabilities of the variants to make the best possible selection.

Optimization in HPLC: Concepts and Strategies, First Edition. Edited by Stavros Kromidas.
© 2021 WILEY-VCH GmbH. Published 2021 by WILEY-VCH GmbH.

4.4.1 Vanquish Core, Flex, and Horizon – Three Different Tiers, all Dedicated to Specific Requirements

The Vanquish (U)HPLC platform comprises systems built on the concept of integrated modularity. The different modules can be combined without limitations; they connect to each other in a system stack based on a rack enclosure with automatically connecting internal system drainage tubes. This rack concept provides a drawer-like capability to slide out individual modules for maintenance, repair, or exchange without the need for dismantling the entire stack. The whole Vanquish platform offers a choice between three tiers (Vanquish Horizon, Vanquish Flex, and Vanquish Core) to match the specific user requirements from routine HPLC up to ultra-high-end ultrahigh performance liquid chromatography (UHPLC) for specialized research labs.

Vanquish Horizon is the rigorous performance-optimized high-end tier of the platform and it pushes the boundaries of UHPLC. It is biocompatible by default, to relieve the user from making a decision on this, and supports maximum pressure levels of 1500 bar, combined with maximum flow-rate capabilities of 5 ml/min, column temperatures up to 120 °C, ultraprecise and ripple-free binary gradient delivery with lowest gradient delay volume and very precise sample volume dosage even in the sub-1 μl range. The Vanquish Horizon light pipe detector is signal to noise optimized for trace-level analysis combined with a wide linear detection range. Vanquish Horizon is the system of choice to maximize peak capacities in a single dimension, as the outstanding pressure range supports the combination of extended column lengths or coupled column chains with small stationary-phase particle sizes to maximize theoretical plate numbers. It is also the system of choice for maximizing throughput, thanks to excellent control of steep gradients and lowest gradient delay volume (GDV). The parallel piston linear drive pump technology with advanced thermal equilibration control (ATEC) and SmartStroke-based choice between minimum mechanical wear with 120-μl stroke volume or intelligent small-step movements to minimize mixing ripples. This accounts for most accurate proportioning, smoothest baselines even with trifluoroacetic acid (TFA)-containing mobile phases or the choice for maximum long-term stability and reliability, depending on the specific method requirements. The use of Vanquish Horizon as a front system to the unique Thermo Scientific Orbitrap-based high-resolution accurate mass spectrometers (HRAMS) also makes a very powerful combination.

The so-called SmartInject technology is a feature of all Vanquish autosamplers of the platform and so is the proprietary maintenance-free long-term use injection valve. SmartInject is based on ultrafast switching of the injection valve combined with an intelligent control mechanism, which enables to dampen the impact of the injection shock to almost zero-pressure fluctuation. This substantially improves retention time precision, column lifetime, and peak integration accuracy of early eluting compounds. The vertically oriented thermostatted column compartment of the Vanquish systems offers the unique choice between still-air and forced-air thermostatting modes, as well as a selection between passive- and active mobile-phase thermostatting, where the active mode allows to set temperatures different from the column compartment temperature in the expert mode. These capabilities are very powerful in method transfer scenarios, when the behavior of the column compartment needs to be very similar to that of an originator system, which is using different technology.

The midtier variant of the platform is named the Vanquish Flex, a yet fully biocompatible UHPLC system that supports a maximum pressure of 1000 bar and a maximum flow rate of 8 ml/min (the latter at a pressure of up to 800 bar). Vanquish Flex features identical column thermostatting specifications and capabilities like the Vanquish Horizon. Unlike Vanquish Horizon, Vanquish Flex offers a choice between a binary high-pressure proportioning gradient pump (HPG), quaternary low-pressure proportioning gradient pump (LPG), and the unique dual gradient pump (DGP). The DGP is a standard-size device with two independent ternary low-pressure proportioning gradient pumps in one enclosure. It is an important contributor to the Vanquish Duo system variants that support automation and productivity workflows based on two flow paths. This can be applied for 2D-LC applications, Tandem column modes for maximizing data generation times of MS detectors, detection multiplexing, and the so-called Dual LC mode for operation as a full-blown 2-channel HPLC at standard system size. The Dual LC mode is mainly enabled through the Vanquish Dual Split Sampler, a standard-size autosampler that incorporates two completely independent injection needles and sample dosage devices that only share a common sample carousel. Vanquish Dual Split Samplers are offered for both the Vanquish Flex and for Vanquish Horizon, supporting up to 1000 bar or up 1500 bar, respectively. The concept of Dual LC is illustrated in Figure 4.4.1. The separate flow paths from the DGP and the Dual Split Sampler are connected to two columns in either one thermostatted compartment or two compartments, if methods running at different column temperatures are combined. The two columns are then connected with two stacked detectors. Vanquish Flex provides many choices in detection, comprising diode array (DAD) and variable-wavelength light absorbance detection (VWD), fluorescence detection (FLD), and the unique charged aerosol detection (CAD) for most uniform response characteristics, independent of analyte molecule structures. The Dual LC set-up can either perform a double-analysis throughput by running the same method on both channels, or provide a complementary characterization of a sample by running two different methods at the same time, e.g. simultaneous analysis of water-soluble and fat-soluble vitamins or two critical quality attributes of a biopharmaceutical compound (see Figure 4.4.1). Unlike two separate (U)HPLC systems with their double-bench space occupation, the Dual LC set-up has the additional advantage of possible injections even from the same sample vial into two different methods without the need for aliquoting.

Thanks to its biocompatible design, the Vanquish Flex is the system of choice for any routine analysis in biopharmaceutical labs. Biocompatibility implies the absence of iron-containing materials in contact with the mobile phase or the sample. This is important when corrosive mobile phases are applied (e.g. high chloride content in ion exchange or hydrophobic interaction chromatography) or when analytes have a strong affinity to iron (e.g. phosphopeptides) resulting in peak distortion or loss of sample in the system. The Vanquish Duo–based configurations

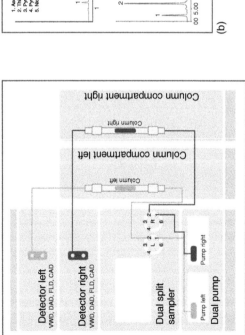
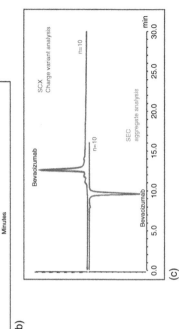

Figure 4.4.1 The Vanquish Duo-based Dual LC concept, (a) Fluidic configuration with two completely independent flow paths, (b) Example of simultaneous gradient separation of water- and fat-soluble vitamins, (c) simultaneous charge-variant analysis based on pH gradient separation on a strong cation exchange column (MAbPac™ SCX-10) and aggregate analysis based on a size exclusion column (MAbPac™ SEC-1) applied to the monoclonal antibody bevacizumab.

offer great capabilities to automate simultaneous multiattribute characterization to boost productivity of biopharma labs. It is also ideal for fast screening of method parameters and thus the ideal tool for method development (see below). The binary (HPG) variant can be configured to low GDV, hence making it a powerful front system for mass spectrometry, where typically lower flow rates in the range of 200–400 µl/min are preferred. Moreover, the wide range of pump and detector variants (including the 1000-bar dual gradient pump) make Vanquish Flex modules also ideally suited for combinations in advanced set-ups to run multidimensional workflows (see Section 4.4.3). The Chromeleon software control supports virtually any possible combination of modules, so that Vanquish Horizon modules can be added to Vanquish Flex modules whenever the lowest GDV and 1500-bar pressure capability is advantageous in a multidimensional setup to address highly complex samples, also in combination with advanced mass spectrometry.

The Vanquish Core is the latest introduction of the Vanquish family. It completes the portfolio at the lower end and was designed with a clear focus on robustness and compliance to meet the typical needs of routine HPLC laboratories. All knowledge and experience, as well as thorough testing protocols for best durability of moving parts from the Vanquish Horizon and Flex development were considered to build an even more robust workhorse for routine applications. Vanquish Core is based on non-biocompatible stainless steel fluidics and supports a maximum pressure level of 700 bar combined with a maximum flow rate of 10 ml/min and a maximum column temperature of 85 °C. The injection volume range has been extended to 1 ml or higher, depending on the sample loop in use. While the autosampler inherited all strong features from Flex and Horizon (SmartInject, maintenance-free valve, sample cooling concept, etc.), there is even more innovation thanks to a new proprietary valve design of the Vanquish Core autosampler. This injection valve can adapt many different positions to support the following specific functions:

- SmartInject control mechanism to dampen the impact of the injection shock
- Metering device idle volume variation (0–230 µl) for stepless GDV tuning independent of V_{Inj}
- Multidraw option to fill sample loops larger than the metering device volume
- Pump and metering device purge into needle wash port
- Automated tightness test of pump
- Automated tightness test of pump and autosampler

This special design combines great injection volume flexibility with outstanding method transfer capabilities thanks to seamless GDV adjustment and powerful automated system diagnostics features. The GDV adjustment (see Figure 4.4.2) is assisted by a software wizard and it is important to note that this technology enables programmed and fully traceably hardware volume adjustment. Unlike manual changes of mixers or capillaries to tune system volumes, this automatic adaption does not affect the instrument qualification status and is thus fully compliant in a regulated environment. Column and mobile-phase thermostatting are another important instrumental parameter in method transfer, because there

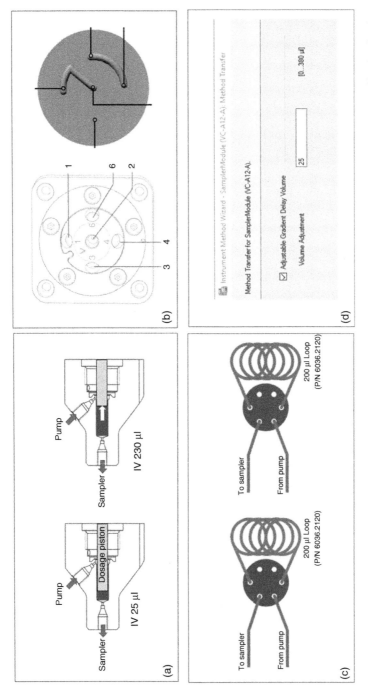

Figure 4.4.2 Vanquish Core features for seamless adjustment of system GDV, (a) autosampler metering device based adjustment of its dwell volume contribution by setting the idle volume (IV) value (piston position when valves switches to INJECT) independent of injection volumes, (b) proprietary injection valve and rotor groove design, here in bypass position to empty metering device into desired position (1: pump, 2: column, 3: needle seat, 4: wash port waste, 6: metering device & loop & needle), (c) optional method transfer kit installed in column compartment to add 200-μl GDV, (d) software assistant to vary autosampler GDV contribution through metering device and switch loop.

is more contribution to the effective internal temperature inside a column than the actual temperature in the surrounding column compartment. The variable fan speed for the forced-air operation and the ability to operate the compartment alternatively in still air mode help to closely simulate the behavior of other column compartments. Eluent preheating is another contributor to instrument specific variability. The Vanquish Core column compartment features a passive preheater by default that should always be used when the column temperature is set to subambient, because this requires precooling of the eluent before entering the column. Optionally, there is also an active preheater available (heat only), which has an independent temperature control and can be set to a temperature value different to the compartment temperature. This enables to mimic insufficient preheating or undesired overheating occurring in the originator system (from which a method is transferred). Combining the GDV and column/mobile phase thermostatting tuning options, the Vanquish Core system sets new standards on method transfer capabilities. A software wizard to holistically address all these instrument parameters and iteratively optimizing them until matching with a target chromatogram is in development.

Vanquish Core offers the largest choice of pump variants, including an isocratic pump in addition to LPG, HPG, and DGP. A wide range of detectors is available with a diode-array technology-based 8-channel multiwavelength detector (MWD), next to the DAD, VWD, and FLD. An integrated stackable refractive index detector (RID) is in development. The Vanquish Core autosampler can be ordered with and without sample thermostatting and a manual injection valve (operating up to 500 bar) is also available. Thanks to the unlimited modular combination possibilities, the Vanquish Core modules can also be combined with detectors from the Horizon and Flex tier. This gives access to the unique charged aerosol detection, light pipe–based diode array detection featuring the 60-mm light path high-sensitivity flow cell (designed for methods based on column diameters of ≥ 3 mm), and single-quadrupole mass spectrometers under full Chromeleon software control single-quadrupole mass spectrometers (ISQ EC, ISQ EM).

The Vanquish Core is the system of choice for all routine applications, whether in regulated or nonregulated environments and is the ideal workhorse when no biocompatibility is required. With a maximum pressure capability of 700 bar, it can run even long columns with particle sizes below 3 µm at high linear velocity for fast separations. When column diameters below 3 mm are targeted, the extra column dispersion contribution will negatively impact on peak resolution and separation will be inferior to that on a Vanquish Flex or Horizon system. This is very critical with short columns and sub-2 µm particles where very small peak volumes are generated, as well as with isocratic separations where the gradient peak compression effects are absent that can correct for precolumn dispersion in the system.

The explicit strength of Vanquish Core lies in the robustness, usability, and compliance aspects for routine analysis of predominantly small molecules in pharma, food & beverage, personal care products, and fine chemicals. The Vanquish Core is an important step into the direction of a smart and self-contained system.

4.4.2 Intelligent and Self-Contained HPLC Devices

With the advent of the era of "Industry 4.0" – the fourth industrial revolution, many industrial laboratories are about to undergo important changes in their operations. A major aspect is to introduce far more automation and related new technologies, such as Internet of Things (IoT), big data analytics, and artificial intelligence. This ultimately aims at improving lab efficiency and/or productivity by lowering human labor and errors and substantially reducing unplanned instrument downtime. Accordingly, the new Thermo Scientific Vanquish Core systems have included intelligent features to facilitate routine tasks in the lab such as the maintenance of instruments and resource allocation. The Vanquish Solvent Monitor (VSM) is an optional module that tracks automatically the volume of eluents and waste in solvent reservoirs and in the waste container. Two types of VSM with either 4 channels or 8 channels (for setups with more fluidic paths) are available. The solvent volumes for all solvent types and mixtures are accurately measured, based on hydrostatic pressure monitoring. An intelligent self-calibration is automatically estimating solvent density based on known volume delivery and known solvent reservoir dimensions. This automated adjustment enables monitoring liquid levels in solvent reservoirs and even in waste containers, where composition and thus density permanently changes under gradient operation. The sequence can be automatically stopped when a predefined or default warning limit is exceeded during analysis. The solvent levels can be monitored via ePanels in Chromeleon, the optional Vanquish User Interface (VUI) and a web browser allows further convenient or even remote monitoring of instrument status. The VUI, consisting of the system controller and the display, allows to monitor and display important parameters of the overall system and modules, as well as instrument parameter data that are collected for maintenance and service purposes.

While these smart tools and features have been first introduced with the Vanquish Core system, they are about to be extended to the entire Vanquish instrument platform and will be further refined in the coming years based on different innovative approaches. One of these is building more intelligence directly into the systems, hence making them more independent from the control of the chromatography data system and the user interference. Another approach is the holistic continuous monitoring and processing of all available data of an HPLC instrument in operation to identify beginning malfunction earlier and predict possible failure. Next to anyway available data from standard operation control (e.g. motor positions or pressure levels) used for background diagnostics, additional sensor devices to monitor parameters in the fluidic path (e.g. pH, conductivity or zones profiles) together with background processed data from chromatograms and spectra (peak symmetry, plate number, resolution, retention time, spectral purity data, Fourier analysis of baselines, etc.) can also be considered to continuously monitor status of instrument, column, and mobile phase. Such automated background processes will be operated in the context of the emerging digital capabilities in Industry 4.0 to enable HPLC systems or HPLC-based analyzers with close to zero unscheduled downtime and never-failing chromatography for utmost productivity.

4.4.3 2D-LC for Analyzing Complex Samples and Further Automation Capabilities (see also Chapter 1.1)

Two-dimensional liquid chromatography (2D-LC) has been the most popular technique for analyzing complex samples since the technique can increase significantly the effective peak capacities (up to the level of 10 000). It also enables to employ orthogonal selectivities in the two dimensions as well as various combinations of desired techniques (IEC × RPLC, RPLC × RPLC, HILIC × RPLC, NPLC × RPLC, SEC × RPLC, etc.). Thanks to resulting combined selectivity and kinetic separation performance, overlapping peaks in complex samples (such as omics samples, samples with large numbers of analytes and/or samples with structurally similar compounds) can be well resolved and the certainty of peak assignments can be enhanced. Basically, 2D-LC separations are performed by collecting fractions (of interest) from the 1D-separation, followed by transferring them to the second separation dimension. There are mainly three types of 2D-LC separations according to the size and feature of fractions transferred to the 2D-part: heart-cut (LC–LC), comprehensive (LC × LC), and selective comprehensive mode (sLC × LC). In heart-cut LC–LC, a single or several fractions from the first separation dimension are cut and then injected to the second dimension. This mode is typically used for semicomplex samples. In case an additional separation or detection technique is needed to obtain additional information that the first 1D-LC cannot provide, the heart-cut LC–LC approach can also be an attractive option. In the comprehensive technique, fractions representing the 1D-separation power holistically are injected to the second dimension. This allows comprehensive analyses for even more complex samples, as the peak capacities from the first and second dimensions are multiplied, resulting in highest total peak capacities. A pre-requisite for successful use of this high theoretical peak capacity is sufficient orthogonality of the following two separation criteria, appropriate population of the whole 2D retention space and sufficient oversampling through the number of transferred fractions. To achieve the latter criterion, the number of fractions must be at least twice as high as the peak capacity of the first dimension. The sLC × LC is a combined tradeoff of the LC–LC and LC × LC approaches, where selected regions of 1D effluent are transferred to the 2D separation with an oversampling rate similar to the LC × LC approach [1, 2].

Another strong incentive for heart-cut LC–LC approaches is given when the 1D eluent is not compatible with MS, but MS-based characterization is required for a given purpose. For example, LC–UV-based impurity methods may require an MS-based peak purity control, or peak identity confirmation, in particular when additional unknown peaks appear in the UV chromatogram. The LC–UV method conditions often apply non-MS-friendly eluent additives (e.g. phosphate) and are often difficult to change in terms of regulatory compliance. Also, it is hardly possible to make changes to an MS-friendly method without altering selectivity, thus complicating correct assignment of peaks from the original LC–UV method. In such cases, it is a great opportunity to heart-cut the respective peaks into an LC–MS run in a second dimension, where the fraction is desalted, possibly further separated, and subjected to MS detection for further characterization with an unambiguous relation to the 1D-LC–UV chromatogram. It is very likely that such instrumental capabilities will experience increasing demand in the near future, due to more strict regulatory requirements for identity confirmation of additional impurity peaks.

In 2D-LC, incompatibility (or solvent mismatch) between the 1D- and 2D-eluents often becomes an important challenge, resulting in peak splitting or peak deformation. To address numerous 2D-LC related challenges, great progress in instrument technology, including the interface (or modulation) between the first and the second separation dimensions, has been made for over the past decade. For more details on the fundamental principles and successful implementation of 2D-LC techniques, including smart interface techniques (such as passive modulation, stationary phase–assisted modulation, active solvent modulation, and vacuum-evaporation modulation), refer to the ref. [1]. Thermo Fisher Scientific offers various online 2D-LC solutions, based on the Vanquish instrument platform and Chromeleon chromatography data system (CDS). The online 2D-LC enables the automatic fraction transfer that is implemented typically via switching valves. The LC–LC approach is the current focus among the three 2D-LC modes (i.e. LC–LC, LC × LC, and sLC × LC) and the 2D-LC solutions using various heart-cut techniques are highlighted below as representative examples [3]. This includes straightforward guidelines for online 2D-LC instrumentation on the Vanquish platform, along with the following major fluidic setups for fraction transfer:

- Loop-based single-heart-cut 2D-LC
- Loop-based multi-heart-cut 2D-LC
- Trap-based single-heart-cut 2D-LC(-MS) for eluent strength reduction
- Trap-based single-heart-cut 2D-LC(-MS) using dual-split sampler

4.4.3.1 Loop-based Single-Heart-Cut 2D-LC

The schematic fluidic setup is shown in Figure 4.4.3. The interface consists of a 2-position 6-port switching valve and a single loop (between ports 3 and 6 in the switching valve). The loop is used to sample and store a desired fraction from the first dimension separation. The switching valve allows to direct the first dimension effluent to either the waste or the loop interface, prior to the 2D analysis.

4.4.3.2 Loop-based Multi-Heart-Cut 2D-LC

The multi-heart-cut setup, as shown in the Figure 4.4.4, consists of two column compartments, a 2-position 6-port switching valve and two 7-port 6-position selection valves in the interface between the two separation dimensions, allowing to transfer several target fractions to the second separation dimension. Maximum six loops can be coupled between two 7-port 6-position selection valves. Similar to the single-heart-cut setup, the upper 2-position 6-port switching valve directs the 1D effluent to each loop or to the waste. The respective target fractions in each loop

4.4.3 2D-LC for Analyzing Complex Samples and Further Automation Capabilities | 365

Figure 4.4.3 Loop-based single-heart-cut 2D-LC setup. The switching valve integrated in the column compartment selects between collecting in the embedded loop and transferring it to the second dimension column.

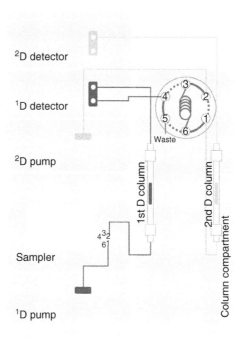

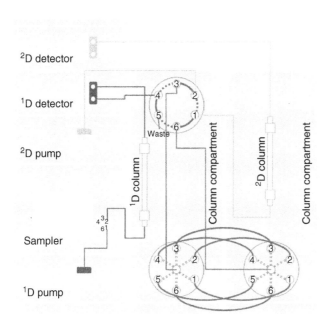

Figure 4.4.4 Loop-based multi-heart-cut 2D-LC setup with the Vanquish platform.

are successively processed when the 2D flow path is connected to the loop, followed by the 2D analysis. One loop can be set as default to flush while no cut is performed, allowing to collect a maximum of five fractions. The Chromeleon software allows straightforward programming of the flow path changes in the upper position valve and the loop selection in the lower valves. Briefly, in the instrument method of the Chromeleon CDS, the position valve 1_2 is set for the complete 1D run time and the position valve 1_6 for the heart-cut and then 2D analysis. With an additional column compartment, the capacity can be extended to 11 fractions. The additional column compartment also allows to cool fractions for improved analyte integrity of stored fractions, independent of the column thermostatting temperature in the other compartment.

4.4.3.3 Trap-based Single-Heart-Cut 2D-LC for Eluent Strength Reduction

The trap-based single-heart-cut 2D-LC setup can be used in the challenging situation mentioned above, where the 1D eluent is not compatible with MS that is required to use the peak purity control in the 2D analysis. The eluent strength in 1D solvent should be reduced before entering the MS detector. Figure 4.4.5 shows the schematic fluidic setup, consisting of two pumps, two (6-port 2-position) switching values, a trap column, and a T-piece. When the 1D separation finishes with the position 1_2 in the upper switching valve, the valve switches the position 6_1. The 1D effluent then reaches at the T-piece, where the 1D effluent was mixed with 2D eluent and then transferred to the trap column in the lower valve. The position 1_2 in the lower valve allows to wash the trap column with the eluent from the second dimension,

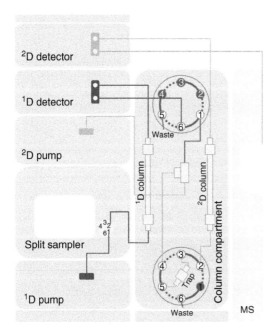

Figure 4.4.5 Trap-based single-heart-cut 2D-LC setup for eluent strength reduction, with the Vanquish platform.

resulting in the overall eluent strength reduction, while fractions are loaded and mixed in the trap column. The lower valve switches to the position 6_1 to direct the fractions to the 2D column and then to start the 2D analysis. It should be noted that the position 6_1 in both valves at the same time should be avoided to prevent the MS detector contamination due to the nonvolatile salts from the 1D eluent. Trap-based fraction transfer also enables to collect fraction volumes that exceed the typical loop sizes and it can be combined with loop-based transfer capabilities, even in the same column compartment, so that the user has the choice to select the mode.

4.4.3.4 Trap-based Single-Heart-Cut 2D LC–MS Using Vanquish Dual Split Sampler

A very advanced setup, with additional column compartment and switching valve, is shown in Figure 4.4.6. This setup has the great advantage that it is utilizing a Vanquish Dual Split Sampler, enabling the separate introduction of samples directly to the second separation dimension. This can facilitate method development in the second dimension for both the separation and MS detection parameters. The injection of the calibration standard directly to the second dimension can also be used for internal calibration, a very powerful option for quantification with MS detection. Lastly, it can help to determine recovery rates of transfer from first to second dimension and can switch the system from a 2D-LC into a two single-channel instrument without manual replumbing.

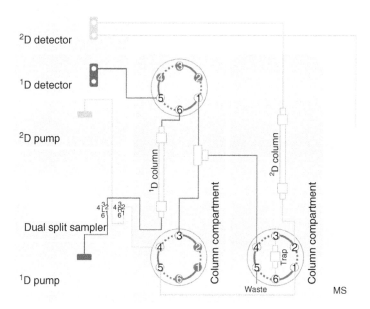

Figure 4.4.6 Trap-based single-heart-cut 2D-LC setup for eluent strength reduction, with the Vanquish platform using Dual Split Sampler.

Additionally, this setup allows to control independently the column temperatures of the two dimensions and avoids the accidental entering of non-MS compatible 1D eluent to the MS in the second dimension, reducing user errors. Similar to the previous setups, the upper switching valve (with the position 6_1) plays the role in transferring the fractions to the trap column via the T-piece, at which 1D eluate is mixed with the 2D flow, reducing the eluent strength. The left lower valve directs the 2D flow to the T-piece and the right lower valve controls if the combined 1D- and 2D flow at the T-piece goes to the trap column or the waste. Both switching valves are located in position 1_2 for the trapping of the fractions and the washing of the trap column, while both lower valves are switched to position 6_1 for the 2D analysis. Furthermore, such setup can be used as two independent 1D-LC systems, by integrating the right sampling unit into the 2D fluidics. This does not need any manual replumbing and also increases the productivity. For the configuration, both lower valves locate at position 6_1, then different samples from the right sampler can be directed to the 2D column via the right lower valve.

The capabilities of the setup from Figure 4.4.6 are illustrated in the application example in Figure 4.4.7 (details described in [3]). The method of the first dimension was optimized for use with UV detection and was not able to separate the two isomeric agrochemical substances, phenmediphan and desmediphan. The switch into a second dimension method enabled both the separation of the isomers and MS-compatibility for accurate mass-based identity confirmation on an Orbitrap MS, proving the two substances are isomeric, both with M+1 of 301.117 Da. In the alternative mode of operation where the system can be switched over automatically, it can be applied as a two-channel device to run both methods simultaneously for complementary characterization of the sample, typically with both channels using UV detection. This flexibility offers various modes for advanced multiparameter sample characterization, and it can be automatically switched between a dual LC for throughput increase or a 2D-LC system.

4.4.4 Software-Assisted Automated Method Development

Method development (MD) effort continues to be another critical bottleneck in HPLC labs and often impedes throughput and productivity. To address this, there is a continuous attempt to lowering involved human labor and reducing the MD time and cost. Automated method scouting is typically performed as a first step, prior to subsequent method optimization, to select basic chromatographic parameters, such as stationary phase and separation mode, as well as mobile-phase components, additives, and appropriate pH ranges. It is also very common in most (bio)pharmaceutical companies to perform column screening as an important step to obtain methods for determination of critical quality attributes (CQA) in new product development processes. Both in small-molecule pharma and biopharma labs, four different stationary phases are typically screened at different mobile-phase compositions and pH, as candidates of orthogonal RPLC methods.

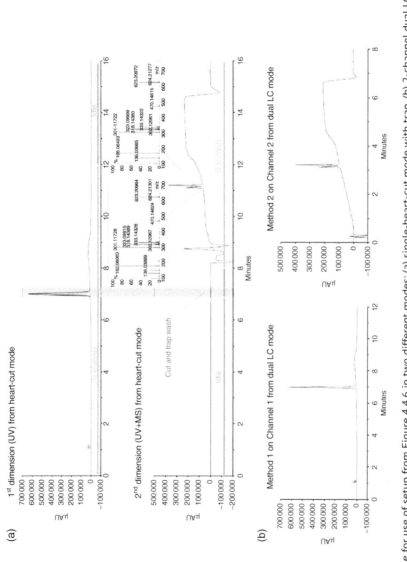

Figure 4.4.7 Example for use of setup from Figure 4.4.6 in two different modes: (a) single-heart-cut mode with trap, (b) 2-channel dual LC mode. Both modes employ identical separation methods, either sequentially in 2 dimensions or simultaneously in 2 channels. Method 1: 0.1% aqueous phosphoric acid to acetonitrile gradient on Thermo Scientific™ Hypersil GOLD™ (2.1 × 100 mm, 1.9 μm) column. Method 2: water to methanol gradient, both at 0.1% formic acid on Thermo Scientific™ Accucore™ PFP (2.1 × 50 mm, 2.6 μm) column. Trap column for heart-cut mode: Thermo Scientific™ Hypersil™ Javelin BDS C18 (3 × 20 mm, 3 μm) guard column. Analytes: phenmediphan and desmediphan (not separated in method 1). UV detection at 210 nm, positive mode ESI–MS detection on Thermo Scientific™ Q Exactive™ Plus Hybrid Quadrupole - Orbitrap™ Mass Spectrometer. *Source*: Gruebner and Greco [3]. © Thermo Fisher Scientific Inc.

The four methods are thoroughly chosen, considering maximal practical peak capacity, individual method's peak capacities, and method orthogonality. The screening step is used as a starting point for further method development and allows a better understanding of process-related impurities [4]. The screening process can be easily performed in unattended manner on a Thermo Scientific Vanquish instrument-based "automated method scouting (AMS) solution." The AMS setup includes two 6-position 7-port switching valves in the column compartment of (U)HPLC system, allowing automatic selection (and switching) of up to six columns. In addition to the column selection valves, a quaternary gradient pump is preferably utilized to facilitate blending of up to four mobile-phase solvents, including additives like buffer components. Additionally, an optional Extension Kit for AMS solution, consisting of a low-pressure solvent selection valve, can be added to one of the solvent channels, allowing automatic variation of up to 13 different mobile-phase conditions (varying different buffer additives and pH). Single-, dual-, or even triple-column compartment setups, depending on target applications or candidate columns, are possible with the Vanquish instrument platform. Two types of fluidic setups (i.e. single- and dual-column compartment setups) of the AMS configuration, with the Vanquish UHPLC system, are shown in Figure 4.4.8.

The AMS is initiated by automated sequence setup with custom variables in the Chromeleon software, which can be downloaded from the AppsLab repository as so-called Chromeleon eWorkflows. Resulting large data sets are automatically evaluated using the intelligent features of Chromeleon software. Specifically, the intelligent run control (IRC) feature is used to check pass/failure of injections, based on predefined criteria (such as minimum resolution between critical peak pairs, the number of detected peaks, and the peak asymmetry). The injection query tool of the software extracts successful injections with successful methods according to the search criteria and the most promising candidates are then listed in a virtual sequence, available in the Chromeleon CDS studio. The Smart Report feature allows to visualize all results, along with the table showing resulting parameters (such as sample number, sample name, elution time of last compound, and the resolution of the critical peak pair). Individual peaks in the best method are assigned by the Chromeleon Component Table Wizard (in the component table of the processing method) and the Spectral Library Screening. Please refer to the Ref. [5] for more details.

The automatic scouting system above, consisting of the Vanquish (U)HPLC instruments and Chromeleon CDS, can be combined with commercially available HPLC method development software packages such as ChromSword Auto, DryLab, and Fusion quality-by-design (QbD). Computer-assisted MD facilitates, by employing retention models, the selection of chromatographic parameters (including columns, mobile-phase buffer and pH, flowrate, temperature, etc.). The systematic workflow using the in silico tools allows to implement QbD practice required in industrial labs, resulting in robust and rugged methods. In addition, the computer-assisted MD supports less experienced users even in developing methods for challenging and complex samples. As an example, ChromSwordAuto® software provides online method scouting and optimization in unattended manner. Briefly,

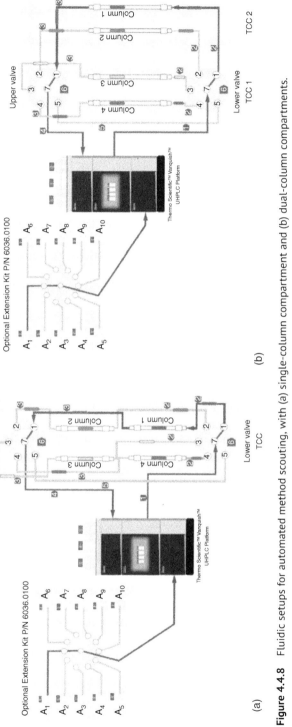

Figure 4.4.8 Fluidic setups for automated method scouting, with (a) single-column compartment and (b) dual-column compartments.

two modules of the ChromSwordAuto® software, which are ChromSwordAuto® Scout and ChromSwordAuto® Developer, were used for automatic method scouting and optimization. This workflow can be applied with all the Vanquish (U)HPLC systems. For the use of the ChromSwordAuto software with (U)HPLC systems, the configuration between a (U)HPLC system in use and ChromSword is required. Similar to the Thermo Scientific AMS system, the ChromSwordAuto Scout can be used to screen automatically the most promising columns, rough mobile-phase conditions (such as organic solvent type, buffer additives, and pH), using the ChromSword switching columns/solvents algorithm. With the ChromSwordAuto Scout, different gradient profiles can be scouted simultaneously with the abovementioned method parameters since different gradient methods can also be programmed in the same sequence. Once rough method parameters (such as a candidate column, buffer additives, buffer pH, organic solvent type [e.g. methanol or acetonitrile]) are determined by the method scouting, the ChromSwordAuto Developer can be used for further method optimization, again in unattended manner. Typically, two types of development tasks (i.e. rapid optimization and fine optimization indicated as "sample profiling – isocratic and gradient optimization") can be used to find the best method for small molecules, in terms of the separation (e.g. with minimum $R_s > 1.5$) and analysis time (i.e. short as much as possible). The selected task type and the starting conditions (i.e. column, solvents, flow rate, injection volume) are together programmed in the Developer software. The optimization is automated by iterative runs, following study runs building retention models for a test sample. New runs are planned based on the processed results of the previous runs, using the optimization search tools such as Monte Carlo, neural network methods, and genetic algorithms. Typically, 3–4 runs are executed for the rapid optimization and more runs (greater than four runs) for the fine optimization [6]. With the ChromSwordAuto® software, the direct injection of a challenging sample, that is catechins in tea, allowed to even more speed up method optimization, following method scouting with standard samples (Figure 4.4.9). Briefly, the method scouting resulted in selecting the most promising column (i.e. Accucore Phenyl-X column) and mobile phase (20 mM ammonium acetate buffer at pH 3.8). The followed rapid- and fine-gradient optimization enabled a baseline separation of a main analyte peak (i.e. epigallocatechin) from sample matrix [7].

The abovementioned automated method scouting still requires analysts' knowledge and experience to choose the four test columns and initial mobile-phase conditions. Since several hundreds of RP columns are available in market, the selection of four orthogonal columns is still challenging and the approach is limited due to the available materials (e.g. columns) and the amount of solvent for the initial experiments. A promising solution can be the utilization of quantitative structure–retention relationship (QSRR) modeling for column scoping, which retention models based on molecular structures of analytes (i.e. QSRR models) are used to predict retention times of unknown compounds and to simulate separations of a test sample. With the QSRR modeling, the method scouting can be performed without any experimentation, again reducing substantially the MD time and cost. A promising example using the QSRR modeling for RPLC column scoping has

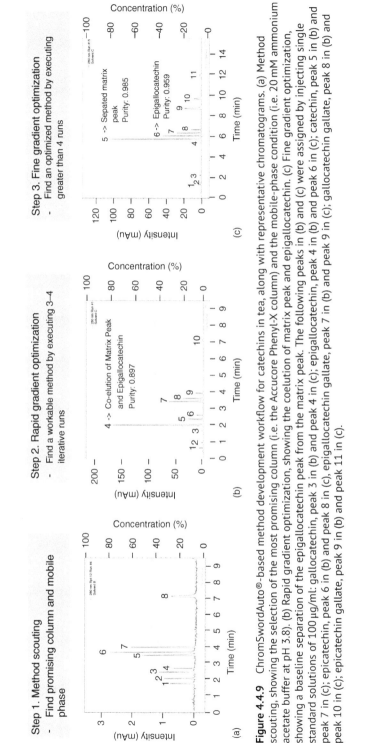

Figure 4.4.9 ChromSwordAuto®-based method development workflow for catechins in tea, along with representative chromatograms. (a) Method scouting, showing the selection of the most promising column (i.e. the Accucore Phenyl-X column) and the mobile-phase condition (i.e. 20 mM ammonium acetate buffer at pH 3.8). (b) Rapid gradient optimization, showing the coelution of matrix peak and epigallocatechin. (c) Fine gradient optimization, showing a baseline separation of the epigallocatechin peak from the matrix peak. The following peaks in (b) and (c) were assigned by injecting single standard solutions of 100 μg/ml: gallocatechin, peak 3 in (b) and peak 4 in (c); epigallocatechin, peak 4 in (b) and peak 6 in (c); catechin, peak 5 in (b) and peak 7 in (c); epicatechin, peak 6 in (b) and peak 8 in (c); epigallocatechin gallate, peak 7 in (b) and peak 9 in (c); gallocatechin gallate, peak 8 in (b) and peak 10 in (c); epicatechin gallate, peak 9 in (b) and peak 11 in (c).

been reported [8]. Predictive QSRR models to model solute parameters (such as η, solute "hydrophobicity" coefficient) in the hydrophobic subtraction model, well known in RPLC, in combination with column parameters to be retrieved from the United States Pharmacopeia (USP) website, successfully proposed the most suitable columns, providing baseline separations of simple test mixtures. The most suitable column was selected by comparing predicted resolution values of critical peak pairs. The QSRR approach can be used on numerous RP columns listed, e.g. on the USP website, providing accurate retention models, and hence, the feasibility of in silico column scoping in RP MD.

Abbreviations

HILIC	hydrophilic interaction chromatography
IEC	ion exchange chromatography
NPLC	normal phase chromatography
RPLC	reversed phase liquid chromatography
SEC	size exclusion chromatography

References

1 Pirok, B.W.J., Stoll, D.R., and Schoenmakers, P.J. (2019). Recent developments in two-dimensional liquid chromatography: fundamental improvements for practical applications. *Analytical Chemistry* 91: 240–263.

2 Iguiniz, M. and Heinisch, S. (2017). Two-dimensional liquid chromatography in pharmaceutical analysis. Instrumental aspects, trends and applications. *Journal of Pharmaceutical and Biomedical Sciences* 145: 482–503.

3 Gruebner, M. and Greco, G., Flexible HPLC instrument setups for double usage as one heart-cut-2D-LC system or two independent 1D-LC systems. TN-73298.

4 Szucs, R., Brunelli, C., Lestremau, F., and Hanna-Brown, M. (2017). Liquid chromatography in the pharmaceutical industry. In: *Liquid Chromatography: Applications* (eds. S. Fanali, P.R. Haddad, C.F. Poole and M.-L. Riekkola), 431–453. Amsterdam: Elsevier.

5 Fabel, S. and Heidorn, M., Fast and easy HPLC method development: Application of an Automated Method Scouting system. TN-161.

6 Galushko, S., Tanchuk, V., Shishkina, I. et al. (2006). ChromSword® software for automated and computer-assisted development of HPLC methods. In: *HPLC Made to Measure* (ed. S. Kromidas), 587–600. Weinheim: Wiley-VCH.

7 Grosse, S., Park, S.H., De Pra, M., and Steiner, F. *Automated Method Development in HPLC for the Quantitative Determination of Catechins in Tea*. Germering, Germany: Thermo Fisher Scientific. AN-72830.

8 Park, S.H., De Pra, M., Haddad, P.R. et al. (2019). Localised quantitative structure-retention relationship modelling for rapid method development in reversed-phase high performance liquid chromatography (2020). *Journal of Chromatography A* 1609: 460508.

4.5

Systematic Method Development with an Analytical Quality-by-Design Approach Supported by Fusion QbD and UPLC–MS

Falk-Thilo Ferse[1], Detlev Kurth[1], Tran N. Pham[2], Fadi L. Alkhateeb[2], and Paul Rainville[2]

[1]Waters GmbH, Helfmann-Park 10, 65760, Eschborn, Germany
[2]Waters Corporation, 34 Marple Street, Milford, MA 01757, USA

Reversed-phased liquid chromatography (LC) separation methods are developed for all kinds of chromatographic assays. Their performance is critical for ensuring the quality, safety, and efficacy of pharmaceutical products; monitoring clinical patient parameters, food contents, and contaminants; screening environmental samples for toxins or any other health-impacting substances; and for characterizing products used in material sciences. The sample itself as well the sample matrices vary in complexity. Method development is costly and time consuming and there is a desire to streamline such processes to bring new products faster to market. Currently, there are several approaches to method development. A traditional, commonly used approach is one factor at a time (OFAT) [1–3]. In the OFAT protocol, only one parameter is varied and its effect on responses is evaluated while others remain constant. When no more improvements are attained from changing this factor, another parameter is then explored [4]. This approach follows the human logic to keep the overview of the experiments. However, it is not very comprehensive because complex relationships between varied parameters cannot be seen or considered.

Analytical Quality-by-Design (AQbD) is a more comprehensive, systematic, and risk-based approach for method development that starts with predefined method performance criteria. In this approach, multiple parameters and settings are explored to provide a broad knowledge about the impact of the studied factors on the method performance. This knowledge is used to establish the method operable design region (MODR) that corresponds to the multidimensional combination of variables that have been verified to meet the method performance criteria.

The result of this approach is a robust and well-designed method that reliably delivers the expected performance [5, 6]. Another key advantage for employing the AQbD approach in method development is the potential for regulatory flexibility with regards to changes to the analytical method [5]. As such, AQbD is a desirable approach to be followed in analytical method development. Before explaining the

Optimization in HPLC: Concepts and Strategies, First Edition. Edited by Stavros Kromidas.
© 2021 WILEY-VCH GmbH. Published 2021 by WILEY-VCH GmbH.

AQbD approach with an example, there are some general considerations and criteria for a successful LC method development. The most important step is to gather as much information about the sample as possible. Examples of important information are the chemical properties, sample solubility, sample matrix properties, ionizable species, polarity of the analytes, pK_a, and molecular weight. In addition, the number of present compounds to be separated and the complexity of the separation need to be evaluated. Furthermore, sample concentration range, quantitative requirements, and detection technique available in the lab (UV, ELS, RI, FL, MS, etc.) need to be assessed.

Peak assignment can be challenging unless an MS detector is used. Due to testing different LC conditions during method development, peak retention times may shift, UV spectra will vary, and coelution can occur. An MS detector helps track the peaks very easily because the peaks are tracked by mass. Because method development is costly and time consuming (not only the run time, but also the equilibration time of the columns must be considered), it is strongly recommended to use an ultra performance liquid chromatography (UPLC) instrument, but it is not a requirement.

The following example demonstrates briefly the analytical potential of the software-assisted AQbD approach for achieving high-performance separations of formoterol, budesonide, and their related compounds [7]. The chemical structures of budesonide and formoterol are shown in Figure 4.5.1. Both compounds are used for treatment of asthma; according to Zetterstrom et al., inhaling the two pharmaceutical ingredients as one dose in combination proved to be more clinically effective [8]. Several key chromatographic parameters are investigated for their effect on the efficiency of the separations and the findings are presented and discussed.

In the process of LC method development, several factors are normally varied to achieve the desired separation goals. Some of these factors, such as the column stationary phase, the strong solvent, and the pH, have strong effects on separations while others like gradient steepness and separation temperature, have weaker effects. The success of any method development is generally evaluated based on the chromatographic separation and the peak shape obtained from varying these factors. The more chromatographic parameters being screened, the more time consuming the method development, especially if multiple data points for each

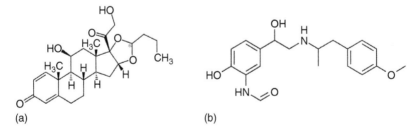

Figure 4.5.1 Chemical structures of budesonide (a) and formoterol fumarate (b). Formoterol is ionizable and therefore its peak shape and retention time are strongly affected by the pH of the mobile phase. *Source:* ©Waters Corporation.

parameter are screened. The advantage for using an AQbD software platform is that it uses statistical sampling approaches to create comprehensive and representative experimental designs. This significantly reduces the number of experiments required for method development because it covers the same design space as a generalized "full factorial" design by generating and modeling a much smaller but representative subset of all possible factor combinations. For example, given a "full factorial" study for two variables at five different study levels and running all possible combinations would require 25 experiments. However, if a method development software, like Fusion QbD, is used to model a comprehensive study to represent all these factors, only 13 experiments will be needed. This significant reduction in the number of experiments is due to the fact the statistical sampling design of Fusion QbD does not run all the combination points, but rather runs some points and use statistical modeling to predict the other points. This gives a comprehensive understanding of all significant factor effects across the entire chromatographic space with fewer actual experiments.

The design of experiment (DOE) is defined by the ICH as "a structured, organized method for determining the relationship between factors affecting a process and the output of that process" [9]. The software, used in this example, is Fusion QbD by S-Matrix. It employs the DOE approach in the following steps to develop the method:

- Initial screening
  - Initial sample workflow
  - Chemistry system screening design and selecting responses
- Optimizing method by adding additional method criteria and additional screenings
- Establishing robustness range

The main purpose of the initial screening step is to find initial chromatographic conditions that can be used as a starting point. These conditions are determined based on the initial method criteria. The peaks are not expected to be well separated or with good shape at this point. They only need to be retained and integrable.

The goal of this step is to perform a comprehensive screening experiment on parameters expected to have the largest effects on method performance. In this experiment, the pH of the mobile phase, gradient time (5–12 minutes), and the chemistry of the stationary phases are the three major chromatographic parameters explored. All other chromatographic parameters, including flow rate, temperature, and injection volume are set as constant. In this example, pH values between 2.0 and 4.2 are explored in half-pH-unit increments. Considering the stationary phases, it is important to choose a set of columns that provide a wide range of selectivities. The plot, shown in Figure 4.5.2, can help to choose a set of columns. A good set of stationary phases are a C18, an embedded polar group, a pentafluorophenyl (PFP), and a phenyl column. Sometimes, it can be beneficial to test a column that gives increased retention for polar analytes. Waters ACQUITY BEH C18, ACQUITY Shield RP18, CORTECS T3, CORTECS Phenyl, and HSS PFP are used for this experiment. All columns were 50 mm long with sub-2-µm particles for rapid gradient analysis and high flow rates.

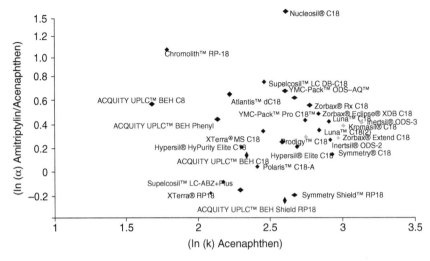

Figure 4.5.2 The plot demonstrates the differences in selectivity of LC columns, x-coordinate: retention of the non-ionizable compound acenaphthene, y-coordinate represents the silanol activity. For method development, it is strongly recommended to select the appropriate column stationary phase with different selectivity. *Source:* ©Waters Corporation.

As mentioned before, if all possible combinations are measured, there would be 125 experiments (five levels of $t_G \times$ five levels of pH $\times$ five columns) but the statistical-modeling software DOE approach by Fusion QbD is highly efficient and reduces the screening experiments to only 44.

The processing of the chromatograms involves integrating only all the peaks of interest, followed by the selection of the "best-looking" chromatogram, which is normally the one with the most visible peaks corresponding to actual analytes and has the most baseline resolved peaks. The data from the initial screening are not optimum and further experiments are needed to explore other chromatographic parameters. For example, as can be seen in Figure 4.5.3, the formoterol peak has very high tailing factor (2.10) and the budesonide-related compounds are not all baseline resolved. For peak assignment, it is very beneficial to use an MS detector. The mass spectrometry information allows to identify and track the peaks of interest despite the changes in their retention times and order of elution from the different runs.

After processing the data, Fusion QbD builds mathematical models and combines to predict the "best overall answer (BOA)." The performance goals from this initial screening stage are total number of peaks, number of peaks with a USP resolution of ≥ 1.50, and other chromatographic parameters. The software uses the measured responses to calculate models for each chromatographic result.

These models calculate a "cumulative desirability result," which is a numerical value that ranges from zero to one and represents the probability that the chromatographic conditions will meet the user-specified performance goals. In this example, it was found that the best combination of conditions to achieve the

Systematic Method Development with an Analytical Quality-by-Design Approach | 379

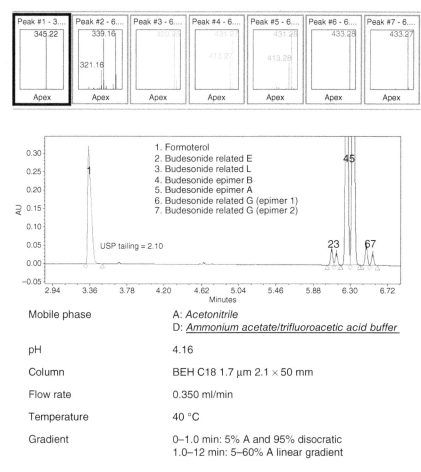

Figure 4.5.3 The "best-looking" chromatogram from the chemistry screening experiment; however, further experiments are needed due to the poor peak shape of formoterol (peak 1). *Source:* ©Waters Corporation.

set performance goals is: BEH C18 column, acetonitrile solvent, pH 4.2, and a gradient time of 12 minutes. However, in this example, the acidic pH is not able to achieve the desired performance goals. Some analytes are not baseline resolved and the formoterol peak is unsymmetrical with tailing factor over 2. Therefore, it is necessary to perform a second screening experiment to see if further improvements can be attained under different chromatographic conditions. Since formoterol has one basic structural functionality, it is ionizable by the pH of the mobile phase and will strongly affect the retention time and the peak shape (Figure 4.5.4). The pH value of the mobile phase has therefore a major influence on the separations.

The second screening experiment under different pH conditions is pursued.

In this example, pH values ranging from 6.7 to 10.7 are screened using the ACQUITY BEH C18 column. Results of this experiment show that the formoterol

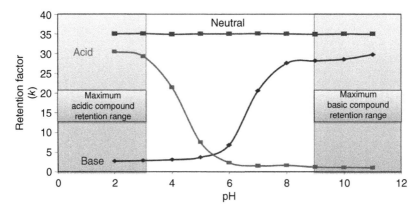

Figure 4.5.4 Influence on peak retention of ionizable compounds by the pH of the mobile phase. *Source:* ©Waters Corporation.

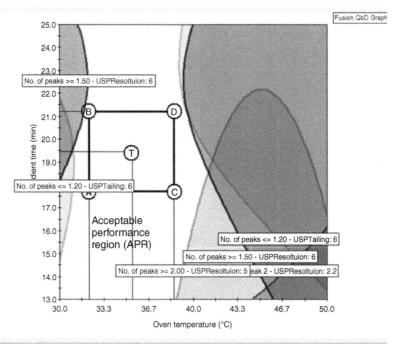

Figure 4.5.5 Fusion QbD graph of the design space and the APR obtained from the optimization experiment. *Source:* ©Waters Corporation.

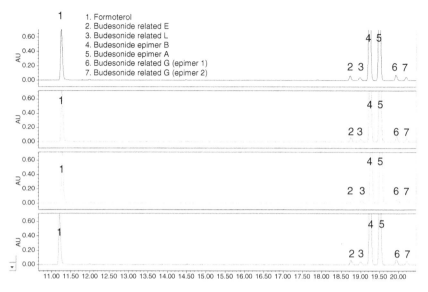

Figure 4.5.6 Four replicate injections of formoterol, budesonide, and its related compounds under the BOA conditions that were obtained from the optimization experiment. These conditions are: pH = 8.2, temperature = 33 °C, and flow rate = 0.350 ml/min. *Source:* ©Waters Corporation.

peak shape significantly improved when high pH mobile phases are used. The tailing for the formoterol peak is less than 1.3 for 10 of the 15 high pH screening runs. The results of these screening runs are processed in Fusion QbD to find the BOA. Fusion QbD models also provide performance maps that show the "acceptable performance region" (APR), an unshaded region around the BOA where the method meets or exceeds the predefined screening performance goals (Figure 4.5.5).

The major performance goals set in this search include baseline resolution for at least four peaks and a target tailing of 1.2 for the formoterol peak. The results of the basic screening after optimization are shown in Figure 4.5.6. The optimization in LC method development is normally done by fine tuning some of the less important parameters such as the temperature, gradient slopes, gradient times, and flow rates. The final gradient conditions are: initial hold of two minutes at 5% acetonitrile and 95% ammonium acetate buffer pH of 8.2 followed by a linear gradient of acetonitrile from 5–60% over 25 minutes. The tailing of formoterol decreases as the pH increases.

In Figure 4.5.5, the rectangle within the final design space illustrates the "acceptable performance region (APR)," a defined region within the design space where if a parameter is adjusted within the region and all other parameters remained constant, the results still meet the method performance goals. These findings indicate that a robust method that achieves all the desired performance goals can be created within the middle of the APR space.

Table 4.5.1 summarizes the individual considerations of a systematic method development.

Table 4.5.1 General method development considerations.

	Method development parameters	What	Result/goal
Preliminary considerations	Sample characteristics (chemical and physical properties)	Chemical structure of main analyte, unknowns, ionizability (neutrals, acids, bases, zwitterionic), solubility, concentration, detectability, sample matrix, sample pretreatment (filtration, protein precipitation, SPE), stability, polarity, isomers, etc.	Gathering as much as possible information about sample
	Desired separation goals	Which separation mode (RP, HILIC, etc.), which desired runtime, detection limits, minimum resolution, linearity, separating all peaks or some like critical components, sample preparation: yes/no (filtration/PPT, SPE, etc.), which mobile phase (organic modifier, buffer compatibility of detectors)	Criteria for intended method
	Potential limitations	System limitations (backpressure or flow rate, available detectors, etc.), automation capabilities (column switching valve), column heater/cooler, available column chemistries, isocratic or gradient (e.g. using RI detection no gradient possible)	Fit available equipment to method needs (no universal detector, isocratic methods more robust, gradient methods more sensitive, etc.)
	Columns dimensions and particle sizes	Choose for UPLC 2.1 × 50 mm columns (sub-2-µm particles), for UHPLC 3.0 × 50 mm (2.5-µm or 3.5-µm particles), for HPLC 3.9 or 4.6 × 100 mm (3.5-µm or 5-µm particles) (column ID and particle sizes depend on system bandspreading and backpressure capabilities)	Increased resolution, sensitivity and speed by using smaller particle sizes and inner diameters
	MS detection consideration	Use a single quadrupole detector	Easy peak tracking during column screening, peak specific standards no longer needed
Determination of method conditions by the analyst			

Table 4.5.1 (Continued)

			Software assistance for creating Design Space, scouting methods, and optimization of the final method — S-Matrix Fusion QbD software
Systematic column screening	Mobile phase	Solvents (ACN, MeOH), modifiers (buffers, acids, bases, salts like formic acid, TFA, ammonium carbonate, etc.)	Consider UV absorption, MS compatibility, fluorescence quenching
	Column chemistries	Use different stationary phases, include high pH stable columns, different ligands like C18, C18 with embedded polar group, C8, etc.	Systematic column screening, covering broadest possible selectivity range
	pH screening	pH is the biggest selectivity driver in RP chromatography. Screen low and high pH (< pH 3 and > pH 10)	For RP: Making sure the analyte is unionized. The pH of the mobile phase should be at least 2 pH units higher or lower than the pK_a value of the analyte
	Gradient method	UPLC/UHPLC: 2–5 minutes linear gradient, HPLC: 7–20 minutes gradient time	Short runtimes for speeding up the systematic screening
Method optimization	Choose best fitting separation and related conditions	Which stationary phase, mobile phase, and pH are getting best separation?	Selection based on your predefined separation criteria (see above)
	Optimize best fitting separation	Modify gradient slope, flow rate, column temperature if necessary	Optimizing final separation conditions
	Method robustness test	Check impact of slightly changed separation conditions on chromatographic results staying within acceptable performance region	Prove method and instrument suitability and transferability between labs and systems
Optional method validation	Reproducibility and robustness of methods	Check the impact of slightly changed separation conditions on chromatographic results, use column method validation kits (different batches)	Validation of the final method for regulatory authorities — Empower Method Validation Manager (Software assistance for method validation)

Notes: Gray underlaid: Software-assisted steps using S-Matrix Fusion QbD software and Empower Validation Manager.

References

1 Dispas, A., Avohou, H.T., Lebrun, P. et al. (2018). Quality by design approach for the analysis of impurities in pharmaceutical drug products and drug substances. *TrAC, Trends in Analytical Chemistry* 101: 24–33.
2 Shao, J., Cao, W., Qu, H. et al. (2018). Novel quality by design approach for developing an HPLC method to analyze herbal extracts: a case study of sugar content analysis. *PLoS One* 13: e0198515.
3 Thakor, N. and Amrutkar, S. (2017). Implementing quality by design (QbD) in chromatography. *Austin Journal of Analytical and Pharmaceutical Chemistry* 4: 1–5.
4 Peraman, R., Bhadraya, K., and Padmanabha Reddy, Y. (2015). Analytical quality by design: a tool for regulatory flexibility and robust analytics. *International Journal of Analytical Chemistry* 2015: 1–9.
5 Reid, G.L., Morgado, J., Barnett, K. et al. (2013). Analytical quality by design (AQbD) in pharmaceutical development. *American Pharmaceutical Review*: 144191.
6 Argentine, M., Barnett, K., Chatfield, M. et al. (2017). Evaluating progress in analytical quality by design. *Pharmaceutical Technology* 41 (4): 52–59.
7 Alkhateeb F. L. and Rainville P., Applying a software-assisted analytical quality-by-design approach for the analysis of formoterol, budesonide, and related compounds by UPLC-MS. Application notes 720006654EN.
8 Zetterstrom, O., Buhl, R., Mellem, H. et al. (2001). Improved asthma control with budesonide/formoterol in a single inhaler, compared with budesonide alone. *The European Respiratory Journal* 18: 262–268.
9 Ranga, S., Jaimini, M., Sharma, S.K. et al. (2014). A review on design of experiments (DOE). *International Journal of Pharmaceutical and Chemical Sciences* 3 (1): 216–224.

Index

a

acetonitrile   29–32, 35, 44, 64, 65, 74, 78, 93, 97, 107, 110, 116, 118, 121, 125, 133, 174, 203, 226, 228, 229, 234, 236, 258, 259, 263, 269, 270, 276, 278, 281, 282, 285, 287, 291–295, 297, 298, 300, 301, 306, 319, 345, 346, 347, 369, 372, 379, 381
acetyl pyridine   147, 148
achiral separations   90–93
acidic functional groups   152–153, 255
acidic pH values   31, 43
acidity constants
  acid constant estimation   263
  influence on peak shape   259–263
  polarity of acidic or alkaline substances   257–259
  UV spectra   259
active pharmaceutical ingredients (API)   108, 255, 278–279
additives   43–44, 52, 54, 65, 76, 81, 83, 87–90, 93, 96–97, 100, 101, 114, 117–119, 123–126, 135, 151, 182, 184, 190, 236, 292, 294, 319, 320, 363, 368, 370, 372
advanced thermal equilibration control (ATEC)   356
Agilent 1100 HPLC system   321
Agilent 1290 Infinity II Flexible Pump   319
Agilent InfinityLab Quick Connect UHPLC Column Fittings   314
Agilent technologies

automation   319–321
  HDR   325–326
  ISET   326–327
  laboratories sample throughput
    alternating column regeneration   316
    dual needle multisampler   316
    StreamSelect   316–317
  modular or stepwise upgrade of existing systems   318
  multimethod and method development systems   321–322
  multiwash   317–318
  online SPE   322–323
  second chromatographic dimension (2D-LC)   323–324
  separation performance
    minimizing the dispersion   314–315
    zero dead-volume fittings   314
  SFC-HPLC hybrid system   324–325
Agilent 1290 Infinity II High Dynamic Range Diode Array Detection (HDR-DAD) Impurity Analyzer Solution   325
alcohols   121
aldehydes, hydrate formation of   146–148
amylose derivatives   131
analytical procedure lifecycle management (APLM) approach   199, 201
analytical quality-by-design (AQbD)
  advantage   375, 377
  Fusion QbD   377

Index

analytical target profile (ATP)   201–202, 217
antibody–drug conjugates (ADCs)   59, 76
APLM Stage 1   202–204
APLM Stage 2   199, 202, 214
assistant AZURA ASM 2.2L   333
automated method development   49, 180–182, 184–188, 196, 243, 368–374
automated method scouting (AMS)   368, 370–372
automatic optimization   180, 181

b

best overall answer (BOA)   378, 381
Bioinert components   340
BlendAssist   320
budesonide   376, 378, 379, 381
building block phase   245, 248
butamirate dihydrogen citrate, cough syrup   279, 283, 284

c

carbonyl compounds   147, 271
cation exchange chromatography (CEX)   74
computer-assisted method development (CAMD)   180
charged aerosol detection (CAD)   291, 357, 361
chemical and enzymatic modifications   57
chemical oxidation   44
chemistry system screening   202, 204–207, 377
chiral chromatography   133
chiral HPLC and SFC   110, 132–135
chiral polysaccharide stationary phases   110–113
chiral separation   88–90, 93, 96, 100, 101, 107–138, 180
chiral stationary phase (CSP)   88, 96, 107
chromatographer   89, 98, 147, 163, 180, 194
chromatographic modes   58, 64, 66, 74, 76, 102, 201
chromatographic optimization parameters   97
chromatographic separation
  conjugated ADC   76
  mobile phase modifiers   76
  monoclonal antibody   74
  protein formulations   74
  released and labeled N-glycans   74
  size exclusion chromatography   76
chromatography data system (CDS)   49, 180, 182, 191, 192, 362, 364
Chromeleon eWorkflows   370
ChromSwordAuto®   180–182, 184, 332, 370, 372, 373
coated CSPs   113
coefficient of variation (VC)   165, 166, 171–175
coinjection   346–348
column bleeding   35, 49, 51
column identification kits   327
column selection   89, 321, 334, 370
conjugated antibody–drug conjugates (ADCs)   59, 76
cosmotropic salinity   33
critical quality attributes (CQA)   189, 192, 357, 368
cumulative desirability result   378
CUP laboratories   239–249
cyproconazole   119

d

data integrity   200, 216
design of experiments (DOE)   87, 96, 99, 102, 180, 201, 217, 377, 378
difluoroacetic acid (DFA)   66
dimethyl formamide (DMF)   148, 149, 256
diode array (DAD)   87, 171, 272, 325, 357, 361
1,4-dioxane   149
distance effects   152
donor–acceptor interactions   142, 146

drug development
　early stages of　185
　late stages of　185
dual LC mode for operation　357, 369

e

electrospray ionization (ESI)　39, 40, 79, 100, 287
Empower Validation Manager　383
enantiomer separation problems　135–136
enantioselective chromatography　108, 137
E-phase, or Execute　245
Erlenmeyer rule　147
ethers, peroxide formation of　150, 151
ethyl acetate　113, 121, 125, 149, 255, 291, 345
2-ethylpyridine-bonded silica stationary phase　90

f

fast chiral separations　136–138
fit for purpose method　109
flow injection analysis (FIA)　44
fluorescence detection (FLD)　74, 78, 304, 357
formoterol　376–379, 381
functional moieties, impact of　142–143
Fusion QbD　377
　BOA　378, 381
　DOE approach　377, 378
　S-Matrix　383

g

generic methods
　active pharmaceutical ingredients　278–280
　butamirate dihydrogen citrate, cough syrup　279
　highly polar compounds　282
　limits of　279
　MS detection improvement　282
　proteins　282
　strongly nonpolar connections　282

UV detection, deep wavelength range　282
gradient delay volume (GDV) adjustment　356, 359–361
gradient methods　95, 96, 167, 184, 185, 269, 283, 295, 372, 382, 383
gradient separations　42, 74, 78, 167, 168, 358

h

Hansen-solubility parameters　160, 162
High-Definition Range (HDR) HPLC　325
high-efficiency stationary phases　92
high performance chromatography　58
high performance liquid chromatography (HPLC)
　acid–base constants　255
　Agilent technologies　313
　avoidance of peak tailing　295–302
　basic functional groups　153–155
　blank samples　287–288
　column dimension and particle sizes　305–308
　content method　242
　dissolved analytes, stability of　255, 272–278
　experimental data　226
　intelligent and self-contained　362
　measurement uncertainty and method design
　　calibration–calibration model　304–305
　　detection　304
　　dilutions　303
　　internal standards　305
　　peak shape/separation　304
　　weighing in or measuring　302–303
　multidimensional separation modes　80
　nonpolar analytes　228
　octanol–water partition coefficient　254, 255, 263–270
　optimizing sensitivity　79–80
　pH value in　151–152

high performance liquid chromatography
(HPLC) (*contd.*)
  polar analytes 228–230
  process control analytics 235–237
  quality control 234–235
  research and development 221, 233–234
  robust approaches 222
  software assisted automated method development 368–374
  solubility 252–257
  speed optimization 77–79
  task, definition of 252
  2D-LC separations 363–368
  UNTIE®process, CUP laboratories 239
  UV absorption 255, 270–272
  UV detection wavelengths 288–291
  Vanquish Core 356–361
  Vanquish Flex 357
  Vanquish Horizon 357
high pH reversed-phase (RP) fractionation 81
high-pressure proportioning gradient pump (HPG) 357
high-resolution accurate mass spectrometers (HRAMS) 47, 356
high-resolution/accurate mass (HR/AM) measurements 47, 356
host-cell protein (HCP) contaminants 81
hybrid modes 7–8
hydrogen bonds 26, 27, 91, 94, 134, 142–146, 148, 150, 154, 156, 158–161, 224
hydroperoxides 150, 151, 156, 157
hydrophilic interaction liquid chromatography (HILIC) 78
  classification 27
  free silica phase 26
  4-hydroxybenzoic acid 25
  logD value 23
  mass spectrometric detection 35
  neutral stationary phases 27
  organic solvent 28–31
  pH value 33–34
  salts 31–33
  silica particles 26
  zwitterionic stationary HILIC phases 28
hydrophobic compounds 76, 91
1-hydroxy- and 4-hydroxybenzoic acid 23–25, 31, 32, 34
hydroxychloroquine (HCQ) 121
hyphenated method 48–49

i

immobilized polysaccharide CSPs 89, 112–127
InfinityLab LC Companion 327
intelligent peak deconvolution analysis (I-PDeA) 351–353
intelligent run control (IRC) 370
Intelligent System Emulation Technology (ISET) 326–327
interactive forces 141, 142
Internet of Things (IoT) 341, 353, 354, 362
ion chromatography 23, 24, 240, 338
ion source parameters 45
isocratic methods 96, 170, 382
isocratic separations 168, 294, 361

k

ketones, hydrate formation of 146–148

l

L-ABGA 123
laboratory information management system (LIMS) 341
limit of quantification (LOQ) 108, 165
lincomycin 154, 155
liquid chromatography (LC) 39, 57, 200, 375
long life and/or easy maintenance pump heads 327
loop-based multi-heart-cut 2D-LC 364–366
loop-based single-heart-cut 2D-LC 364, 365
low-pressure proportioning gradient pump (LPG) 357

low wavelengths
　acids and buffer additives 292–294
　drift at solvent gradients 294–295
　solvents 291–292

m

macrocyclic glycopeptide CSPs 89
MarvinSketch 263, 270
mass accuracy 47
mass spectrometry (MS) 3, 7, 12, 14, 15, 23–36, 39, 40, 42, 43, 50–54, 58, 76, 81, 87, 100, 137, 200, 234, 272, 352, 359, 378
　control software 48
　detection 12–14, 41–43, 47–49, 52, 73, 79, 81–83, 282, 301, 364, 367, 369, 382
methanol 29, 30, 44, 93–95, 97, 101, 110, 113, 116, 118, 121, 125, 133, 135, 148, 153, 167, 226, 227, 254–256, 270, 274, 278, 281, 287, 291–294, 301, 319, 325, 345, 348, 369, 372
method development (MD) 3–18, 28, 29, 39–54, 88, 96, 101, 102, 109, 179–182, 184–188, 192, 194, 196, 199–217, 221–231, 235, 236, 239, 241–245, 248, 249, 252, 253, 272, 318, 320–322, 337, 350, 351, 359, 367–383
method operable design region (MODR) 199, 208, 213, 217, 375
method scouting 243–245, 322, 368, 370–373
mobile-phase composition (MP) 13, 87, 90, 93, 95–99, 101, 213, 368
mobile-phase modifiers 76, 127
mobile phase production
　buffer solutions preparation of 285
　degassing 287
　filtration of solvents and buffer 285–287
　reagents 283–285
　solvent measurement 284–285
　vessels and fluid bottles 284
mobile-phase viscosity 78, 94

monoclonal antibodies (mAbs) 57, 74, 78, 81, 82, 179–196, 200, 358
monovalent ions 52
Monte Carlo simulation 210–213
Monte Carlo optimization procedure 186
multidimensional separation modes 80–81
multivariate curve resolution-alternating least squares (MCR-ALS) 351

n

neutral stationary phases 27
Nexera system 243, 342, 349
noise reduction 171
normal-phase (NP) separation 9, 109, 113, 325

o

octanol–water partition coefficient 142, 157–160, 162, 254, 255, 263–270
one factor at a time (OFAT) 99, 375
1D method 11, 12
Online SPE 318, 322–323
optimization, HPLC
　assistant AZURA ASM 2.2L 333
　automatic method optimization and column screening 334–335
　Bioinert components 340
　catalogue of measures 331
　chemical resistance of wetted components 338–340
　device level 332–333
　equipment level 330
　fractionation 335–336
　individual optimizations, assistant 334
　Internet of Things (IoT) 341
　material certification 340
　method levels 330
　outcomes and improvements 329
　peak recycling 335–336
　personnel level 331–332
　purification 336–337
　sample preparation 335–336

optimization, HPLC (contd.)
  software 340–341
  specification sheet 329–331
  time-consuming method optimization 332
  wetted vs. dry components 337–338
optimizing mean performance 207–210
organic solvents 4, 9, 14, 15, 25, 28–31, 33, 36, 43, 65, 74, 76, 81, 82, 112, 116, 118, 125, 136, 191, 229, 243, 255, 256, 263, 274, 276, 281, 282, 287, 302, 338, 345, 346, 372
out-of-specification (OOS) 202

p

paracetamol 137, 160–162
particle size 12, 40–42, 59, 73, 75, 78, 79, 82, 83, 88, 92, 98, 99, 109, 123, 136, 166, 204, 280, 305–308, 323, 356, 361, 382
peak recycling 335–336
peak-to-noise ratio 165, 171, 172
PeakTracker 208
pharmacopoeia methods 165, 166
pH dependency 151, 152
photodiode array (PDA) detector 343, 348–349, 351
pH value 23, 25, 28, 31, 33–34, 36, 43, 46, 52, 142, 151–152, 155, 156, 167, 221, 257, 258, 260, 261, 263, 272, 273, 283, 285, 298, 301, 338, 377, 379
Pirok compatibility table 9
polar organic mode 110, 113–116, 124
polarity tables 141
polynuclear nitrogenous aromatics 256
polysaccharide-based stationary phases 88
polysaccharide CSPs 88, 110, 127, 129, 133, 135
POPLC 222, 223
protein biopharmaceuticals
  high performance chromatography 58–62
  microheterogeneity 57
  multidimensional LC approaches 64–66
  nondenaturing LC modes 62–64
  therapeutic proteins and chromatographic methods 58
protein oxidation 183
proven acceptable range (PAR) 381
pumping system 87, 100

q

quality-by-design concepts 49
quantitative structure–retention relationship (QSRR) modeling 372, 374
quaternary pumps 203, 204, 319, 320, 323

r

rac-Norketotifen 107, 108
recombinant monoclonal antibodies (mAbs) 179
replication strategy 199, 202, 214, 217
reversed-phase HPLC methods
  automated method development 184–185
  columns screening 185–186
  fine optimization and sample profiling 188
  interaction with instruments 181–182
  performance improvement 194, 196
  rapid optimization mode 186–188
  robustness test 188–189
  sample preparation and HPLC analysis 183–184
  software tools 180, 181
  variables selection 189–190
reversed phase liquid chromatography (RPLC) 23, 24, 57, 74, 98
rule of thumb 80, 82, 96, 222, 252, 278, 325

s

sample diluents 74, 97
sample injector 87

screening coated and immobilized
polysaccharide CSPs
in normal phase and polar organic
mode 113–116
polar organic supercritical fluid
chromatography conditions
121–125
in reversed-phase mode 116–119
screening immobilized polysaccharide
CSPs
in medium-polarity mode 119–122
in medium-polarity supercritical fluid
chromatography conditions
125–127
second chromatographic dimension
(2D-LC) 323–324
selective comprehensive mode (sLCxLC)
363
semipolar phases 26, 27
separation, HPLC
hyphenated method 48–49
ion source conditions 44–47
mass spectrometry 50–51
MS-compatible mobile phase 43
MS detection 47–48
rapid screening separations 42
sample throughput and separation
quality 42
sensitivity and limit of detection
40–41
software-based parameter variation
49–50
SFC–HPLC hybrid system 325
Shimadzu LabSolutions Chromatography
Software (CDS) 347
Shimadzu Nexera Method Scouting
System 245
Shimadzu Scouting System 246
simulated moving bed (SMB)
chromatography 336, 337
single-/multi-heartcut system 52–54
size exclusion chromatography (SEC)
57, 76, 81, 82
slit width 343, 348–349
SmartInject technology 356

SmartStroke 356
S-Matrix Fusion QbD® Software 199, 383
application to chromatographic
separation modes 200
non-LC method development
procedures 200–201
small and large molecule applications
200
analytical target profile 201–202
APLM Stage 1 202–204
APLM Stage 2 214–215
chemistry system screening 204–207
experimental design and data modeling
201
Monte Carlo simulation 210–213
optimizing mean performance
207–210
USP <1210> 214–216
software-assisted automated method
development 368–374
software optimization 332, 340–341
solid phase extraction (SPE) 79
speed optimization 77–79
Stage-Gate®model 241
stationary phase-optimized-selectivity
(SOS) 92
stationary phases (SP) 9, 25–34, 36, 44,
51, 52, 59, 61, 64, 66, 75, 76, 87–93,
95–98, 100–102, 107, 109–113, 124,
127, 129, 132, 135, 136, 138, 141,
151, 157, 166, 167, 204–208, 221,
222, 224, 226, 228, 234–236, 255,
257, 263, 269, 278, 280, 282, 284,
295, 297, 298, 301, 302, 305, 324,
347, 356, 364, 368, 376–378, 383
statistical process control (SPC) 211
statistical test planning 49
StreamSelect 316
streptomycin 155, 156, 159
supercritical fluid chromatography (SFC)
mobile phases 93
optimization parameter 97
polysaccharide-based stationary phases
88
proportion of cosolvent 94

supercritical fluid chromatography (SFC) (contd.)
  SFC–MS coupling 100–101
  synthetic polymer CSP 89
  two-dimensional chromatography 102
synthetic polymer CSP 89

t

target measurement uncertainty (TMU) 202
Tefzel® 338, 339
Thermo Scientific Vanquish Core systems 362
time-consuming method optimization 332, 352
tolerance interval (TI) 199, 202, 214–217
toluene 23, 24, 111, 161, 162, 256
trap-based single heart-cut 2D-LC for eluent strength reduction 366–367
trap-based single heart-cut 2D LC–MS using Vanquish Dual Split Sampler 367–368
trifluoroacetic acid (TFA) 43, 61, 76, 153, 186, 281, 282, 284, 285, 293–295, 297, 298, 301, 356, 379
two-dimensional liquid chromatography (2D-LC) 13, 363
  loop-based multi-heart-cut 2D-LC 364–366
  loop-based single-heart-cut 2D-LC 364
  trap-based single heart-cut 2D-LC for eluent strength reduction 366–367
  trap-based single heart-cut 2DLC-MS using Vanquish Dual Split Sampler 367–368
two-dimensional separation 80
  choice of mode 5
  complex samples 4
  comprehensive 2D-LC methods 13
  difficult-to-separate samples 3
  fixed first dimension conditions 11–13
  four 2D separation modes 5–7

hybrid modes 7–8
LC–LC method 14
rules of thumb 13–14
separation goals 4
separation types/mechanisms 8–11
2D-LC separation of surfactants 14–17

u

(U)HPLC performance
  advanced peak processing 350–353
  auto-diagnostics and auto-recovery 349–350
  compressibility settings 345
    mixture of water and organic solvent 345, 346
    pressure and pressure fluctuations 343
    single solvent 345
  Internet of Things 353
  photodiode array detector, silt width 348–349
  solvent composition and injection volume 346–348
  systems 168
ultra-high-performance supercritical fluid chromatography (UHPSFC) 92
Ultra Low Dispersion Kit 314
unique dual gradient pump (DGP) 357
United States Pharmacopoeia 241
UNTIE® process, CUP laboratories 248
  development and optimization method 243–245
  existing method, test of 242–243
  understanding customer needs 241–242
  validation 245–248
UNTIE pyramid 240
USP <1210> 199, 214–216
UV cut-off 291, 301
UV or diode-array detection 87

v

Vanquish Core 356–362
Vanquish Duo-based Dual LC concept 358

Vanquish Flex   356–361
Vanquish Horizon   356–361
Vanquish (U)HPLC platform   355, 356, 370, 372
Vanquish Solvent Monitor (VSM)   362
Vanquish User Interface (VUI)   362
variable-wavelength light absorbance detection (VWD)   357
volatile ionic detergents   44

z

zero dead-volume fittings   314
zwitterionic stationary HILIC phases   28